series *Lecture Notes in Electrical Engineering* (LNEE) publishes the
...lopments in Electrical Engineering - quickly, informally and in high
...hile original research reported in proceedings and monographs has
...y formed the core of LNEE, we also encourage authors to submit books
... supporting student education and professional training in the various
...applications areas of electrical engineering. The series cover classical and
...opics concerning:

...unication Engineering, Information Theory and Networks
...nics Engineering and Microelectronics
... Image and Speech Processing
...ss and Mobile Communication
...s and Systems
... Systems, Power Electronics and Electrical Machines
...-optical Engineering
...nentation Engineering
...ics Engineering
...l Systems
...t-of-Things and Cybersecurity
...dical Devices, MEMS and NEMS

...al information about this book series, comments or suggestions, please
...ontina.dicecco@springer.com.
...omit a proposal or request further information, please contact the
...g Editor in your country:

...Dou, Associate Editor (jasmine.dou@springer.com)

...pan, Rest of Asia

...eherishi, Executive Editor (Swati.Meherishi@springer.com)

...st Asia, Australia, New Zealand

...Nath Premnath, Editor (ramesh.premnath@springernature.com)

...nada:

... Luby, Senior Editor (michael.luby@springer.com)

...r Countries:

... Di Cecco, Senior Editor (leontina.dicecco@springer.com)

...xing: Indexed by Scopus. **

...formation about this series at http://www.springer.com/series/7818

Lecture Notes in Electrical

Volume 710

Series Editors

Leopoldo Angrisani, Department of Electrical and Information T
Federico II, Naples, Italy
Marco Arteaga, Departament de Control y Robótica, Universida
Mexico
Bijaya Ketan Panigrahi, Electrical Engineering, Indian Institute o
Samarjit Chakraborty, Fakultät für Elektrotechnik und Informatic
Jiming Chen, Zhejiang University, Hangzhou, Zhejiang, China
Shanben Chen, Materials Science and Engineering, Shanghai Jiad
Tan Kay Chen, Department of Electrical and Computer Engineer
Singapore, Singapore
Rüdiger Dillmann, Humanoids and Intelligent Systems Laborator
Karlsruhe, Germany
Haibin Duan, Beijing University of Aeronautics and Astronautics
Gianluigi Ferrari, Università di Parma, Parma, Italy
Manuel Ferre, Centre for Automation and Robotics CAR (UPM-C
Madrid, Spain
Sandra Hirche, Department of Electrical Engineering and Informat
München, Munich, Germany
Faryar Jabbari, Department of Mechanical and Aerospace Enginee
CA, USA
Limin Jia, State Key Laboratory of Rail Traffic Control and Safety,
Janusz Kacprzyk, Systems Research Institute, Polish Academy of S
Alaa Khamis, German University in Egypt El Tagamoa El Khames
Torsten Kroeger, Stanford University, Stanford, CA, USA
Qilian Liang, Department of Electrical Engineering, University of T
Ferran Martín, Departament d'Enginyeria Electrònica, Universitat A
Barcelona, Spain
Tan Cher Ming, College of Engineering, Nanyang Technological U
Wolfgang Minker, Institute of Information Technology, University c
Pradeep Misra, Department of Electrical Engineering, Wright State
Sebastian Möller, Quality and Usability Laboratory, TU Berlin, Berl
Subhas Mukhopadhyay, School of Engineering & Advanced Techno
Palmerston North, Manawatu-Wanganui, New Zealand
Cun-Zheng Ning, Electrical Engineering, Arizona State University, T
Toyoaki Nishida, Graduate School of Informatics, Kyoto University,
Federica Pascucci, Dipartimento di Ingegneria, Università degli Studi
Yong Qin, State Key Laboratory of Rail Traffic Control and Safety, Be
Gan Woon Seng, School of Electrical & Electronic Engineering, Nan
Singapore, Singapore
Joachim Speidel, Institute of Telecommunications, Universität Stuttga
Germano Veiga, Campus da FEUP, INESC Porto, Porto, Portugal
Haitao Wu, Academy of Opto-electronics, Chinese Academy of Scien
Junjie James Zhang, Charlotte, NC, USA

Arun Kumar Singh · Manoj Tripathy
Editors

Control Applications in Modern Power System

Select Proceedings of EPREC 2020

Editors
Arun Kumar Singh
National Institute of Technology Jamshedpur
Jamshedpur, India

Manoj Tripathy
Indian Institute of Technology Roorkee
Roorkee, India

ISSN 1876-1100 ISSN 1876-1119 (electronic)
Lecture Notes in Electrical Engineering
ISBN 978-981-15-8814-3 ISBN 978-981-15-8815-0 (eBook)
https://doi.org/10.1007/978-981-15-8815-0

© The Editor(s) (if applicable) and The Author(s), under exclusive license to Springer Nature Singapore Pte Ltd. 2021
This work is subject to copyright. All rights are solely and exclusively licensed by the Publisher, whether the whole or part of the material is concerned, specifically the rights of translation, reprinting, reuse of illustrations, recitation, broadcasting, reproduction on microfilms or in any other physical way, and transmission or information storage and retrieval, electronic adaptation, computer software, or by similar or dissimilar methodology now known or hereafter developed.
The use of general descriptive names, registered names, trademarks, service marks, etc. in this publication does not imply, even in the absence of a specific statement, that such names are exempt from the relevant protective laws and regulations and therefore free for general use.
The publisher, the authors and the editors are safe to assume that the advice and information in this book are believed to be true and accurate at the date of publication. Neither the publisher nor the authors or the editors give a warranty, expressed or implied, with respect to the material contained herein or for any errors or omissions that may have been made. The publisher remains neutral with regard to jurisdictional claims in published maps and institutional affiliations.

This Springer imprint is published by the registered company Springer Nature Singapore Pte Ltd.
The registered company address is: 152 Beach Road, #21-01/04 Gateway East, Singapore 189721, Singapore

Preface

Originally, the Electric Power and Renewable Energy Conference (EPREC 2020) was to be organized onsite by the Department of Electrical Engineering, National Institute of Technology Jamshedpur, India, during May 29–30, 2020. However, due to a global pandemic COVID-19, the conference was organized in an online mode and attracted national and international audience from different countries like Canada, Brazil, USA, *etc.* The editors are thankful to all the contributors for submitting the papers of high standard and making EPREC 2020 a huge success. Out of total 351 valid submissions, only 142 were selected for publication in three different volumes, i.e., an acceptance rate of nearly 40%. This volume, i.e., Control Applications in Modern Power System, is one of the three volumes to be published by Springer in the book series "Lecture Notes in Electrical Engineering (LNEE)." It contains 47 high-quality papers which provide an overview of recent developments in control system applications in modern power systems.

We thank all the organizing committee members, technical program committee members, reviewers, and student coordinators for their valuable support and volunteer work. We also appreciate the role of session chairs/co-chairs. We also thank the series editors of LNEE and Ms. Priya Vyas and Dr. Akash Chakraborty, Associate Editors, Applied Sciences and Engineering, Springer, for their help and quick responses during the preparation of the volume.

The editors hope that this volume will provide the readers relevant information on the latest trends in control system area.

Jamshedpur, India	Arun Kumar Singh
Roorkee, India	Manoj Tripathy

Contents

1. **Closed Loop Control of Non-ideal Buck Converter with Type-III Compensator** 1
   Abhishek Kumar, Durgesh Chandra Nautiyal, and Prakash Dwivedi

2. **Vector Error Correction Model for Distribution Dynamic State Estimation** 15
   C. M. Thasnimol and R. Rajathy

3. **Optimal Battery Charging Forecasting Algorithms for Domestic Applications and Electric Vehicles by Comprehending Sustainable Energy** 29
   S. Parvathy, Nita R. Patne, and T. Safni Usman

4. **Performance Index-Based Coordinated Control Strategy for Simultaneous Frequency and Voltage Stabilization of Multi-area Interconnected System** 45
   Ch. Naga Sai Kalyan and G. Sambasiva Rao

5. **Load Frequency Control of Two-Area Power System by Using 2 Degree of Freedom PID Controller Designed with the Help of Firefly Algorithm** 57
   Neelesh Kumar Gupta, Manoj Kumar Kar, and Arun Kumar Singh

6. **Stabilizing Frequency and Voltage in Combined LFC and AVR System with Coordinated Performance of SMES and TCSC** 65
   Ch. Naga Sai Kalyan and G. Sambasiva Rao

7. **Optimal Fuzzy-PID Controller Design by Grey Wolf Optimization for Renewable Energy-Hybrid Power System** 77
   Himanshu Garg and Jyoti Yadav

8. **A New Classical Method of Reduced-Order Modelling and AVR System Control Design** 89
   Nafees Ahamad, Afzal Sikander, and Gagan Singh

9	**Unscented Kalman Filter Based Dynamic State Estimation in Power Systems Using Complex Synchronized PMU Measurements** Shubhrajyoti Kundu, Anil Kumar, Mehebub Alam, Biman Kumar Saha Roy, and Siddhartha Sankar Thakur	99
10	**Real-Time Electric Vehicle Collision Avoidance System Under Foggy Environment Using Raspberry Pi Controller and Image Processing Algorithm** Arvind R. Yadav, Jayendra Kumar, Roshan Kumar, Shivam Kumar, Priyanshi Singh, and Rishabh Soni	111
11	**Modified Sine Cosine Algorithm Optimized Fractional-Order PD Type SSSC Controller Design** Preeti Ranjan Sahu, Rajesh Kumar Lenka, and Satyajit Panigrahy	119
12	**Performance Enhancement of Optimally Tuned PI Controller for Harmonic Minimization** Anish Pratap Vishwakarma and Ksh. Milan Singh	131
13	**Design and Performance Analysis of Second-Order Process Using Various MRAC Technique** Saibal Manna and Ashok Kumar Akella	143
14	**Investigation Analysis of Dual Loop Controller for Grid Integrated Solar Photovoltaic Generation Systems** Aditi Chatterjee and Kishor Thakre	153
15	**Damping Enhancement of DFIG Integrated Power System by Coordinated Controllers Tuning Using Marine Predators Algorithm** Akanksha Shukla and Abhilash Kumar Gupta	165
16	**IoT-Integrated Voltage Monitoring System** Himanshu Narendra Sen, Ashish Srivastava, Mucha Vijay Reddy, and Varsha Singh	177
17	**Intrinsic Time Decomposition Based Adaptive Reclosing Technique for Microgrid System** Shubham Ghore, Pinku Das, and Monalisa Biswal	187
18	**Design of Energy Management System for Hybrid Power Sources** Akanksha Sharma, Geeta Kumari, H. P. Singh, R. K. Viral, S. K. Sinha, and Naqui Anwer	197
19	**Control and Coordination Issues in Community Microgrid: A Review** Seema Magadum, N. V. Archana, and Santoshkumar Hampannavar	217

20 Optimal Generation Sizing for Jharkhand Remote Rural Area by Employing Integrated Renewable Energy Models Opting Energy Management 229
Nishant Kumar and Kumari Namrata

21 IoT-integrated Smart Grid Using PLC and NodeMCU 241
Kumari Namrata, Abhishek Dayal, Dhanesh Tolia, Kalaga Arun, and Ayush Ranjan

22 Fuzzy Model for Efficiency Estimation of Solar PV Based Hydrogen Generation Electrolyser 251
Sandhya Prajapati and Eugene Fernandez

23 Temperature-Dependent Economical and Technical Aspect of Solar Photo Voltaic Power Plant........................ 261
Subhash Chandra

24 An Efficient Optimization Approach for Coordination of Network Reconfiguration and PV Generation on Performance Improvement of Distribution System........................ 269
Sachin Sharma, Khaleequr Rehman Niazi, Kusum Verma, and Tanuj Rawat

25 Enhancement of Hybrid PV-Wind System by Ingenious Neural Network Technique Indeed Noble DVR System 279
Roopal Pancholi and Sunita Chahar

26 Modified Particle Swarm Optimization Technique for Dynamic Economic Dispatch Including Valve Point Effect 311
Gaurav Kumar Gupta and Mayank Goyal

27 Electromagnetic Compatibility of Electric Energy Meters in the Presence of Directional Contactless Electromagnetic Interference... 325
Illia Diahovchenko and Bystrík Dolník

28 Improvement of Small-Signal Stability with the Incorporation of FACTS and PSS 335
Prasenjit Dey, Anulekha Saha, Sourav Mitra, Bishwajit Dey, Aniruddha Bhattacharya, and Boonruang Marungsri

29 Optimal Threshold Identification of Fault Detector Using Teaching and Learning-Based Optimization Algorithm 345
Ch. Durga Prasad and Monalisa Biswal

30 A Novel MTCMOS Stacking Approach to Reduce Mode Transition Energy and Leakage Current in CMOS Full Adder Circuit 355
Anjan Kumar and Sangeeta Singh

31 A Backward/Forward Method for Solving Load Flows
    in Droop-Controlled Microgrids 367
    Rahul Raj and P. Suresh Babu

32 Monte Carlo Simulation Application in Composite Power System
    Reliability Analysis 379
    Atul Kumar Yadav, Soumya Mudgal, and Vasundhara Mahajan

33 Enhancement of Static Voltage Stability Margin Using
    STATCOM in Grid-Connected Solar Farms 387
    S. Venkateswarlu and T. S. Kishore

34 Energy and Economic Analysis of Grid-Type Roof-Top
    Photovoltaic (GRPV) System 399
    Abhinav Kumar Babul, Saurabh Kumar Rajput, Himmat Singh,
    and Ramesh C. Yadaw

35 Automation of Public Transportation (Bus Stands) 409
    Gaurav Yadav, Archit, Parth Dutt, and Sankalp Sharma

36 Analysis of Different Aspects of Smart Buildings and Its Harmful
    Effects on the Ecosystem 419
    D. K. Chaturvedi and Boudhayan Bandyopadhyay

37 Survey and Analysis of Content-Based Image Retrieval
    Systems 427
    Biswajit Jena, Gopal Krishna Nayak, and Sanjay Saxena

38 A Prototype Model of Multi-utility Mist Vehicle for Firefighting
    in Confined Areas 435
    Sasidhar Krishna Varma, Pankaj Bhagath,
    and Nadakuditi Gouthamkumar

39 Security Analysis of System Network Based on Contingency
    Ranking of Severe Line Using TCSC 445
    Kumari Gita and Atul Kumar

40 A Renewable Energy-Based Task Consolidation Algorithm
    for Cloud Computing 453
    Sanjib Kumar Nayak, Sanjaya Kumar Panda, Satyabrata Das,
    and Sohan Kumar Pande

41 Multi-objective Optimization for Hybrid Microgrid Utility
    with Energy Storage 465
    Kapil Gandhi and S. K. Gupta

42 Damaged Cell Location on Lithium-Ion Batteries Using Artificial
    Neural Networks 477
    Mateus Moro Lumertz, Felipe Gozzi da Cruz,
    Rubisson Duarte Lamperti, Leandro Antonio Pasa, and Diogo Marujo

43	**Fuzzy Logic-Based Solar Generation Tracking** .... Anish Agrawal and Anadi Shankar Jha	487
44	**Airborne Manoeuvre Tracking Device for Kite-based Wind Power Generation** .... Roystan Vijay Castelino and Yashwant Kashyap	497
45	**Integration of Electronic Engine and Comparative Analysis Between Electronic and Mechanical Engine** .... Sapna Chaudhary and Rishi Pal Chauhan	509
46	**Monitoring Cyber-Physical Layer of Smart Grid Using Graph Theory Approach** .... Neeraj Kumar Singh, Praveen Kumar Gupta, Vasundhara Mahajan, Atul Kumar Yadav, and Soumya Mudgal	519
47	**New Active-Only Impedance Multiplier Using VDBAs** .... A. Hari Prakash Reddy, R. N. P. S. S. Charan, and Mayank Srivastava	527

About the Editors

Dr. Arun Kumar Singh is currently a Professor at the Department of Electrical Engineering, National Institute of Technology Jamshedpur, India. He received his B.Sc. (Engg.) from Kurukshetra University, Kurukshetra; M.Tech from the Banaras Hindu University, Varanasi; and PhD degree in electrical engineering from the Indian Institute of Technology Kharagpur, India. He has over 30 years of teaching experience and taught subjects like Control System, Control & Instrumentation, Power System Operation and Control, Non-Conventional Energy, Circuit & Network Theory. His major areas of research interests include control system, control system applications in different areas & non-conventional energy. Prof. Singh is a Life Member of ISTE and a life fellow member of Institute of Engineers. He has been the Dean of Student Welfare at National Institute of Technology.

Dr. Manoj Tripathy received his B.E. degree in electrical engineering from Nagpur University, Nagpur, India, in 1999; M.Tech degree in instrumentation and control from Aligarh Muslim University, Aligarh, India, in 2002; and PhD degree from the Indian Institute of Technology Roorkee, Roorkee, India, in 2008. He is currently working as Associate Professor in the Department of Electrical Engineering, Indian Institute of Technology Roorkee, Uttarakhand, India. His fields of interest are wavelets, neural network, optimization techniques, content-based image retrieval, digital instrumentation, digital protective relays and digital speech processing. Dr Tripathy is a reviewer for various international journals in the area of power systems and speech.

Chapter 1
Closed Loop Control of Non-ideal Buck Converter with Type-III Compensator

Abhishek Kumar, Durgesh Chandra Nautiyal, and Prakash Dwivedi

Introduction

In the electrical and electronics industry, almost every product uses several components that are designed to operate under a constant voltage source. Buck converter supplies step-down voltage from its input to output and a class of switch mode power supply (SMPS) with low weight and maximum efficiency. Buck converter has nonlinear operation [1–3] due to its switching behaviour but at the same time, it is linear during its particular switching state as ON and OFF.

Therefore, by employing the averaging method, it is possible to exchange a nonlinear system with a linear one. Now, due to the variation in the supply and load, the transient and steady-state response of open loop buck converter is not coming satisfactory as many applications require the steady-state error below 5 percent. In order to obtain stable and fast transient response with small steady-state error for buck converter, type-III compensator [4–7] is used here. A type-III compensator can boost the phase theoretically up to 180° but practically limited to 160°. In this paper, to obtain stable and fast transient and small steady-state error for buck converter, the design and closed loop operation of type-III compensator will be performed. The presented compensator takes good care against the system non-line arties and load variations to make the overall system "a source of regulated power supply".

A. Kumar (✉) · D. C. Nautiyal · P. Dwivedi
National Institute of Technology, Jamshedpur, India
e-mail: abhishek100895@gmail.com

D. C. Nautiyal
e-mail: durgeshdc@gmail.com

P. Dwivedi
e-mail: prakashdwivedi001@gmail.com

© The Editor(s) (if applicable) and The Author(s), under exclusive license to Springer Nature Singapore Pte Ltd. 2021
A. K. Singh and M. Tripathy (eds.), *Control Applications in Modern Power System*, Lecture Notes in Electrical Engineering 710,
https://doi.org/10.1007/978-981-15-8815-0_1

The outline of the paper is described as in Sect. 1.2, the principle and operation along with the mathematical model by using state space approach has been discussed and small-signal analysis of buck converter has been done to derive its small-signal transfer function. In Sect. 1.3, brief introduction to the generalized type-III controller and design consideration for the elements used in the compensator has been presented. The computer simulation results are presented with complete discussion in Sect. 1.5, and finally, Section 1.6 concludes this paper.

Buck Converter: Principle and Operation

Buck converter is also known as step-down chopper (DC-DC converter) in which output voltage needs to be lower than the input voltage. It is one of the basic switch mode power supply (SMPS). The input of the buck converter must be DC derived from any DC supplies or rectified AC. It is popularly used where electrical isolation is not needed between switching circuits and output. Switch (SW) used here is IGBT and MOSFET, and for higher rating applications, we can use GTO and thyristor also. In order to reduce the harmonics and improve the quality of the output waveform, we must use filters.

In the above circuit, i.e. Figure 1.1, inductor and capacitor are used as filter circuit. By introducing the filter circuit, the non-ideality r_l and ESR of L and C, respectively, will come into the account and their effect in the performances of the buck converter is considered in this paper.

There are two operational mode: (i) continuous conduction mode (CCM) in which current through inductor remains always positive and (ii) discontinuous conduction mode (DCM) in which current through inductor returns to zero in each period. As per our application, we are considering the CCM mode. Now when switch SW is ON and diode D_1 is OFF, the current through inductor starts rising linearly and capacitor starts charging. And when switch SW is OFF and diode D_1 is ON, the current through inductor starts decreasing linearly in the same direction as the inductor current

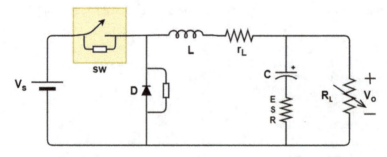

Fig. 1.1 Non-ideal buck converter

cannot change instantaneously. However, the voltage polarity across the inductor L is reversed to maintain the current in the same given direction.

Mathematical Model [8]

Case I When switch SW is ON, diode D remains in OFF condition as reverse voltage is directly applied by the source. The current through inductor starts rising linearly and capacitor starts charging. The equivalent circuit during ON time is given in Fig. 1.2. Now, by applying KVL, KCL in the circuit shown in Fig. 1.2 and further rearranging, we get Eqs. (1.1), (1.2) and ((1.3)).

$$\frac{di_L}{dt} = -i_L\left(r_L + \frac{R_L r_C}{R_L + r_C}\right)\frac{1}{L} - \frac{V_c}{L} + \frac{V_s}{L} \quad (1.1)$$

$$\frac{dV_c}{dt} = \left(\frac{i_L R_L}{R_L + r_C}\right)\frac{1}{C} - \frac{V_c}{R_L C} \quad (1.2)$$

$$V_o = i_L\left(\frac{R_L r_C}{R_L + r_C}\right) + V_c \quad (1.3)$$

By using Eqs. (1.1)–(1.3), the state space model for the case I operation is given in Eqs. (1.4) and (1.5).

$$\begin{bmatrix} \dot{i}_L \\ \dot{V}_c \end{bmatrix} = \begin{bmatrix} -\left(r_L + \frac{R_L r_C}{R_L + r_C}\right)\frac{1}{L} & -\frac{1}{L} \\ \left(\frac{R_L}{R_L + r_C}\right)\frac{1}{C} & -\frac{1}{R_L C} \end{bmatrix}\begin{bmatrix} i_L \\ V_c \end{bmatrix} + \begin{bmatrix} \frac{1}{L} \\ 0 \end{bmatrix}V_s \quad (1.4)$$

$$V_o = \left[\left(\frac{R_L r_C}{R_L + r_C}\right) r_C 1\right]\begin{bmatrix} i_L \\ V_c \end{bmatrix} \quad (1.5)$$

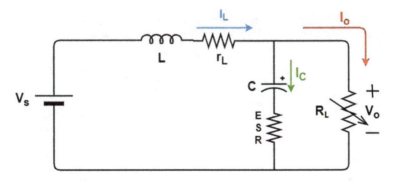

Fig. 1.2 Equivalent circuit when SW is ON

The system, input and output matrix identified from the Eqs. (1.4) and (1.5) is given in Eqs. (1.6), (1.7) and (1.8), respectively.

$$A_1 = \begin{bmatrix} -\left(r_L + \frac{R_L r_C}{R_L + r_C}\right)\frac{1}{L} & -\frac{1}{L} \\ \left(\frac{R_L}{R_L + r_C}\right)\frac{1}{C} & -\frac{1}{R_L C} \end{bmatrix} \tag{1.6}$$

$$B_1 = \begin{bmatrix} \frac{1}{L} \\ 0 \end{bmatrix} \tag{1.7}$$

$$C_1 = \left[\left(\frac{R_L r_C}{R_L + r_C}\right) r_C 1\right] \tag{1.8}$$

Case II When switch SW is OFF and diode D is ON, the current through inductor starts decreasing linearly in the same direction as the inductor current cannot change instantaneously. However, the voltage polarity of the inductor (L) is reversed to maintain the current in the same given direction. The equivalent circuit during OFF time is given in Fig. 1.3. Now, by applying KVL, KCL in the circuit shown in Fig. 1.3 and further rearranging, we get Eqs. (1.9), (1.10) and (1.11)

$$\frac{di_L}{dt} = -i_L\left(r_L + \frac{R_L r_C}{R_L + r_C}\right)\frac{1}{L} - \frac{V_c}{L} \tag{1.9}$$

$$\frac{dV_c}{dt} = \left(\frac{i_L R_L}{R_L + r_C}\right)\frac{1}{C} - \frac{V_c}{R_L C} \tag{1.10}$$

$$V_o = i_L\left(\frac{R_L r_C}{R_L + r_C}\right) + V_c \tag{1.11}$$

By using Eqs. (1.9)–(1.11), the state space model for the case II operation is given in Eqs. (1.12) and (1.13).

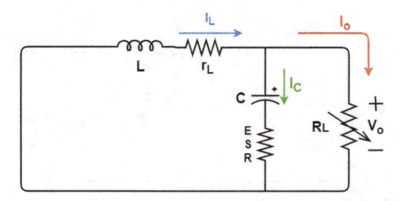

Fig. 1.3 Equivalent circuit when SW is OFF

$$\begin{bmatrix} \dot{i}_L \\ \dot{V}_c \end{bmatrix} = \begin{bmatrix} -\left(r_L + \frac{R_L r_C}{R_L + r_C}\right)\frac{1}{L} & -\frac{1}{L} \\ \left(\frac{R_L}{R_L + r_C}\right)\frac{1}{C} & -\frac{V_c}{R_L C} \end{bmatrix} \begin{bmatrix} i_L \\ V_c \end{bmatrix} \quad (1.12)$$

$$V_o = \begin{bmatrix} \left(\frac{R_L}{R_L + r_C}\right) r_C 1 \end{bmatrix} \begin{bmatrix} i_L \\ V_c \end{bmatrix} \quad (1.13)$$

The system, input and output matrix is identified from Eqs. (1.12) and (1.13) is given in Eqs. (1.14), (1.15) and (1.16), respectively.

$$A_2 = \begin{bmatrix} -\left(r_L + \frac{R_L r_C}{R_L + r_C}\right)\frac{1}{L} & -\frac{1}{L} \\ \left(\frac{R_L}{R_L + r_C}\right)\frac{1}{C} & -\frac{1}{R_L C} \end{bmatrix} \quad (1.14)$$

$$B_2 = \begin{bmatrix} 0 \\ 0 \end{bmatrix} \quad (1.15)$$

$$C_2 = \begin{bmatrix} \left(\frac{R_L r_C}{R_L + r_C}\right) r_C 1 \end{bmatrix} \quad (1.16)$$

Averaging and Small-Signal (Small AC Disturbance) Analysis

To find resultant state space model of the converter using system, input and output matrices obtained in both the cases, i.e. from Eqs. (1.6) to (1.8) and from Eqs. (1.14) to (1.16), averaging is required. So this section cope with the averaging and small-signal analysis of the converter. The average system, input and output matrices are given in Eqs. (1.17), (1.18) and (1.19), respectively.

$$A = A_1 D + A_2(1 - D) \quad (1.17)$$

$$B = (A_1 - A_2)X + (B_1 - B_2)V_s \quad (1.18)$$

$$C = C_1 D + C_2(1 - D) \quad (1.19)$$

The final averaged state space equation of the given converter is given in Eqs. (1.20) and (1.21).

$$\dot{X} = AX + BV_s \quad (1.20)$$

$$V_o = CX \quad (1.21)$$

The averaged system, input and output matrices obtained in Eqs. (1.20) and (1.21) are concerned with steady-state response. The disturbances, i.e. variations present around the steady-state value, are not considered in the Eq. (1.17), (1.18) and (1.19). The small-signal (small ac disturbance) analysis is a way to include the variations in the state space modelling. On adding variations as $d = D+\hat{d}$, $x = X+\hat{x}$, $v_s = V_s+\hat{v}_s$ and rearranging Eq. (1.20) and re-written as Eq. (1.22).

$$\hat{x} = A\hat{x} + B\hat{v}_s + [(A_1 - A_2)X + (B_1 - B_2)V_s]\hat{d} \tag{1.22}$$

After substituting the values of A_1, B_1, C_1, A_2, B_2, C_2 from Eqs. (1.6) to (1.8), (1.14), to (1.16) respectively and putting $\hat{v}_s = 0$, the Eq. (1.22) is reduced to Eq. (1.23)

$$\hat{x} = A\hat{x} + BV_s\hat{d} \tag{1.23}$$

After applying Laplace transform in Eq. (1.23), it becomes Eq. (1.24)

$$\hat{x}(s) = (sI - A)^{-1}BV_s\hat{d}(s) \tag{1.24}$$

Similarly adding deviations in D and X and after applying Laplace transform in Eq. (1.21), results into Eq. (1.25).

$$\hat{v}_o(s) = C\hat{x}(s) \tag{1.25}$$

On putting the value of $\hat{x}(s)\hat{x}(s)$ from Eq. (1.24) in Eq. (1.25) yields Eq. (1.26)

$$\hat{v}_o(s) = C(sI - A)^{-1}BV_s\hat{d}(s) \tag{1.26}$$

So the resultant transfer function of output voltage with respect to duty cycle perturbation is shown in Eq. (1.27)

$$\frac{\hat{v}_o(s)}{\hat{d}(s)} = C(sI - A)^{-1}BV_s \tag{1.27}$$

Substituting the values of average system, input and output matrices from Eqs. (1.17) to (1.19) in Eq. (1.27) yields Eq. (1.28)

$$\frac{\hat{v}_o(s)}{\hat{d}(s)} = \frac{\left(s + \frac{1}{R_LC}\right)\left(\frac{R_Lr_C}{R_L+r_C}\right)\frac{V_s}{L} + \left(\frac{R_L}{R_L+r_C}\right)\frac{V_s}{LC}}{s^2 + s\left(\left(r_L + \frac{R_Lr_C}{R_L+r_C}\right)\frac{1}{L} + \frac{1}{R_LC}\right) + \left(\frac{r_L+R_L}{R_LLC}\right)} \tag{1.28}$$

Type-III Compensator

An open loop non-ideal buck converter cannot maintain its output voltage constant due to variation in supply voltage and load. This problem can be resolved by using compensator that is used to produce constant output voltage [9, 10]. In this paper, we present type-III analog compensator (refer Fig. 1.4) for non-ideal buck converter.

The small-signal transfer function of type-III compensator is expressed in Eq. (1.29) as:

$$G(j\omega) = -\frac{(j\omega + \omega_{z_1})(j\omega + \omega_{z_2})}{R_3 C_2 j\omega (j\omega + \omega_{p_2})(j\omega + \omega_{p_3})} \tag{1.29}$$

The zeros and poles of the above transfer function are given in Eqs. (1.30)–(1.34).

$$\omega_{z_1} = \frac{1}{R_2 C_2} \tag{1.30}$$

$$\omega_{z_2} = \frac{1}{R_1 C_3} \tag{1.31}$$

$$\omega_{p_1} = 0 \tag{1.32}$$

$$\omega_{p_2} = \frac{1}{R_2 C_2} \tag{1.33}$$

$$\omega_{p_3} = \frac{1}{R_3 C_3} \tag{1.34}$$

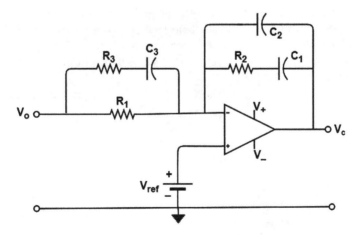

Fig. 1.4 Circuit layout of type-III compensator

The phase offered by the compensator is given in Eq. (1.35)

$$\theta_{comp} = -270^0 + \tan^{-1}\frac{\omega}{\omega_{z_1}} + \tan^{-1}\frac{\omega}{\omega_{z_2}}$$
$$- \tan^{-1}\frac{\omega}{\omega_{p_2}} - \tan^{-1}\frac{\omega}{\omega_{p_3}} \quad (1.35)$$

($-180°$ is from the negative sign and $-90°$ is from the pole at the origin.)

Controller Parameters Design

The K factor method can be selected for determining the values of component associated with the type-III compensator. Also, the corresponding value of cross over frequency (f_{co}) should be considered first. Generally, the cross over frequency can be taken of the range of 1/5–1/10 of switching frequency to avoid the switching noises.

The K factor for type-III compensator is given in Eq. (1.36)

$$K = \left[\tan\left(\frac{\theta_{comp} + 90}{4}\right)\right]^2 \quad (36)$$

where θ_{comp} is that phase which compensator has to be provided and is given in Eq. (1.37).

$$\theta_{comp} = \theta_{desired} - \theta_{converter} \quad (37)$$

By using K factor method for type-III compensator, we can theoretically boost the phase upto 0°–180° but practically upto 160° only. It is because, if 180° is used to boost up the phase, the K factor will become undefined. The components of the compensator can be calculated by setting R_1 it self and by following equations given from Eq. (1.38) to (1.42):

$$R_2 = \frac{|G(j\omega_{co})|R_1}{\sqrt{K}} \quad (1.38)$$

$$C_1 = \frac{\sqrt{K}}{\omega_{co} R_2} \quad (1.39)$$

$$C_2 = \frac{1}{\omega_{co} R_2 \sqrt{K}} \quad (1.40)$$

$$C_3 = \frac{\sqrt{K}}{\omega_{co} R_1} \quad (1.41)$$

1 Closed Loop Control of Non-ideal Buck Converter …

$$R_3 = \frac{1}{\omega_{co}\sqrt{K}C_3} \quad (1.42)$$

Simulation Work

The specifications of the buck converter are tabulated in Table 1.1.

The specifications of the buck converter are tabulated in Table 1.1. Now, after putting the specification values in Eq. (1.28), the transfer function of converter is given in Eq. (1.43):

$$\text{TF} = \frac{1.4545 * 10^4 s + 10^9}{s^2 + 0.3939 * 10^4 s + 0.5151 * 10^8} \quad (1.43)$$

The above Fig. 1.5 shows the bode plot of the open loop buck converter. We have chosen the crossover frequency as $f_{co} = 10\text{KHz}$. So, the magnitude and angle of converter at cross over frequency are -9.19 dB and $-133°$, respectively (refer Fig. 1.5).

The PWM gain is calculated and given in Eq. (1.44)

$$\text{PWM (gain)} = 20\log\frac{1}{3} = -9.54\,\text{bB} \quad (1.44)$$

Therefore, the total gain has to be compensated by the compensator at f_{co} is given in Eq. (1.45)

$$\text{Total Gain} = -9.19 - 9.54 = -18.7\,\text{dB} \quad (1.45)$$

Table 1.1 Specifications of the buck converter

Description	Parameter	Nominal value
Input voltage	V_s	20 V
Capacitance	C	500 μF
ESR	r_c	30 mΩ
Inductance	L	40 μH
Inductor resistance	r_l	0.1 Ω
Switching frequency	f_s	1 MHz
Load resistance	R_L	3 Ω
Peak gain of PWM	V_P	3 V
Desired phase	$\theta_{desired}$	45°
Desired output voltage	V_{od}	10 V

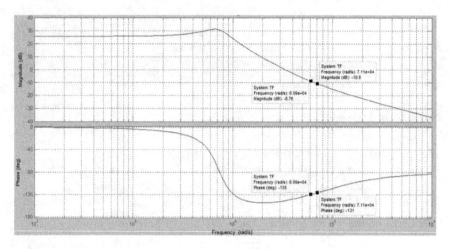

Fig. 1.5 Open loop bode plot of non-ideal buck converter

Now, by using Eq. (1.37), the phase required to compensate by the compensator is given in Eq. (1.46), where $\theta_{\text{desired}} = 45°$ and $\theta_{\text{converter}} = 133°$.

$$\theta_{\text{comp}} = 45 - 133 = 178° \tag{1.46}$$

Putting the value of θ_{comp} from Eq. (1.46) to Eq. (1.36), we get the the value of K as and given in Eq. (1.47)

$$K = 5.55 \tag{1.47}$$

By using the above value of K and by setting $R_1 = 1\,\text{k}\Omega$, the other parameters of the compensator can be obtained from Eqs. (1.38)–(1.42):

$$R_2 = 5.55\,\text{k}\Omega,\ C_1 = 10.3\,\text{nF},\ C_2 = 1.85\,\text{nF},$$
$$C_3 = 37.5\,\text{nF and } R_3 = 180\,\Omega$$

By putting above calculated values in Eq. (1.29), we get the small-signal transfer function of type-III compensator in Eq. (1.48)

$$G(j\omega) = -\frac{1.78 * 10^5 s^2 + 7.01 * 10^9 s + 6.91 * 10^{15}}{s^3 + 2.6 * 10^5 s^2 + 1.7 * 10^{10} s} \tag{1.48}$$

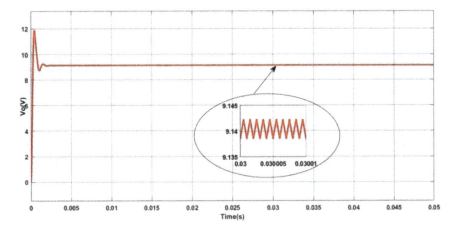

Fig. 1.6 Output voltage of open loop configuration

Simulation Result

In this section, the time response analysis of the buck converter is done. The comparison between the open loop response and closed loop response has been shown. Also the effect of load and ESR(s) variations has been presented through the MATLAB simulation results.

Comparison: Time Response Analysis

The output voltage response for the open loop configuration is shown in Fig. 1.6, in which the output voltage is settled with an average value of 9.14 V but the required or desired output voltage is 10 V.

Therefore, there is a significant error of 8.6% in the output. Whereas, the output voltage response of the closed loop configuration is settled at desired value of 10 V with settling time of 4 ms and with steady-state error of only 0.3% which is significantly low (Fig. 1.7).

Load Variation

Figure 1.8 shows the variation of output voltage with respect to the variation in load. At $t = 0.025$ s, the load changes from 3 to 6 Ω and the output voltage settled within 0.004 s at 10 V with 0.3% of ripple, which makes the closed loop configuration robust and a source of regulated power supply

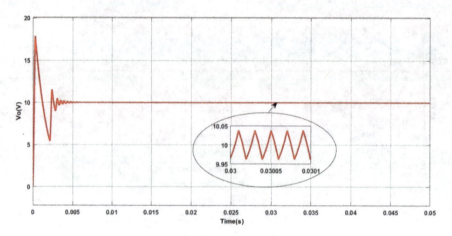

Fig. 1.7 Output voltage of closed loop configuration

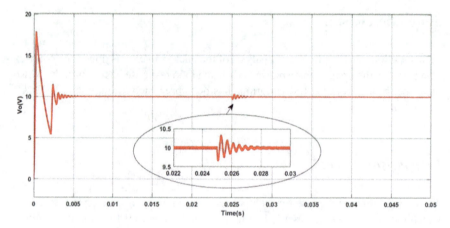

Fig. 1.8 Output voltage variation with respect to load changes

Conclusion

It can be seen from the simulation results that the type-III compensator can effectively take care the non-ideality present in buck converter. The type-III compensator is being used with non-ideal buck converter to obtain a stable (phase margin of closed loop configuration comes out to be 52°) with significantly small steady-state error than open loop configuration. Also, the presented converter can take good care against the load variation which makes it robust controller. And as the output voltage is less affected by the load variation makes the overall system "a source of regulated power supply".

References

1. Mohan N, Undeland TM, Robbins WP (2003) Power electronics: converters, applications, and design, 3rd edn. Wiley
2. Kazimierczuk MK (2015) Pulse-width modulated DC-DC power converters, 2nd edn. Wiley
3. Chan KF, Lam CS, Zeng WL, Zheng WM, Sin SW, Wong MC (2015, Nov)Generalized Type III controller design interface for DC-DC converters. In: TENCON 2015–2015 IEEE region 10 conference. IEEE, pp 1–6
4. Hwu KI, Shieh JJ, Jiang WZ (2018, May) Analysis and design of type 3 compensator for the buck converter based on PSIM.In: 2018 13th IEEE conference on industrial electronics and applications (ICIEA). IEEE, pp 1010–1015
5. Salimi M, Siami S (2015, Nov)Closed-loop control of DC-DC buck converters based on exact feedback linearization. In: 2015 4th international conference on electric power and energy conversion systems (EPECS). IEEE, pp 1–4
6. Salimi M, Soltani J, Markadeh GA, Abjadi NR (2013) In-direct output voltage regulation of DCDC buck/boost converter operating in continuous and discontinuous conduction modes using adaptive backstepping approach.IET Power Electron 6(4): 732–741
7. Rim C, Joung GB, Cho GH (1988) A state space modelling of non-ideal DC-DC converters. In: PESC 88 Record 19th annual IEEE power electronics specialists conference. IEEE, pp 943–950
8. Garg MM, Pathak MK (2018) Performance comparison of non-ideal and ideal models of DC-DC buck converter.In: 2018 8th IEEE India international conference on power electronics (IICPE). IEEE, pp 1–6
9. Peretz MM, Yaakov SB (2011) Time-domain design of digital compensators for PWM DC-DC converters. IEEE Trans Power Electron 27(1):284–293
10. Wu PY, Tsui SYS, Mok PKT (2010) Area-and power-efficient monolithic buck converters with pseudo-type III compensation. IEEE J Solid-State Circuits 45(8):1446–1455

Chapter 2
Vector Error Correction Model for Distribution Dynamic State Estimation

C. M. Thasnimol and R. Rajathy

Introduction

From the period 1960s, state estimation is being widely applied by transmission system operators for the foolproof operation of transmission systems. One-way power flow and passive nature of the distribution system are the major obstacles for the popularity of state estimation in the distribution systems. State estimation is highly essential for the execution of distribution automation applications because of the increased proliferation of distributed energy resources (DER) and demand response programs. Due to the absence of a real-time monitoring system in the conventional distribution system, the distribution network cannot be automated. Accurate state estimation demands complete observability of the system, which is now possible with the advent of micro-phasor measurement units (Micro-PMU).

Most of the researches concentrate on static state estimation of the distribution system. Major problems with static state estimation are measurement availability and measurement intervals. It is highly essential to have corrective and preventive control actions and regular watch of the system states for a proper and secure distribution system operation. The state estimated by the static state estimator actually represents the past state of the system. But this will not be an issue in the traditional distribution system where the states will not change considerably during the time frame of state estimation. At present, there is a high degree of intervention of distributed energy resources (DER) in the distribution network. This will result in a change in sub-second levels of system states. Electric vehicles and microgrids are also causing

---

C. M. Thasnimol (✉) · R. Rajathy
Pondicherry Engineering College, Puducherry, India
e-mail: cmthasni@pec.edu

R. Rajathy
e-mail: rajathy@pec.edu

© The Editor(s) (if applicable) and The Author(s), under exclusive license to Springer Nature Singapore Pte Ltd. 2021
A. K. Singh and M. Tripathy (eds.), *Control Applications in Modern Power System*, Lecture Notes in Electrical Engineering 710,
https://doi.org/10.1007/978-981-15-8815-0_2

uncertainty in the distribution system. All these factors highlight the importance of forecasting capabilities of state estimator.

Time series forecasting techniques can be broadly classified into univariate and multivariate forecasting. The various statistical uni-variate modeling techniques include Auto-Regression (AR), Naive2, Auto-Regression Integrated moving Average (ARIMA), Seasonal ARIMA (SARIMA), Exponential Smoothing, etc. Vector Auto-Regression (VAR) [1], Vector Auto-Regression Moving Average (VARMA), and Vector Error Correction Model (VECM) are examples of multivariate time series analysis problems. A VAR(1) model is proposed in [2] to capture the temporal and spatial correlation among multiple nodal voltage time series. VAR model is determined based on the characteristics of the data, and there is no need for theoretical information about the relationships that exist between the variables. The only assumption that has to be taken while modeling the VAR model is the stationarity of the variables. If the variables are not stationary, it should be converted into stationary by taking the difference. VAR model can be set up employing the statistical properties of the data. If there exists a co-integrating relation among the time series, the VECM model can be established from the VAR model [3].

This work is motivated by the challenges of growing proliferation of DER which will remarkably increase the risk of the sudden change in grid states within a small time interval to anomalous levels. This chapter is organized as follows. Section 'Introduction' gives a brief introduction about dynamic state estimation and different time series methods employed for power system state estimation. Section 'Problem Formulation' presents the test system and discusses the training data generation employed in this chapter. Formulation of Vector Auto-Regression (VAR) model is discussed in Sect. 'Vector Auto-Regression (VAR)' and state forecasting using the VECM model is presented in Sect. 'VECM Based State Forcasting'. Section 'Conclusion' presents the results of the study and thereafter conclusion.

Problem Formulation

The Distribution Dynamic State Estimation (DDSE) is modeled as a multivariate time series forecasting problem. The objective is to predict y, a stalk of multiple voltage time series.

Test System

Test system considered in this study is the standard IEEE 13 bus system. IEEE 13 bus system is an unbalanced system consisting of 13 buses and 40 nodes. The state matrix of the system consists of 80 variables including 40 variables representing voltage magnitudes and another 40 variables representing voltage angles. For the present study, we have considered only the forecasting of voltage magnitude; hence, total variables to be forecasted at any instant of time are 40 for this system.

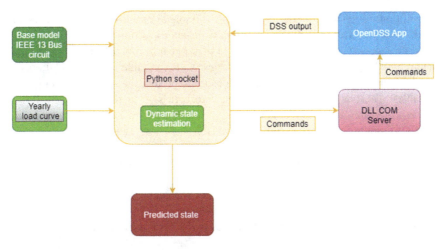

Fig. 2.1 Simulation setup for dynamic state estimation

Training Data Generation

The training data is created through simulation which is done in OpenDSS through a Python COM interface. Simulations are done using real load data taken from the New York Independent System Operator's (ISO New York) database [4]. Real load data archived by ISO New York is fed as the load curve for our test system, and the state variables are estimated for all the nodes. We have used 20 months load demand data starting from 1-1-2018 till 9-10-2019. The first 18 months data are used for training the model, and the last two months data are used for validation purpose. The ISO New York's data contains load data of nine buses reported at an interval of 5 min. A total of 123456 data points are there in the load data of each bus. The training data generated from OpenDSS load flow studies are in the form of time series data with time labels. Figure 2.1 illustrates the dynamic state estimation process using python OpenDSS COM interface.

Vector Auto-Regression (VAR)

Vector Auto-Regression (VAR) is first proposed by Sims in 1980 [5]. VAR is a multivariate time series forecasting method which is suitable when the relationship between time series are bi-directional. It is an autoregressive method similar to AR, ARMA, and ARIMA in the sense that the future values can be forecasted based on the past or time-delayed variables. The difference between the basic autoregressive method is its bi-directional nature. The future values of the series not only depend on its own past values but also on the delayed variables of the other time series. The basic assumption in the VAR model is that each variable at any instant of time depends on its own past values and also on the past instances of other related time series. Therefore, each variable can be expressed as a linear combination of its own past values and the past values of other time series. VAR can understand the relationship between variables and can use this relationship to predict the future values of these variables. The basic representation of an autoregressive model is given in Eq. 1, where a is a constant, b_n is the coefficient of nth delay term Y_{t-n}, and e is the error which is assumed as the Gaussian white noise.

$$Y_t = a + b_1 * Y_{t-1} + b_2 * Y_{t-2} + b_3 * Y_{t-3} + \cdots + b_n * Y_{t-n} + e \qquad (1)$$

If there are multiple time series that are influencing each other, then the current and future values of each time series in the system depends on its own time lags and also on the lagged values of the other time series in the system as shown in Eq. 2

$$\begin{aligned} Y_{1t} &= a_1 + b_{11,1} * Y_{1_{t-1}} + b_{12,1} * Y_{1_{t-1}} + b_{13,1} * Y_{1_{t-1}} \\ Y_{2t} &= a_2 + b_{21,1} * Y_{2_{t-1}} + b_{22,1} * Y_{2_{t-1}} + b_{23,1} * Y_{3_{t-1}} \\ Y_{3t} &= a_3 + b_{31,1} * Y_{4_{t-1}} + b_{32,1} * Y_{4_{t-1}} + b_{33,1} * Y_{4_{t-1}} \end{aligned} \qquad (2)$$

Equation 2 shows the VAR(n) model for a multivariate time series. In general, a standard n variable VAR model with lag p can be represented as in Eq. 3.

$$Y_t = a + \sum_{i=1}^{p} b_i * Y_{t-i} \qquad (3)$$

where Y_t is the vector of multivariate time series of order $k \times 1$, a is the vector of intercepts, and b_i is the vector of coefficient vectors.

Stationarity of the data should be checked before feeding into the model, and if found non-stationary should be converted into stationary by taking the first difference of the data or first difference of the logarithm of the data. For applying the VAR model, the first difference or the first difference of the logarithm of data should be stationary. Further, if the first difference is co-integrating, we can employ Vector Error Correction Model (VECM) which is a modified version of VAR.

Co-integration and Vector Error Correction Model (VECM)

The order of integration $I(d)$ is the minimum number of differencing operations required to make a time series stationary [6]. Co-integration is a statistical property of a set of time series. If the order of integration of the linear combination of a set of time series is less than that of the individual series, then the set of time series are said to be co-integrated. If a set of time series is co-integrated, then they may exhibit a long run statistically stable equilibrium although they are exhibiting dynamical relationship in the short run [7]. If the first difference is found co-integrated, then we can apply Vector Error Correction Model (VECM) which is a modified version of VAR.

VECM [8, 9] is a restricted VAR [10], in which the time series to be forecasted are co-integrated [11]. The authors of [12] compared the forecasts generated from the VECM model with that from the VAR model and found out that VECM is giving better forecast with smaller squared forecast error. The co-integrating relationship between the time series and the co-integration rank is estimated using a VAR model fitted with the data. The estimated co-integrating rank from the VAR model should be given as one of the parameters for establishing the VECM model. The co-integrating rank is equal to the difference between the number of variables or the number of time series in the system and the number of co-integrating vectors [13]. Subtracting Y_{t-1} from both sides of Eq. 2 gives a VECM model for multivariate time series [14] as given in Eq. 4.

$$\Delta Y_t = a + \sum_{i=1}^{p} \Gamma_i Y_{t-i} + \Pi Y_{t-i} \qquad (4)$$

VECM Based State Forcasting

Various steps involved in VECM based state forecasting is shown in Fig. 2.2 and is explained in detail in the following subsections.

Understand the patterns in the time series and time series decomposition The goal of statistical time series analysis is to find out the patterns in the series and to find out the relationship between various time series. This information is used for predicting the outcome over time. Time series can be stationary or non-stationary. The properties of a stationary time series do not depend on the time of origin of

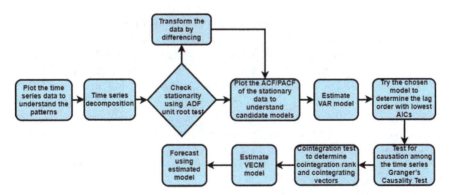

Fig. 2.2 Flow diagram of VECM

the data, whereas a non-stationary time series data is dependant on the time of observation. The time-series data may exhibit seasonal variations and increasing or decreasing trends. A time series with seasonal properties or trends are not stationary as the observed values vary with respect to time. Before giving in to the algorithm for training, it is important to analyze the time-series data to find out the presence of seasonality or trends.

Auto-correlation and cross-correlation   Correlation coefficient gives a measure of similarity between two different time series data or it gives the similarity between two data points of the same series at different times. Auto-correlation plot shows the correlation between the present values of the time series and the past values. Cross-correlation is a measure of the correlation between different time series.

Split the series into training and testing data   The time series should be in chronological order before the splitting operation. In normal series, we can do the random splitting of data, but in time series data, the order of the series is important, and hence, the chronological order should be considered while splitting the series into train and test series.

Check for stationarity   A stationary time series is the one who's mean and variance wil not vary with respect to time. Most of the machine learning methods assume a stationary time series as the input, and hence, it is important to check the stationarity before giving to the model for training. There is a collection of test called unit-root tests for checking the stationarity of time series. The most popular tests are the augmented Dickey–Fuller test (ADF Test), KPSS test, and Philip–Perron test. We have employed the augmented Dickey–Fuller test (ADF Test) for checking the stationarity. If the series is found to be non-stationary, it should be converted into stationary series before fed into the algorithm.

Transform the series to make it stationary   There are two methods for converting a given series into stationary series. The first method is to difference the series once and check for stationarity. If it is still non-stationary, do differencing again and check again. The second method is to take the logarithm of the series. We have employed the second method for making the series stationary. Differencing operation reduces the length of the differenced time series by one. Since all-time series in the system should be having the same length, it is better to do differencing for all the time series in the system if anyone is found to be non-stationary.

Test for causation among the time series—Granger's causality test   An important assumption in multivariate time series analysis is the interdependency between the different time series in the system. Future values of a time series will be influenced by the present and past values of the same series and present and past values of the other time series in the system. Granger's causality test [15, 16] helps to determine the influence of each time series on one another. Granger's causality test assumes a null hypothesis that the different time series in the system does not cause each other. The basic principle of Granger causality analysis is to test whether past values of a variable X (the driving variable) help to explain current values of another variable Y (the response variable) [17].

Estimation of VAR model   The auto-correlation function ACF plot shows how the present values are influenced by past values. Multicollinearity issues arise due to excessive correlation between independent variables. The multicollinearity problem should be removed before doing any auto-regression modeling. We can get the relevant features or an approximate order of AR process using partial autocorrelation function PACF plots as it will remove the features which are already defined by earlier lags. The lag order selection can be done more accurately by employing a VAR model fitted with variables at the level. Four information criteria are available for deciding the lag order. The common procedure to find out the lag order is to examine the different information criteria using different VAR models and choose the one which will give the minimum value. But, out of the four methods for lag order selection, AIC, is recommended for forecasting purpose [18], and it selects the best model which is having the smallest squared forecast error.

Cointegration test   Co-integration test helps to find out the statistically significant connection between the time series in the system. If the linear combination of the time series in the system has an order of integration less than that of individual time series, then system of time series is said to be co-integrated. Order of integration is the number of differencing operations needed to transform a non-stationary time series into a stationary one. Johansen test for co-integration [19, 20] can be used to test co-integration between any number of time series up to 12.

Estimation of VECM model   VECM is a restricted VAR in which the time series to be forecasted are co-integrated. For specifying a VECM model, we need the lag order and the co-integrating rank of the system [21]. The co-integrating relationship between the time series and the co-integration rank is estimated using a VAR model fitted with the data. The co-integrating rank is equal to the difference between the number of variables or the number of time series in the system and the number of co-integrating vectors.

Roll back the transformations   After training the model using the transformed train dataset, the model should be validated using the test dataset. Since the forecasted output of the test data will be on the scale of the differenced data, it should be converted back into the original level. This is done by taking de-differencing and anti-log transformation of the data.

Results and Discussion

Time series patterns and time series decomposition   Figure 2.3 shows the patterns in the voltage data of four nodes in the system. Figure 2.4 shows the trend, seasonality, and residual component of the time series data. It is clear from the graph that our time series data contain a remarkable seasonal component. Also it does not show any increasing or decreasing trend. For making forcast, the first 18 months state vectors are used for training purpose, and the remaining two months data are used for validation.

Augmented Dickey–Fuller test (ADF Test)   We have employed the augmented Dickey–Fuller test (ADF Test) for checking the stationarity. For ADF test,

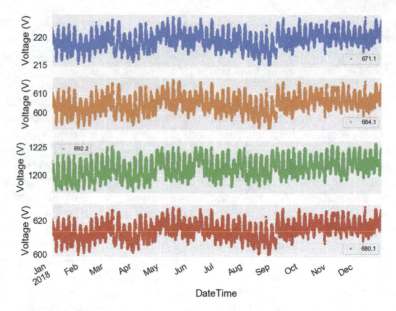

Fig. 2.3 Time series pattern

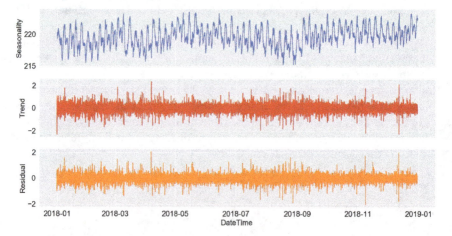

Fig. 2.4 Time series decomposition

significance level is assumed as 0.05; 1% critical value is taken as -3.431; for 5%, it is -2.862; and for 10%, it is -2.567. Result of ADF test on bus 684.3 is given below. Series is not stationary. Therefore, we have to convert it into a stationary series before fed into the algorithm. ADF test on bus 684.3 before and after log transformation and first differencing is given in Table 2.1. The series is non-stationary before applying transformation and differencing operation, but became stationary after doing differencing of the log-transformed data. Differencing

Table 2.1 Augmented Dickey–Fuller test

Node	Before log transformation and differencing				After log transformation and differencing			
	Test statistic	No. lags chosen	P-value	Inference	Test statistic	No. lags chosen	P-value	Inference
650.2	−4.0829	36	0.001	Stationary	−5.7765	35	0	Stationary
633.1	−2.0537	24	0.2635	Non-stationary	−23.2294	23	0	Stationary
634.1	−2.4531	36	0.1273	Non-stationary	−12.4413	35	0	Stationary
671.1	−2.2102	36	0.2026	Non-stationary	−16.3226	35	0	Stationary
645.2	−3.3651	36	0.0122	Stationary	−22.4445	35	0	Stationary
646.2	−3.2234	36	0.0187	Stationary	−16.3226	35	0	Stationary
692.1	−3.0618	36	0.0295.	Stationary	−22.4445	35	0	Stationary
675.1	−2.1971	36	0.2073	Non-stationary	−16.3037	35	0	Stationary
611.3	−2.0534	18	0.2637	Non-stationary	−30.579	17	0	Stationary
632.1	−2.5298	30	0.1084	Non-stationary	−23.6629	29	0	Stationary
680.1	−3.1012	24	0.0265	Stationary	−15.9049	35	0	Stationary
684.1	−2.1888	36	0.2104	Non-stationary	−16.3344	35	0	Stationary

operation reduces the length of the differenced time series by one. Since all-time series in the system should be having the same length, it is better to do differencing for all the time series in the system if anyone is found to be non-stationary.

Granger's causality matrix Figure 2.5 shows the plot of Granger's casuality matrix. The value given in each cell is the p-value or the casuality between the corresponding variables. If a given p-value is < significance level (0.05), then the corresponding driving variable (x-axis) causes the response variable (y-axis). It is clear from Fig. 2.5 that the voltages in each bus are greatly influenced by the other bus voltages. Therefore, we can reject the null hypothesis assumed by Granger's causality test, which implies a strong interdependency between different time series in our system.

Estimation of VAR Model

Auto-correlation function (ACF) Figure 2.6a shows the ACF plot of the node '650.1'. The ACF plot shows how the present values are influenced by past values. The light blue shaded area represents the confidence interval. From the ACF plot, it is clear that the node '650.1' has very good positive correlation up to a lag of 280, the point where it cuts the upper confidence interval. But we cannot use all

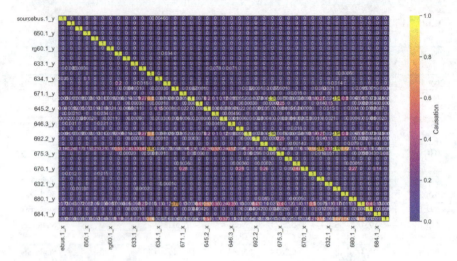

Fig. 2.5 Granger's casuality matrix

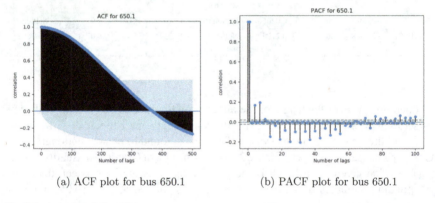

(a) ACF plot for bus 650.1   (b) PACF plot for bus 650.1

Fig. 2.6 Auto-correlation plots

of these 280 lags for modeling the AR process, as it will create multicollinearity issues.

Partial auto-correlation function (PACF)   We can get the relevant features or the order of the AR process using PACF plots as it will remove the features which are already defined by earlier lags. Figure 2.6b shows the PACF plot for the node '650.1'. In the PACF plot, the upper and lower dotted lines represent confidence boundaries based on 95% confidence interval. From the plot, it is clear that the optimum order or the 'p' value of our AR process is '5'. This is the point where the PACF plot cuts the upper confidence interval. We can model our time series forecasting problem as an AR process having an order of 5, which means that the future values of the series or the target value can be expressed as a linear combination of the past five values in the series.

Table 2.2 Lag order selection

Lag order	AIC	BIC	FPE	HQIC
1	−623.65110	−622.2817	1.41829e−271	−623.1849
2	−634.0839	−631.3776	4.1767e−276	−633.1626
3	−635.694	−631.65085	8.3448e−277	−634.31795
4	−639.66143	−634.28025	1.579945e−278	−637.8295
5	−628.7662	−622.04723	8.5202e−274	−626.4788
6	−648.0617	−640.0047	3.553e−282	−645.3188
7	−657.998	−648.6032	1.7189e−286	−654.800
8	−647.7825	−637.0486	4.70130e−282	−644.1283
9	−650.219	−638.1464	4.11383e−283	−646.1092

Choosing the lag order   Table 2.2 shows the results of the lag order selection using four different methods. From the table, it is clear that 7th lag, all the four criteria are giving the lowest value. Therefore, the optimum lag order can be chosen as 7.

Cointegration test   Since our problem contains more than 12-time series, Johansen test is first conducted between the first 12 time series, and the process is repeated to get the co-integration between the other variables also. It is found out that there are 40 co-integrating vectors in the system. The rank of co-integration found from Johansen test for co-integration is 1.

Estimation of VECM model   In our case, we have 41 nodes in our system; hence, there are 41 voltage-time series, and the number of co-integrating vectors is found to be 40 from Johansen co-integration test. Therefore, the co-integrating rank of our model is 1. Figure 2.7 shows the error plot between original and VECM forecasted data. Table 2.3 shows the performance evaluation of VECM.

Conclusion

Dynamic state estimation of the distribution system has been modeled as a multivariate time series problem. Training data for state forecasting is generated through simulations done in Python OpenDSS COM interface. VAR model is estimated first, and various statistical analyses have been done on the fitted VAR model to identify the underlining patterns and the characteristics of the voltage-time series data. Finally, vector error correction model is employed for multivariate voltage forecasting problem. It is found that the performance of VECM is satisfactory with respect to all the performance factors considered.

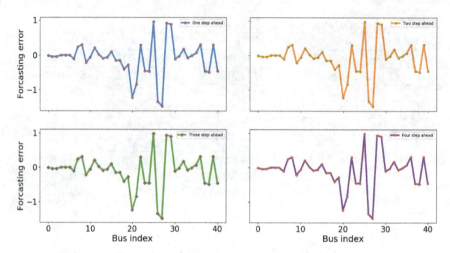

Fig. 2.7 Forcasting error

Table 2.3 Performance evaluation of VECM

Method	Mean absolute error	Mean squared error	Root mean squared error
VECM	0.06089116	0.0037082187	0.06089514

References

1. Zivot E, Wang J (2006) Vector autoregressive models for multivariate time series. In: Modeling financial time series with S-plus®, pp 385–429
2. Hassanzadeh M, Evrenosoğlu CY, Mili L (2015) A short-term nodal voltage phasor forecasting method using temporal and spatial correlation. IEEE Trans Power Syst 31(5):3881–3890
3. Attari MIJ, Taha R, Farooq MI (2014) Tax revenue, stock market and economic growth of Pakistan. Acta Universitatis Danubius, Œconomica 10(5)
4. NIS OPERATOR, Newyork independant system operator, real time load data, 24
5. Sims CA (1980) Macroeconomics and reality. Econometrica J Econ Soc 1–48
6. Tkacz G (2001) Estimating the fractional order of integration of interest rates using a wavelet ols estimator. Stud Nonlinear Dyn Econ 5(1)
7. Hylleberg S, Engle RF, Granger CW, Yoo BS (1990) Seasonal integration and cointegration. J Econ 44(1–2):215–238
8. Engle RF, Granger CW (1987) Co-integration and error correction: representation, estimation, and testing. Econometrica J Econ Soc 251–276
9. Hoffman DL, Rasche RH (2002) A vector error-correction forecasting model of the US economy. J Macroecon Citeseer
10. Purna FP, Mulyo PP, Bima MRA (2016) Exchange rate fluctuation in indonesia: vector error correction model approach. Jurnal Ekonomi & Studi Pembangunan 17(2):143–156
11. Maysami RC, Koh TS (2000) A vector error correction model of the singapore stock market. Int Rev Econ Fin 9(1):79–96
12. Engle RF, Yoo BS (1987) Forecasting and testing in co-integrated systems. J Econ 35(1):143–159

13. Zou X (2018) VECM model analysis of carbon emissions, GDP, and international crude oil prices. Discrete Dyn Nat Soc
14. Kazanas T (2017) A vector error correction forecasting model of the Greek economy. Hellenic Fiscal Council 2:2–6
15. Engsted T, Johansen S (1997) Granger's representation theorem and multicointegration
16. Johansen S (1995) Likelihood-based inference in cointegrated vector autoregressive models. Oxford University Press on Demand
17. Papana A, Kyrtsou C, Kugiumtzis D, Diks C et al (2014) Identifying causal relationships in case of non-stationary time series. Department of Economics of the University of Macedonia, Thessaloniki
18. Athanasopoulos G, de Carvalho Guillén OT, Issler JV, Vahid F (2011) Model selection, estimation and forecasting in var models with short-run and long-run restrictions. J Econ 164(1):116–129
19. Johansen S (1998) Likelihood-based inference in cointegrated vector autoregressive models. Econ Theory 14(4):517–524
20. Hansen PR, Johansen S (1998) Workbook on cointegration. Oxford University Press on Demand
21. Liang C, Schienle M (2019) Determination of vector error correction models in high dimensions. J Econ 208(2):418–441

Chapter 3
Optimal Battery Charging Forecasting Algorithms for Domestic Applications and Electric Vehicles by Comprehending Sustainable Energy

S. Parvathy, Nita R. Patne, and T. Safni Usman

Introduction

Plug-in hybrid electric vehicles are driving popularity owing to various grounds. They are commodious, silent, visually appealing and produce less air pollution in the environment. PHEVs have the inherent capacity to cut down the fossil energy intake resulting in reduced greenhouse gas emissions [1]. PHEVs enhance the coupling of sustainable energy sources like wind energy and solar energy [2].

The national grid in its present position has not yet completely inclined to the large degree of PHEV penetration [3]. There exist quite a number of problems in the charging process of PHEVs, where there is an inconsistency related to the number of batteries and the energy demand put forward by each battery which in turn leads to a predicament in DSM [4]. Uncoordinated charging process may lead to various economic and technical problems such as extensive voltage fluctuations, depraved system efficiency and increased probability of blackouts due to network overloads [5].

Researchers have developed algorithms for the charging of multiple PHEVs in the smart grid based on minimization of total cost of generation and grid energy losses, though the inclusion of renewable energy resources was not considered [6–8]. Moreover, the charging priority was assigned based on the payment made by

---

S. Parvathy · T. Safni Usman (✉)
Robert Bosch GmbH, Bangalore, India
e-mail: safniusman@gmail.com

S. Parvathy
e-mail: parvathy2202sobha@gmail.com

N. R. Patne
VNIT, Nagpur, India
e-mail: nrpatne.eee@vnit.ac.in

© The Editor(s) (if applicable) and The Author(s), under exclusive license to Springer Nature Singapore Pte Ltd. 2021
A. K. Singh and M. Tripathy (eds.), *Control Applications in Modern Power System*, Lecture Notes in Electrical Engineering 710,
https://doi.org/10.1007/978-981-15-8815-0_3

the owner of the PHEV rather than considering the SOC of the battery or the time duration for which the PHEV will be available for charging [9–11].

Also, the existing literature gives less importance to the idea of DSM. The major advantage that DSM put forwards is cost effectiveness and will have a reasonable impact on the load rather than installing storage devices or constructing a power plant [12]. DSM handles the customer demand with almost efficiency. In the proposed research, an attempt has been made to improve the efficiency of battery charging algorithm by employing DSM techniques and day ahead pricing [13, 14]. The algorithms developed in this paper consider load shifting technique which shifts the load from peak to off-peak time of day.

The paper is arranged into two main sections. First section discusses an algorithm for the day ahead scheduling of charging time of a residential inverter battery in the house hold, where grid is the energy supplier. Second section discusses a real-time algorithm for a parking garage charging system for PHEVs at a work place where solar energy is the main energy supplier along with grid.

A. Residential inverter battery charging schedule algorithm

Methodology

The proposed algorithm schedules the charging slots for the charging of household inverter battery depending on day ahead price [15]. The day ahead pricing used for billing has been taken from Ameren Corporation, Illinois [16, 17]. The algorithm optimizes the energy consumption by charging the battery when the demand is low. By load shifting technique, the peak load on the utility reduces which cuts down the consumer energy cost. The day ahead price for the demand is obtained online and offline analysis is done. The major part of the algorithm deals with the classification of day ahead prices into high, medium and low ranges. Three strategies are proposed based on which the algorithms are developed further.

I. Classifying algorithms
1. Elementary algorithm: The day ahead prices are obtained for the next 24 h of the day and are classified into equally sized sets of low (L), medium (M) and high (H) categories.
2. Statistical algorithm: The day ahead prices obtained are classified into three categories based on statistical parameters of the data, e.g. mean, variance, standard deviation, etc. When classification is done based on this strategy, it is not mandatory that all the three classes (i.e. L, M and H) will be equally sized but may vary depending on the variance of the data.
3. Clustering algorithm: Cluster analysis is an unsupervised classification of objects which creates groups or clusters of data. Objects are classified such that the same cluster will hold very similar objects and objects in different clusters will be distinct.

Clustering analysis involves three major steps [18].

1. Representing pattern
2. Measuring pattern proximity
3. Grouping or clustering.

Representing the pattern is the important task in pattern recognition. It refers to the available scale, number and type of features, number of available patterns, and number of classes. It may include feature selection, i.e. identifying the most appropriate subset among original features to be used in clustering. Pattern proximity is generally measured by distance functions defined on pattern pairs. Diverse techniques like Euclidean distance, probabilistic distance, or another metric are employed for measuring the distance, i.e. similarity between two patterns. Grouping step can be completed in a variety of ways. The clusters developed may be hard (i.e. the data being classified) or fuzzy (i.e. the patterns will have a variable degree of freedom in the developed clusters). In this paper, clustering analysis is done with two different methods.

1. k-means clustering [18, 19].
2. Hierarchical clustering [18–20].

Based on the above two techniques, an algorithm is proposed for scheduling the charging time of household inverter battery. The low-price slot is filled at priority followed by medium. High-price slots are filled at least priority. Figure 3.1 shows steps involved in charging algorithm.

Results and Discussion

Day ahead price (dap) of energy varies with load on the utility, i.e. the demand. The consumer demand depends on time of day as well as season of the year. Figure 3.2 shows the 24 h variation of dap with four seasons of the year. Generally in the evenings when the demand is high, dap rises, while at night dap lowers as the demand falls, during daytime dap holds moderate values, but in winter due to excessive heating loads connected, demand shoots even in the morning so as the dap. k-means clustering algorithm classifies dap into three groups, L, M and H. Figure 3.3 shows the classification, and centroid of each cluster is marked. It also shows the range of the each cluster to which data is classified.

Figure 3.4 shows the dendrogram developed by hierarchical clustering algorithm. The numbers along the horizontal axis represent the indices of the objects in the original data set. The links between objects are presented as upside-down U-shaped lines. The height of the U shows the distance between the objects. For example, the link representing the cluster containing objects 11 and 21 has a height of 0.00203. The link representing the cluster that group the above cluster together with objects 6, 9, 18, 10, 8 and 20 has a height of 0.00483. The link representing the cluster that

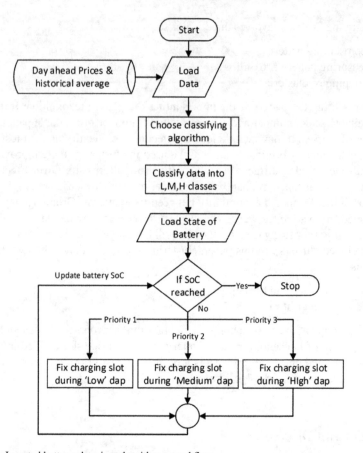

Fig. 3.1 Inverted battery charging algorithm—workflow

groups all the objects has a height of 0.01358. The height represents the distance computed between objects.

Here, the clusters are formed by natural grouping. the cophenetic relation for a hierarchical cluster tree is the linear correlation coefficient between the cophenetic distances obtained from the tree, and the original distances (or dissimilarities) are used to develop the tree. Hence, cophenetic correlation determines the accuracy involved in finding dissimilarities among observations.

The cophenetic distance between two observations is represented in a dendrogram by the height of the link at which those two observations are first joined. That height is the distance between the two subclusters that are merged by that link. The cophenetic correlation coefficient value should be very close to 1 for a high-quality solution. Here, the value obtained is 0.9281.

Dap of one particular day is considered for analysis. Table 3.1 shows the results of classification of dap into low, medium and high prices by all the strategies mentioned. The size of each cluster varies with the method employed. In elementary algorithm,

3 Optimal Battery Charging Forecasting Algorithms … 33

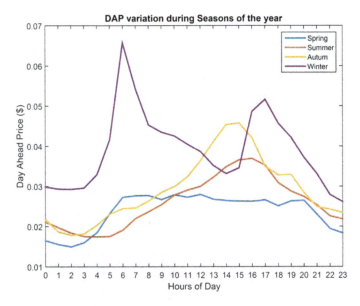

Fig. 3.2 Seasonal dap variation

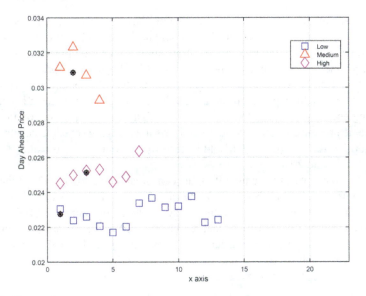

Fig. 3.3 Clustered groups

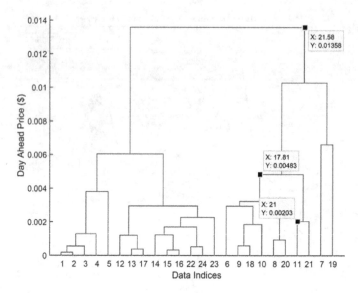

Fig. 3.4 Dendrogram developed by hierarchical clustering

24 h slots are divided into three equal slots having 8 h of low, medium and high energy prices. By statistical approach, the clusters hold unequal number of time slots. Low-price cluster holds 9 h, medium-price cluster holds 9 h and high-price cluster holds 6 h. With k-means algorithm, low-price group holds 13 h, medium-price group holds 7 h and high-price group holds 4 h. Hierarchical algorithm puts 8 h in low-price cluster, 9 h in medium-price cluster and 7 h in high-price cluster. The advantage of grouping energy prices into low, medium and high slots is the reduction in billing price gained by the consumer. Figure 3.5 shows the charging schedule developed by the algorithm for charging house hold battery, by analysing three possible charging times of 8, 15 and 20 h.

Figure 3.6 compares the daily bill of the consumer owing to the battery charging. Appreciable savings are obtained via the introduction of DSM strategy. The load being shifted to off-peak time resulted in savings to the consumer. Table 3.2 shows the possible yearly energy savings that can be obtained by the consumer by adapting the load shifting technique. For different charging times, the obtained energy savings are included. Analysis shows that the charging algorithm is capable of providing 2–5% savings depending on the charging duration on inverter alone.

Ideally the cost of energy consumption remains same for all methods employed. The advantage of one method over the other comes in the rate of charging. For instance, when the medium-price cluster is distributed in various slots along the whole day, the time taken for completing the charging process will be stretched across the day. On the other hand, if the medium-priced slots are concentrated at one part rather than being distributed, the possibility of charging getting completed quickly is high, i.e. the system need not wait for the farther medium slot to reach but it can move on to the next higher price. However, there should not be much compromise

Table 3.1 Clusters formed

Time	Dap ($)	Algorithm			
		Elementary	Statistical	k-means	Hierarchical
00:00–01:00	0.02305	L	L	L	L
01:00–02:00	0.0224	L	L	L	L
02:00–03:00	0.02261	L	L	L	L
03:00–04:00	0.02206	L	L	L	L
04:00–05:00	0.02171	L	L	L	L
05:00–06:00	0.02204	L	L	L	L
06:00–07:00	0.02338	M	M	L	M
07:00–08:00	0.0245	M	M	M	M
08:00–09:00	0.02499	H	M	M	M
09:00–10:00	0.02527	H	M	M	H
10:00–11:00	0.02531	H	H	M	H
11:00–12:00	0.02461	M	M	M	M
12:00–13:00	0.02369	M	M	L	M
13:00–14:00	0.02314	M	L	L	M
14:00–15:00	0.02321	M	M	L	M
15:00–16:00	0.02489	M	M	M	M
16:00–17:00	0.03116	H	H	H	H
17:00–18:00	0.03232	H	H	H	H
18:00–19:00	0.03071	H	H	H	H
19:00–20:00	0.02927	H	H	H	H
20:00–21:00	0.02637	H	H	M	H
21:00–22:00	0.02376	M	M	L	M
22:00–23:00	0.02227	L	L	L	L
23:00–00:00	0.02244	L	L	L	L

on the billing cost, i.e. for the system to be optimized it should give a hand-in-hand solution with low pricing and reduced charging time. From analysing dap of different days of various seasons, it is observed that k-means clustering algorithm comes out with the most optimized solution of reduced time and price.

A. PHEV Parking Garage Charging System

Methodology

In this section, an optimizing algorithm is proposed for scheduling the optimal charging of PHEV in a parking garage. The parking garage under consideration is in a company located in Bangalore, India and has three working shifts (06:00

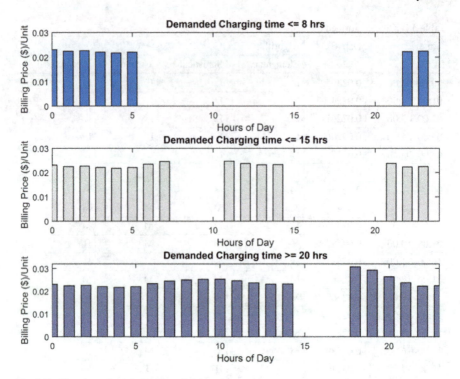

Fig. 3.5 Charging schedule for household battery

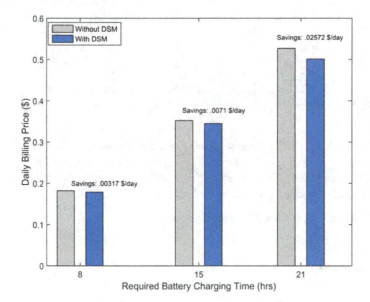

Fig. 3.6 Comparison of daily billing price

3 Optimal Battery Charging Forecasting Algorithms ...

Table 3.2 Estimated savings in energy price

Charge hours	Energy price ($)		Savings ($)		
	Without DSM	With DSM	Per day	Per month	Per year
8	0.18175	0.17858	0.00317	0.0951	1.15705
15	0.35197	0.34487	0.0071	0.213	2.5915
21	0.52669	0.50097	0.02572	0.7716	9.3878

Table 3.3 Data collected

S. No.	Data	Frequency	Source
1	Solar radiation at location (historical data)	Hourly	IMD
2	No. of vehicles parked	Shiftwise	Site survey
3	Battery capacity of vehicles	One time	Site survey
4	Time duration available for charging	Shiftwise	Site survey
5	Battery SoC pattern	Daily	Site survey
6	Distance travelled by vehicles	Daily	Site survey

AM–02:00 PM, 02:00 PM–10:00 PM, 10:00 PM–06:00 AM). The data required is collected through site survey as given in Table 3.3 and algorithm recommends the specification of solar panel required to be installed for maximum utilization of available solar radiation.

Solar PV Output Power Forecasting Model

For the operation of PHEV parking garage in real time, the solar PV power output needs to be predicted accurately. One year historical solar radiation data of the location is employed to predict the day ahead solar generation [21–23]. The power output of solar PV panel is predicted with forecasted solar radiation data and specifications for solar PV panel are recommended. Charging losses are considered while prediction.

PHEV Power Demand Forecasting Model

To develop an authentic algorithm for scheduling charging slot of PHEVs, the general pattern followed in the number of vehicles, battery SOC, etc., need to be familiarized thoroughly and these data need to be depicted for analysis. Hence, the probability distribution function of these random variables is a prerequisite.

Here, the probability distribution function is developed using log-normal distribution, which is a continuous distribution of a random variable whose logarithm is normally distributed, i.e. when X, the random variable is distributed log—normally then, $Y = \log(X)$ has a normal distribution.

Probability density functions of a random variable x is determined as,

$$N(\ln x; \mu : \sigma) = \frac{1}{\sigma\sqrt{2\pi}} \exp\left[\frac{-(\ln x - \mu)^2}{2\sigma^2}\right] \quad (3.1)$$

A change of variables must conserve differential probability.

$$N(\ln x)\mathrm{d}(\ln x) = N(\ln x)\frac{\mathrm{d}\ln x}{\mathrm{d}x}\mathrm{d}x$$

$$= N(\ln x)\frac{\mathrm{d}x}{x} = \ln N(x)\mathrm{d}x \quad (3.2)$$

$$\ln N(x; \mu:\sigma) = \frac{1}{x\sigma\sqrt{2\pi}} \exp\left[\frac{-(\ln x - \mu)^2}{2\sigma^2}\right], x > 0 \quad (3.3)$$

The required random variables include:

1. Parking duration: the distribution of PHEVs daily parking time duration.
2. Net miles driven: based on the site survey and analysing the driving pattern statistics, the daily travel distance of phevs are known
3. Power consumed by PHEV: power consumed by PHEV when plugged into the parking garage charging system.

With a view to head off severe equipment causality, the PHEV batteries should not be over discharged. The battery SOC of PHEVs should remain within 10–80% of the total rated capacity. As PHEVs have the ability to use both fossil fuel energy and electrical energy, it stops discharging when the SOC is below 10% of the rated capacity. The power demand of PHEV, i.e. PPHEV, can be obtained with support from probability distribution function (PDF) of daily parking duration and daily travel distance.

The constant charging power demand of PPHEV when energy consumption is less than 80% of battery capacity is given by

$$P_{\text{PHEV}} = \frac{M_d * E_m}{O_t - I_t} \quad (3.4)$$

The constant charging power demand of PPHEV when energy consumption is equal to or more than 80% of the battery capacity is given by,

$$P_{\text{PHEV}} = \frac{80\% * B_c}{O_t - I_t} \quad (3.5)$$

where

M_d PHEV's daily travel distance.
I_t PHEV's in time.
O_t PHEV's out time.
E_m PHEV's energy consumption per mile.
B_c cPHEV's battery capacity.

Step 1   Preliminary analysis is carried out on the data collected regarding the availability of solar radiation and expected solar panel output at the site in order to make proper recommendations regarding solar PV panel installation.

Step 2   Proposed algorithm schedules charging time and slot for PHEVs being parked in the garage based on data collected regarding the nature of PHEVs and solar panel output. Required PHEV data is loaded from system details and power demand of PHEV, PPHEV is calculated by Eq. (3.5) and Fig. 3.7 shows the work flow of the optimizing algorithm.

Results and Discussion

The parking garage in a work place is under consideration here. The building has a roof area of 5000 m² for the installation of panels. The garage holds 150 car parking slots where an average of 100 vehicles will be parking daily during each shift. Analysis is done considering morning shift. Figure 3.8 shows the forecasted PV panel output for a day. Maximum and minimum radiations obtained are found to be 0.554 MJ/m² and 0, respectively. As the conversion factor is low, the corresponding PV panel output, considering 15% efficient system (including efficiency of panel and battery charging efficiency) is 0.0831 MJ/m², which will efficiently meet the demand of parked PHEVs.

Figure 3.9 shows the constant power requirement of PHEVs parked. Due to similar daily routine observed by drivers, the power demand of PHEVs is comparable. Fixing SoC variation between 10 and 80% of the total battery capacity, maximum power demand on a regular working day is 4.3 kW with 0.0352 kW being the minimum load demand.

Figure 3.10 shows battery SOC of the PHEVs after completing the charging duration. Comparison shows the efficiency of algorithm in maintaining SOC at 80% of the battery charge, though SOC requirement is met in both the cases.

After analysing the PHEV power demand for different days (i.e. normal working days, working days after holiday, etc.), the maximum demand on the system comes up to 175 kW, hence solar PV panels of 200 kW is suggested. With the basic thumb rule of solar PV generation of 10 m² for 1 kW, 2000 m² area will be required for installation, which is readily available [21–23]. Considering the electricity tariff, for charging 100 vehicles during first shift with a minimum duration of 2 h, the net billing amount is around 50,000 Rs which is hundred percent savings with installation of solar PV panels.

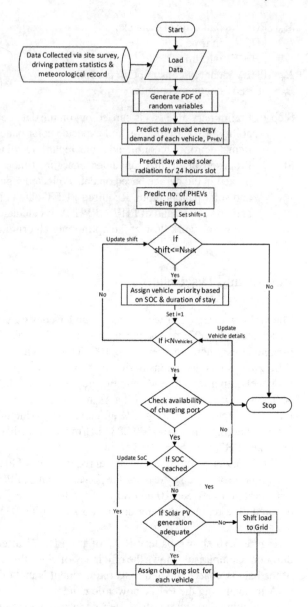

Fig. 3.7 Flow chart of optimizing algorithm

Conclusion

The first part of the paper has proposed an algorithm for scheduling the charging time of residential inverter battery. Day ahead price of electrical energy varies with variation in demand. Consumer demand varies widely within a day and is also prone to seasonal changes. Hence, the load scenario is difficult to predict. However, demand-side management still has scope with flexible loads, when the consumer is willing

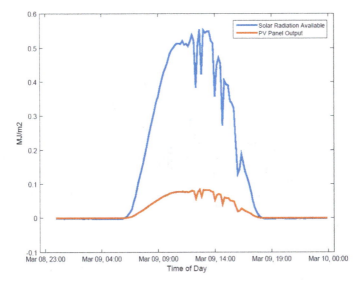

Fig. 3.8 Forecasted solar radiation and PV panel output

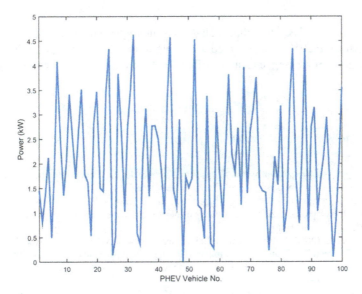

Fig. 3.9 PHEV power demand

to shift the load from peak to off-peak, it reduces the demand on the grid along with reduced energy cost for the consumer. The developed algorithm assists in scheduling the charging time of batteries by choosing the lowest electricity price slots and thus resulting in economic savings to the customer. The paper also discussed an optimizing algorithm for PHEV battery charging. The major part of the load is supplied by

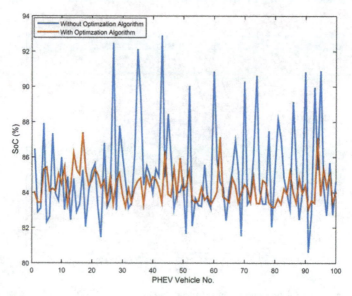

Fig. 3.10 SOC without and with optimizing algorithm while leaving garage

solar PV panels. The algorithm includes an intelligent charge scheduling system that classifies PHEVs to be charged according to the state of charge of the battery, time duration of parking, etc., and attributes optimized time slots for charging these PHEVs.

References

1. Saber AY, Venayagamoorthy GK (2011) Plug-in vehicles and renewable energy sources for cost and emission reductions. IEEE Trans Ind Electron 58(4):1229–1238
2. Ipakchi A, Albuyeh F (2009) Grid of the future. IEEE Power Energy Mag 7(2):52–62
3. Tomic J, Kempton W (2007) Using fleets of electric drive vehicles for grid support. J. Power Sources 168(2):459–468
4. Masters GM (2004) Renewable and efficient electric power systems. Wiley, Hoboken, NJ
5. Acha S, Green TC, Shah N (2009) Impacts of plug-in hybrid vehicles and combined heat and power technologies on electric and gas distribution network losses. Proc IEEE Conf Sustain Alternative Energy: 1–7
6. Deilami S, Masoum AS, Moses PS, Masoum MAS (2011) Realtime coordination of plug-in electric vehicle charging in smart grids to minimize power losses and improve voltage profile. IEEE Trans Smart Grid 2(3):456–467
7. Masoum AS, Deilami S, Moses PS, Masoum MAS, Abu-Siada A (2011) Smart load management of plug-in electric vehicles in distribution and residential networks with charging stations for peak shaving and loss minimisation considering voltage regulation. IET Gener Transmiss Distrib 5(8):877–888
8. Mitra P, Venayagamoorthy GK, Corzine KA (2011) Smartpark as a virtual STATCOM. IEEE Trans Smart Grid 2(3): 445–455

9. Hutson C, Venayagamoorthy GK, Corzine KA (2008) Intelligent scheduling of hybrid and electric vehicle storage capacity in a parking lot for profit maximization in grid power transactions. In: Proceedings of IEEE Energy 2030 Conference, Nov 2008, pp 1–8
10. A Mohamed M Elshaer O Mohammed 2011 Bi-Directional AC-DC/DC-AC converter for power sharing of hybrid AC/DC systems. In: Proceedings of IEEE PES General Meeting. Detroit, MI, USA, Jul 2011, pp 1–8
11. Gellings CW, Chamberlin JH (1993) Demand side management: concepts and methods, 2nd edn. PennWell Books, Tulsa, OK
12. Mohsenian-Rad AH, Leon-Garcia A (2010) Optimal residential loadcontrol with price prediction in real-time electricity pricing environments. IEEE Trans Smart Grid 1(2):120–133
13. Triki C, Violi A (2009) Dynamic pricing of electricty in retail markets. Q J Oper Res 7(1):21–36
14. Bringing residential real-time pricing to scale in Illinois: Policy recommendations technical report CNT Energy (2010)
15. Gan G, Ma C, Wu J (2010) Data clustering: theory, algorithms and applications
16. Bakirci K (2009) Models of solar radiation with hours of bright sunshine: a review. Renew Sustain Energy Rev 13:2580–2588
17. Batlles FJ, Rubio MA, Tovar J, Olmo FJ, Alados-Arboledas L (2000) Empirical modeling of hourly direct irradiance by means of hourly global irradiance. Energy 25:675–88
18. Karakoti I, Pande B, Pandey K (2011) Evaluation of different diffuse radiation models for Indian stations and predicting the best fit model. Renew Sustain Energy Rev 15:2378–2384
19. Myers DR (2005) Solar radiation modeling and measurements for renewable energy applications: data and model quality. Energy 30:1517–1531
20. Kaldellis JK, Zafirakis D, Kondili E (2010) Optimum sizing of photovoltaic- energy storage systems for autonomous small islands. Int J Electr Power Energy Syst 32:24–36
21. Patel MR Wind and solar power systems: design, analysis, and operation
22. Ibrahim IA, Hossain MJ, Duck BC (2020) An optimized offline random forests-based model for ultra-short-term prediction of PV characteristics. IEEE Trans Industr Inf 16(1):202–214. https://doi.org/10.1109/TII.2019.2916566
23. Kawanobe A, Wakao S, Taima T, Kanno N, Mabuchi H (2019) Prediction of representative waveform of surplus PV power based on cluster analysis of power demand and PV output. In: 2019 IEEE 46th photovoltaic specialists conference (PVSC) Chicago, IL, USA, pp 2121–2125. https://doi.org/10.1109/PVSC40753.2019.8980571

Chapter 4
Performance Index-Based Coordinated Control Strategy for Simultaneous Frequency and Voltage Stabilization of Multi-area Interconnected System

Ch. Naga Sai Kalyan and G. Sambasiva Rao

Introduction

The stability of interconnected system is substantiated only by controlling the deviations in both frequency and voltage. These deviations can be reduced by LFC loop by making use of speed governing action and AVR loop through generator excitation. AVR is coupled to the LFC loop through cross-coupling coefficients [1]. A lot of research work is available on LFC and AVR separately. AVR coupling is desired to be considered in order to investigate the system in a more practical approach. However, the excitation system is coupled to generation control loop through week coupling coefficients, and the effect of voltage control loop on frequency control loop is prominent. A few investigations are available on combined LFC and AVR model. In [2], analysis is carried out on combined LFC and AVR model, but limited to single area only. Authors in [3] investigated the combined model of multi-area with conventional generation sources, but incorporation of renewable energy sources is not considered as they are gaining momentum nowadays. Authors in [4] examine the multi-area combined system incorporating solar thermal power plant, but wind energy penetration was not considered. This motivates the author to investigate the combined system consisting of both conventional and renewable energy sources.

The deviations in frequency and voltage are mitigated by the secondary controller in LFC and AVR loops, respectively. Classical controllers like (I/PI/PID) [5], fractional-order (FO) [8] and intelligent fuzzy controllers [6, 9] are proposed by the authors in their work in LFC domain. However, finding the optimum parameters

---

Ch. Naga Sai Kalyan (✉)
Department of EEE, Acharya Nagarjuna University, Guntur, India
e-mail: kalyanchallapalli@gmail.com

G. Sambasiva Rao
Department of EEE, RVR & JC College of Engineering, Chowdavaram, Guntur 522019, India

© The Editor(s) (if applicable) and The Author(s), under exclusive license to Springer Nature Singapore Pte Ltd. 2021
A. K. Singh and M. Tripathy (eds.), *Control Applications in Modern Power System*, Lecture Notes in Electrical Engineering 710,
https://doi.org/10.1007/978-981-15-8815-0_4

for the controller using optimization algorithm is a key task. Optimization algorithms like particle swarm optimization (PSO) [10], lightening search algorithm (LSA) [4], gray wolf optimization (GWO) [6], simulated annealing (SA) [3], etc. are reported in literature. But most of these algorithms are suffering from slow convergence as well as tendency of getting easily trapped into local minima. So, in this work, a new DE-AEFA algorithm is presented to optimize the controller gains which outrage all the aforementioned disadvantages by maintaining the average balance between exploration and exploitation. Moreover, under large disturbances, secondary controllers are not enough to maintain system stability. So, IPFC and RFBs coordinated control strategy is implemented in addition to DE-AEFA optimized secondary controller to smoothen the frequency and voltage deviations effectively.

In view of the above literature, the major contributions of this work are as follows:

(a) The two-area combined system with multi-type generation units is constructed in MATLAB/Simulink.
(b) The parameter gains of the controller are optimized with DE-AEFA algorithm under various performance indices.
(c) The system responses under various performance indices are compared to obtain the best objective function.
(d) The superiority of DE-AEFA algorithm over others is demonstrated.
(e) The system dynamic behavior under IPFC-RFBs coordinated control strategy is studied.

Power System Model

The combined LFC and AVR model consisting of hybrid generating units is considered for investigation which is depicted in Fig. 4.1. The considered system has two unequal capacities of ratio 2:1. Area-1 consists of conventional generation units with appropriate generation rate constraints (GRCs), whereas area-2 consists of diesel power plant along with solar photovoltaic (PV) and wind power generation unit. The time and gain constant parameters of generation units in Fig. 4.1 are considered from [4, 7, 8]. AVR is coupled to the LFC loop through excitation coupling coefficients and power system synchronizer and is depicted in Fig. 4.2.

Controller Design and Optimization Algorithm

Controller and Objective Functions

In both LFC and AVR loops, classical PID controller is opted as secondary controller because of its design simplicity, operational efficiency and widespread usage in industries. However, PID controller had the capability to execute even in the situations

4 Performance Index-Based Coordinated Control Strategy … 47

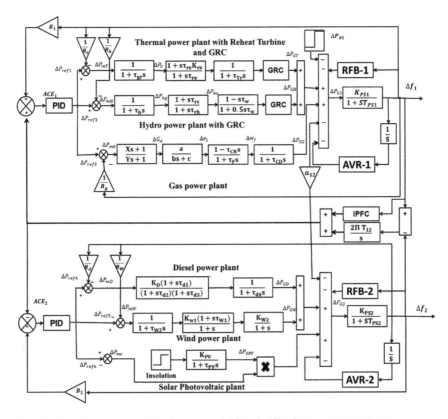

Fig. 4.1 Transfer function model of combined LFC and AVR of two-area power system

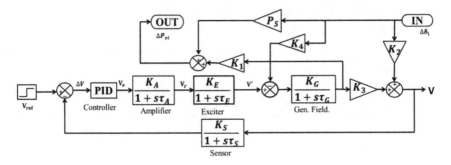

Fig. 4.2 Automatic voltage regulator

of parameter uncertainties. The parameters of the controller are tuned with proposed DE-AEFA algorithm in the range of [0–5] under the following performance indices to choose the best objective function.

$$J_1 = \text{ITSE} = \int_0^{T_{\text{sim}}} t \cdot (\Delta f_1^2 + \Delta f_2^2 + \Delta P_{\text{tie}_{12}}^2) dt \quad (4.1)$$

$$J_2 = \text{ITAE} = \int_0^{T_{\text{sim}}} t \cdot (\Delta f_1 + \Delta f_2 + \Delta P_{\text{tie}_{12}}) dt \quad (4.2)$$

$$J_3 = \text{ISE} = \int_0^{T_{\text{sim}}} (\Delta f_1^2 + \Delta f_2^2 + \Delta P_{\text{tie}_{12}}^2) dt \quad (4.3)$$

$$J_4 = \text{IAE} = \int_0^{T_{\text{sim}}} (\Delta f_1 + \Delta f_2 + \Delta P_{\text{tie}_{12}}) dt \quad (4.4)$$

$$J_5 = J_1 + J_2 \quad (4.5)$$

$$J_6 = J_2 + J_3 \quad (4.6)$$

$$J_7 = J_3 + J_4 \quad (4.7)$$

$$J_8 = J_1 + J_4 \quad (4.8)$$

Differential Evolution-Artificial Electric Field Algorithm (DE-AEFA)

In this work, a new DE-AEFA algorithm is presented, which makes use of evolutionary concept of DE and charged particles based on particles searching strategy of AEFA algorithm. Every algorithm has its own benefits and drawbacks. DE algorithm is potential in diverting the population towards the best solution and can utilize the search space effectively, but having the tendency of slow and premature convergence limits the application boundary. On the other hand, AEFA [10] can locate near-optimal solutions with high speed of convergence and can maintain better exploration and exploitation when compared to other optimization algorithms. But the procedure involved in adjusting the step size in particles position and velocity upgradation leads to untimely convergence. The DE and AEFA algorithms have complementary performance. The principal idea behind this proposed algorithm is to make use of the benefits of both algorithms in a collective manner and to overrule the disadvantages of individual algorithms to find optimal solutions with accuracy and high speed of convergence. During searching process in DE-AEFA, half of individuals in total population find the solution with DE strategy and the others with AEFA strategy and the total information of each population is shared among every individual agent. After that, the individual with the best fitness value acquires the chance of entering next generation's optimization. Hence, this proposed approach inherits the efficiency of searching procedure and also assures the global convergence. Moreover, the superiority of the DE-AEFA algorithm is examined on various benchmark standard functions and is demonstrated in [11]. The step-by-step procedure involved in implementation of DE-AEFA algorithm is referred from [11], and flowchart is depicted in Fig. 4.3.

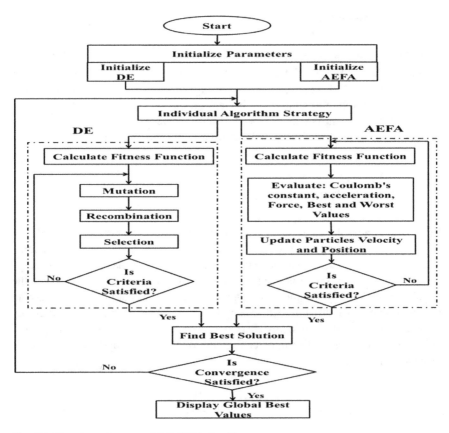

Fig. 4.3 Flowchart of proposed DE-AEFA algorithm

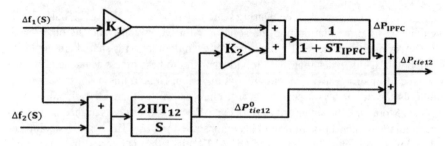

Fig. 4.4 Structure of IPFC as damping controller

Interline Power Flow Controller

IPFC incorporates a variety of voltage source converters (VSCs) provided with a common DC link. Each VSC is facilitated to provide series compensation for transmission lines it is connected. IPFC manages the power flow in multiple lines by injecting appropriate voltages in tie-lines. Thus, the performance of the entire interconnected system is enhanced with the incorporation of IPFC. The damping controller structure of IPFC employed in this work is shown in Fig. 4.4.

Redox Flow Batteries (RFBs)

RFBs come under the category of electrochemical energy storage devices (ESDs). RFBs have a reactor tank consisting of two compartments separated by a membrane. The reactor tank is filled with sulphuric acid containing vanadium ions acted as electrolytic solution. Pump is installed in each compartment which facilitates the circulation of electrolytic solution through battery cell. The charging and discharging process is to be done through oxidation–reduction reaction. However, the efficiency of RFBs depends on charging/discharging cycle period. Lesser the cycle period, higher is the efficiency. The transfer function of RFBs employed in this work is given as follows [11]:

$$G_{RFB} = \frac{K_{RFB}}{1 + ST_{RFB}} \tag{4.9}$$

Simulation Results and Analysis

The analysis performed on considered system shown in Fig. 4.1 is done by applying 1% SLP applied in area-1. The optimization algorithms employed in this work are executed by considering 100 populations for a maximum of 50 iterations.

Responses of the Combined Model Optimized with DE-AEFA Algorithm Under Various Objective Functions

The combined LFC and AVR model shown in Fig. 4.1 is simulated by applying 1%SLP applied in area-1, and the proposed DE-AEFA optimized PID controller is opted as controller under various performance indices discussed in Sect. 3.1. The system dynamic responses optimized using various performance indices are put under comparison to select the best objective function and are rendered in Fig. 4.5. On observing Fig. 4.5, it is evident that the objective function ISE is effective among all the objective functions in terms of mitigating oscillations and bringing back the system to the steady state in faster manner. The objective function values and the settling time (T_s) of the responses depicted in Fig. 4.5 are noted in Table 4.1.

Responses of the Combined Model Optimized with Different Optimization Algorithms

In this subsection, the secondary PID controller is optimized with optimization algorithms like DE, AEFA and DE-AEFA algorithms one at a time for 1%SLP applied in area-1 to corroborate the effectiveness of presented DE-AEFA algorithm. The responses of the model are depicted in Fig. 4.6, and the corresponding numerical results and controller gains are noted in Tables 4.2 and 4.3, respectively. On observing Fig. 4.6 and Table 4.2, it is clear that the system responses under proposed optimization algorithm are improved in terms of oscillations and responses are settled much faster compared to DE and AEFA algorithms.

Responses of the Combined Model Under Coordinated Control Strategy

In practice, the secondary controllers are not sufficient to dampen the system disturbances under large disturbances. So, additional control devices are necessary to maintain stability of the system under large load varying conditions. For this purpose, in this work the system responses are analyzed with and without considering IPFC

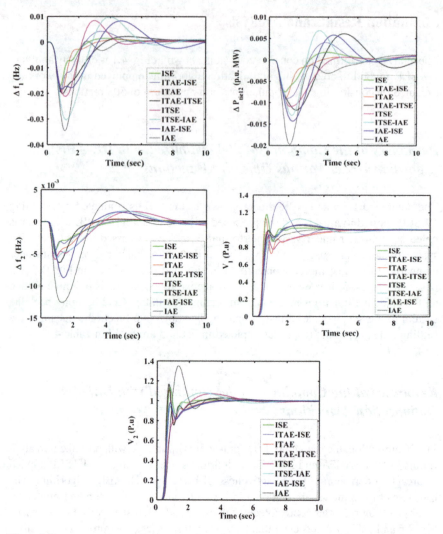

Fig. 4.5 Responses of combined model for 1% SLP applied in area-1 with DE-AEFA-based optimization under various objective functions Δf_1, ΔP_{tie12}, Δf_2, V_1, V_2

and RFBs along with DE-AEFA optimized PID controller. Figure 4.7 depicts the dynamic responses of the system with considering RFBs in both area-1 and area-2, with only IPFC placed in the tie-line with series and with both IPFC and RFBs. The T_S of the responses is mentioned in Table 4.4, and optimal gain and time constants of IPFC are $K_1 = 0.9247$, $K_2 = 0.9524$, $T_{IPFC} = 0.8562$ and for RFBs $K_{RFB} = 0.9441$, $T_{RFB} = 0.997$ employed in this work.

4 Performance Index-Based Coordinated Control Strategy …

Table 4.1 Settling time of system responses with DE-AEFA optimization under different indices

Parameters	Objective function							
	ISE	ITAE-ISE	ITAE	ITAE-ITSE	ITSE	ITSE-IAE	IAE-ISE	IAE
Δf_1	8.78	9.99	11.14	12.18	13.87	15.19	23.67	16.64
Δf_2	8.91	9.43	9.69	11.10	16.53	14.43	19.08	12.67
ΔP_{tie}	7.21	12.29	9.97	15.36	12.77	11.20	16.15	15.31
V_1	4.21	5.02	5.66	5.59	5.76	6.93	6.77	5.21
V_2	4.13	5.72	5.69	6.07	8.70	7.21	5.94	6.38
Function Value	12.72	17.57	21.32	25.45	29.90	36.80	39.14	48.87

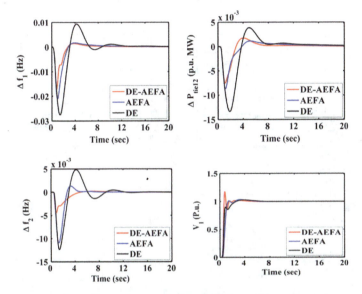

Fig. 4.6 Responses of combined model for 1% SLP applied in area-1 with various optimization methods. Δf_1, ΔP_{tie12}, Δf_2, $V_1.\Delta f_2 V_2$

Table 4.2 Settling time of combined system responses with different algorithms

Optimized controller	Settling time T_S (s)					ISE
	Δf_1	Δf_2	ΔP_{tie12}	V_1	V_2	
DE:PID	19.31	17.63	21.51	8.12	7.81	25.92
AEFA:PID	15.68	11.86	18.17	8.02	7.17	21.48
DE-AEFA:PID	8.78	8.91	7.21	4.21	4.13	12.72

Table 4.3 Optimum parameters of the controller

Controller	Area-1		Area-2	
	LFC Loop	AVR Loop	LFC Loop	AVR Loop
DE:PID	$K_P = 3.2736$ $K_I = 2.9147$ $K_D = 1.2070$	$K_P = 2.0184$ $K_I = 1.8013$ $K_D = 0.9305$	$K_P = 3.4091$ $K_I = 2.6550$ $K_D = 1.9866$	$K_P = 1.9597$ $K_I = 1.6106$ $K_D = 0.9861$
AEFA:PID	$K_P = 3.9925$ $K_I = 2.5499$ $K_D = 1.8142$	$K_P = 2.5109$ $K_I = 1.5490$ $K_D = 1.0090$	$K_P = 3.8716$ $K_I = 2.9595$ $K_D = 1.7653$	$K_P = 2.0467$ $K_I = 1.9119$ $K_D = 0.9415$
DE-AEFA:PID	$K_P = 4.0786$ $K_I = 2.8716$ $K_D = 2.0504$	$K_P = 2.4091$ $K_I = 1.6655$ $K_D = 1.2021$	$K_P = 3.7061$ $K_I = 3.3209$ $K_D = 2.2700$	$K_P = 2.0561$ $K_I = 2.0974$ $K_D = 1.7924$

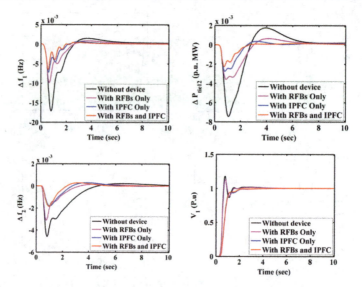

Fig. 4.7 Responses of combined model for 1% SLP applied in area-1 with coordinated control strategy, Δf_1, ΔP_{tie}, Δf_2, V_1. Δf_2 V_2

Table 4.4 Settling time of system responses without and with considering IPFC-RFBs

DE-AEFA:PID	Settling time T_S (s)					ISE
	Δf_1	Δf_2	ΔP_{tie12}	V_1	V_2	
Without devices	8.786	8.919	7.215	4.210	4.135	12.728
With RFBs	7.706	7.873	6.723	3.882	3.683	9.0392
With IPFC	5.575	6.501	5.027	3.552	3.513	3.9624
With IPFC-RFBs	4.789	5.547	3.153	2.899	2.540	0.0079

Conclusion

This paper presented the simultaneous control of voltage and frequency of multi-area system with multi-type generation units using DE-AEFA optimized PID controller. The combined system is analyzed by applying 1%SLP applied in area-1 under DE-AEFA optimized PID controller with various objective functions. It is observed that the responses with ISE objective function are better than that of responses of other objective functions in view of settling time and oscillations. The performance of proposed algorithm is also compared with other algorithms to validate the superiority. Finally, the impact of IPFC and RFBs coordinated control strategy is studied. From this, it is cleared that the system performance is predominantly improved by placing RFBs in both the areas and IPFC with tie-line in series. It is concluded that secondary controller of combined model is not ample enough to maintain system stable when the system is subjected to large disturbances. Therefore, by implementing this coordinated control strategy system can bring back to stable condition even under sudden large disturbances.

References

1. Saadat H (1999) Power system analysis. McGraw-Hill, USA
2. Gupta A, Chauhan A, Khanna R (2014) Design of AVR and ALFC for single area power system including damping control. In: Recent advances in engineering and computational sciences (RAECS). Chandigarh, pp 1–5
3. Vijaya Chandrakala KRM, Balamurugan S (2016) Simulated annealing based optimal frequency and terminal voltage control of multi-source multi area system. Electr Power Energy Syst 78:823–829
4. Rajbongshi R, Saikia LC (2018) Combined voltage and frequency control of a multi-area multi-source system incorporating dish-stirling solar thermal and HVDC link. IET Renew Power Gener 12(3):323–334
5. Lal DK, Barisal AK (2019) Combined load frequency and terminal voltage control of power systems using moth flame optimization algorithm. J Electr Syst Inf Technol 6(1):8
6. Lal DK, Barisal AK (2017) Comparative performances evaluation of FACTS devices on AGC with diverse sources of energy generation and SMES. Cogent Eng 4:1318466
7. Shankar R, Bhushan R, Kalyan C (2016) Small-signal stability analysis for two-area interconnected power system with load frequency controller in coordination with FACTS and energy storage device. Ain Shams Eng J 7(2):603–612
8. Tasnin W, Saikia LC (2018) Comparative performance of different energy storage devices in AGC of multi-source system including geothermal power plant. J Renew Sustain Energy. https://doi.org/10.1063/1.5016596
9. Mukherjee V, Ghoshal SP (2007) Comparison of intelligent fuzzy based AGC coordinated PID controlled and PSS controlled AVR system. Electr Power Energy Syst 29:679–689
10. Yadav A (2019) Artificial electric field algorithm for global optimization. Swarm and evolutionary computation base data. https://doi.org/10.1016/j.swevo.2019.03.013
11. Naga Sai Kalyan C, Sambasivarao G (2020) Automatic generation control of multi-area hybrid system considering communication time delays. J Adv Res Dyn Control Syst 12(02):2071–2083

Chapter 5
Load Frequency Control of Two-Area Power System by Using 2 Degree of Freedom PID Controller Designed with the Help of Firefly Algorithm

Neelesh Kumar Gupta, Manoj Kumar Kar, and Arun Kumar Singh

Introduction

Power system is substantial and complex electrical system which are combination of three different entity which are generation system, transmission system, and distribution system where load on the system is keeping changing every moment as the requirement of the customer. The generating station generated the power how much it required which depends on consumer demands. The generated power has two part which is real (active) and reactive power. The frequency of system varies according to active power and on the base of the reactive power voltage of system change. So in power system, there are two section of control, one for frequency and other for voltage. The frequency control along with active power is called load frequency control (LFC) so with the variation of load of the system, the frequency is also changed. So legitimately compose controller is required for the direction of the system varieties keeping in the mind the end goal to keep up the steadiness of the power system and additionally ensure its solid activity. The quick development of the enterprises has additionally provoked the expanded power system as intricacy system due to which frequency changes every time.

There are hundreds of number of studies have been done for the LFC, Elgerd and Fosa are first to give their view for the robust design of controller for AGC. Various controllers for LFC are tested in [1]. Among the industries, PID controller is well-known controller due to its robustness, simplification and its onsite adjustment facilities. The utilization of combined intelligence approach for AGC interconnected power system is discussed in [1, 2]. The use of artificial bee colony (ABC) algorithm to modulated and regulated of AGC controller which has good capabilities of local

---

N. K. Gupta (✉) · M. K. Kar · A. K. Singh
Department of Electrical Engineering, NIT Jamshedpur, Jamshedpur, Jharkhand, India
e-mail: gptneelesh91@gmail.com

© The Editor(s) (if applicable) and The Author(s), under exclusive license to Springer Nature Singapore Pte Ltd. 2021
A. K. Singh and M. Tripathy (eds.), *Control Applications in Modern Power System*, Lecture Notes in Electrical Engineering 710, https://doi.org/10.1007/978-981-15-8815-0_5

and global search is shown in [3]. The use PID and PI controller in LFC of interconnected power system is shown in [4] where controller is regulated with help of differential evolution (DE) algorithm. Particle swarm optimization (PSO) algorithm has many application in the solving of engineering problem and it is also used for the optimization in LFC [5]. Use of 2-DOF-PID controller in LFC presented in [6] in TLBO algorithm is also used. The information about the firefly algorithm is given in [7, 8]. From the above literature review, here 2 DOF controller will be used for a two-area system whose parameter will be optimized by the firefly algorithm.

System Under Study

In the power system, for the understanding of LFC, a system is required to investigate. Here that investigated system is two-area power whose structural diagram is shown in Fig. 5.1. In this system, each area contains speed governor block, turbine block,

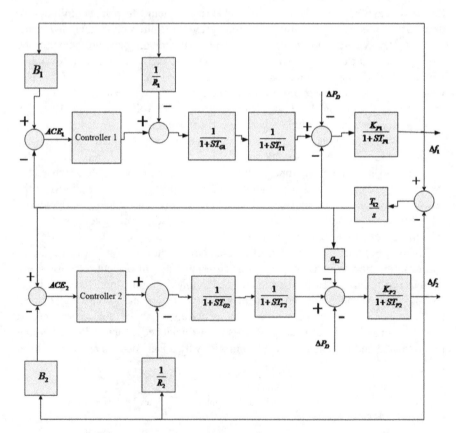

Fig. 5.1 Two-area system's structural diagram

and generator and load block. Here, an identical area has been considered for the study.

Area control error is indispensable piece of LFC which enclose the information about modification in tie-line power and aberration in frequency Δf. In a multi-area system, all area has their own AEC. Whenever there is any disturbance or load variation in any of the area of the system, there will be some controller action

$$ACE_1 = \Delta P_{12} + b_1 \Delta f_1 \tag{5.1}$$

$$ACE_2 = \Delta P_{12} + b_2 \Delta f_2 \tag{5.2}$$

2-DOF-PID Controller

In the control system, DOF of any controller can be define as how many individual close loop it can adopt. In control system, generally the major ambition of designing is to solve the various features problem. So, in solving various features problem, 2 DOF control system can commonly be obtained more accurately then one degree of freedom system or denoted as 1 DOF system.

A 2-DOF-PID controller is able to detect rapid shock or disturbance without symbolic boost of overshoot in setpoint tracking. It will be advantageous to use 2-DOF-PID controller when we want the effect of reference signal is less on the control system.

Figure 5.2 shows a classical control architecture using a 2-DOF-PID controller. Here y represents output and its two input r and y can be shown in parallel form.

The below equation shows transfer function of 2-DOF-PID controller which can be given as

$$u = K_p(br - y) + \frac{K_i}{S}(r - y) + \frac{K_d S}{NS + 1}(cr - y) \tag{5.3}$$

In the above equation K_p, is proportional gain, b is proportional term's set point weight, K_i, is integrator gain, K_d is derivative gain, c is derivative term's set point weight. While N is filter gain in the above equation.

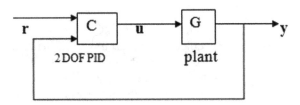

Fig. 5.2 Control architecture using a 2-DOF-PID Controller

Firefly Algorithm

As this algorithm is inspired by the nature of firefly due to which it is called as firefly algorithm. This algorithm is a swarm-based algorithm. Algorithm performance depends upon the two-factor intensification and diversification. Intensification looks into the local region where the best solution can be possible in that region, and it also chose the best solution. On the other hand, diversification tries to examine the other search space more skilfull to yield the solution which will have higher diversity. In traditional gradient-based method, we optimized the gradient of function which has enough information for quick finding for the particular problem and optimized solution is found but in the case where the system has some necessary condition and it is very much nonlinear, non-differentiable, it usually trapped in local search and it faces difficulty on convergence. The firefly algorithm is a new improved technique for finding global search and local search. When the system is nonlinear non-differentiable, it will produce a very good result as this algorithm is based on non-gradient and has a simple objective function.

Xin-she Yang developed this nature stimulated metaheuristic algorithm which is founded on the array of blinking and nature of the firefly [7]. Three basic rules are utilized by this latest algorithm which is as follows

- Firefly is unisexual. They approach more appealing and flashing firefly regardless of their sex.
- Due to their unisexual nature, two firefly attracted each other depending upon the brightness, and depending upon the distance between each other, they will see the brightness. Among the firefly, the less bright will attract toward brighter one. They will move in a random direction if there is no brighter among them.
- The objective function of a system is proportional to the brightness of firefly. Hence, brightness of a firefly is essential to determine the form of objective function.

The mathematical equation for the attractiveness can be given as

$$\beta_r = \beta_0 e^{-\gamma r^m} \text{ with } m\,1 \tag{4}$$

where

β_0 Initial value of attractiveness at $r = o$.

γ Absorption coefficient.

r Distance between the two fireflies.

Absorption coefficient controls the decrement in the light intensity. The value of γ can vary between the 0 to ∞. But in practical, its value depends on the characteristic length of the system. Normally, the range of γ is taken as 0.1–10.

If a firefly 'i' travel toward the brighter firefly 'j,' then a relation between two can be given as

$$x_i^{t+1} = x_i^t + \beta_0 e^{-\gamma r_{ij}^2}\left(x_j^t - x_i^t\right) + \alpha \varepsilon_i^t \tag{5}$$

In the above equation, there are three term. The location of the ith firefly can be known by first term, the middle term gives the information about attraction between firefly, and third term is randomization with α, ϵ denoting randomization parameter and random number which are generated uniformly in range (0, 1).

Results and Discussion

In this paper, compression has been done between the result of simulation done for drift in the frequency and also in the tie-line power when a step change is applied in system of two-area power system when their secondary loop is connected firstly with PID controller and for the second time with 2-DOF-PID controller. To get the optimized value of parameters of the both controller, an algorithm called firefly algorithm is been used.

Here, an identical two-area system is analyzed for the study of LFC. In this case, two scenarios have been considered. In the first scenario, load deviation is happening in area one only while it is happing in the both areas in the second scenario. The parameter will remain same for both area as identical area system has been considered. The schedule value of frequency is 50 Hz for this system. Table 5.1 shows different parameters and its value of the investigate system. 200 MW is the total load of the system which is shared between the two areas.

Case-A. Load Disturbance Given in Only One Area

In this scenario, load deviation (step disturbance of 2%) is happening in only area 1. The value of both controller's parameters will be same for both areas as the system is identical. Both controller different gains for this case are shown in Table 5.2.

With these parameters, the following graph is come as output (Figs. 5.3, 5.4 and 5.5).

Table. 5.1 Two-area system parameters

System parameter	Value
T_{sg}	0.3 s
T_t	0.5 s
K_{ps}	105
T_{ps}	22 s
$1/R$	0.4
B	0.326
T_{12}	0.08

Table. 5.2 Gain of both controller

Controller	K_p	K_i	K_d	N	b	c
2DOF PID	3.345077	4.94234	5	5	4.269	0.06790
PID	1.651	1.651	1.5309	–	–	–

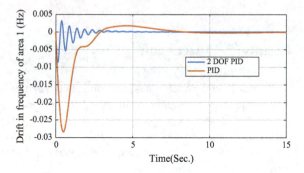

Fig. 5.3 Drift in area 1 frequency

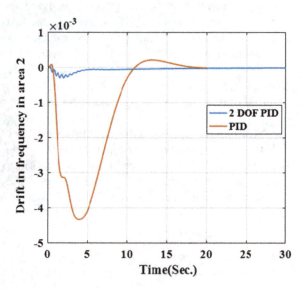

Fig. 5.4 Drift in area 2 frequency

Case-B. Disturbance in Both Areas

In this case, both areas are triggered by load deviation (step of strength 2%). Controllers parameters will have same value for both area as system is an identical system. The controllers gain for this case is shown in Table 5.3.

With these parameters, the following graph is come as output (Figs. 5.6 and 5.7).

Fig. 5.5 Drift in tie-line power

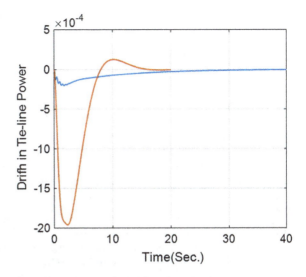

Table. 5.3 Gain of both controller

Controller	K_p	K_i	K_d	N	b	c
2DOF PID	3.35524	4.99984	5	173	4.19	0.03867
PID	3.069	3.7052	5.9647	–	–	–

Fig. 5.6 Drift in area 1 frequency in scenario 2

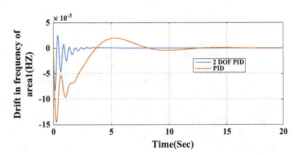

Fig. 5.7 Drift in area 2 frequency in scenario 2

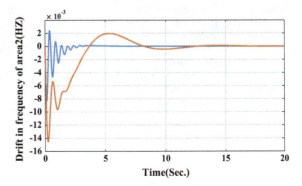

Conclusion

Here, in this paper, load frequency control(LFC) has been study, and for this, a two-area system has been interrogated. The frequency of the system will get deviated from its original value whenever there is load deviation and this also triggers change in the tie-line power among the areas of a multi-area system. To eliminate these changes which are happened due to load variation, controller is used, and to find the optimized value of parameter of controllers, some techniques are required. Hence, two different controller, named PID and 2-DOF-PID controllers, are used in the secondary loop for controlling purpose. Here, for the optimization of controller's parameters, a new meta-heuristic algorithm which works on the nature and flashing patterns of firefly called firefly algorithm is used. Two scenarios have been discussed in which in the first case disturbance will be in one only while there will be disturbance in both areas in the second case. In both scenarios, effect of load disturbance on frequency and tie-line is seen. A comparative study is done between the result of the PID controller and 2-DOF-PID controller which is optimized by FA. It is seen that in every case, 2-DOF-PID controller optimized by firefly algorithm proves its superiority over the PID controller. It concludes that this controller eliminated deviation in frequency in the least time also settling time and oscillation are trim and peak overshoot minimalized.

References

1. Chang CS, Fu W, Wen F (1998) Load frequency control using genetic algorithm based fuzzy gain scheduling of pi controllers.Electr Mach Power Syst 26(1):39
2. Elgerd OI, Fosha CE (1970) Optimum megawatt-frequency control of multi-area electric energy systems. IEEE Trans Power Apparatus Syst PAS 89(4):556–563
3. Gozde H, Cengiz Taplamacioglu M, Kocaarslan L (2012) Comparative performance analysis of artificial bee colony algorithm in automatic generation control for interconnected reheat thermal power system. Int J Electr Power Energy Syst 42(1):167–178
4. Mohanty B, Panda S, Hota PK (2014) Differential evolution algorithm based automatic generation control for interconnected power systems with non-linearity. Alexandria Eng J 53(3):537–552
5. Ghoshal S (2004) Optimizations of PID gains by particle swarm optimizations in fuzzy based automatic generation control. Electr Power Syst Res 72(3):203–212
6. Sahu RK, Panda S, Rout UK, Sahoo DK (2016) Teaching learning based optimization algorithm for automatic generation control of power system using 2-DOF PID controller. Int J Electr Power Energy Syst 77:287–301
7. Yang XS (2009) Firefly algorithms for multimodal optimization stochastic algorithms: foundations and applications, vol 5792. ISBN: 978-3-642-04943-9
8. https://en.wikipedia.org/wiki/Firefly-algorithm

Chapter 6
Stabilizing Frequency and Voltage in Combined LFC and AVR System with Coordinated Performance of SMES and TCSC

Ch. Naga Sai Kalyan and G. Sambasiva Rao

Introduction

In today's world, demand for electrical power is rapidly increasing. The generating stations must cope up with the variable load demand to maintain power reliability. The power system is an interconnected one, which schedules power from surplus generating areas to deficit areas. However, maintaining power supply with quality has an equal importance with power reliability. Quality power results in upholding standard frequency and terminal voltage. The frequency variation is controlled by the speed governor-turbine action and terminal voltage is controlled by varying the generator field excitation system [1]. So, the power system is governed by two loops: LFC loop and AVR loop. LFC loop controls active power and frequency, whereas AVR loop deals with reactive power and terminal voltage. The time constants of LFC and AVR systems are alike, and AVR is much faster than LFC. Hence, a lot of researchers had given a less priority to the incorporation of AVR with LFC. However, a few are available on combined effect.

Though in [2], combined analysis is carried out but limited to single area only. In [3, 4], analysis is broadened to two-area system with conventional generation units only. Authors in [5, 6] investigated the combined effect on three-area system comprises of diesel and thermal plants along with the incorporation of solar thermal power plant. On reviewing the literature, it is concluded that authors had concentrated on analyzing the combined LFC and AVR system by considering two diverse

---

Ch. Naga Sai Kalyan (✉)
Department of EEE, Acharya Nagarjuna University, Guntur, India
e-mail: kalyanchallapalli@gmail.com

G. Sambasiva Rao
Department of EEE, RVR & JC College of Engineering, Chowdavaram, Guntur 522019, India

© The Editor(s) (if applicable) and The Author(s), under exclusive license to Springer Nature Singapore Pte Ltd. 2021
A. K. Singh and M. Tripathy (eds.), *Control Applications in Modern Power System*, Lecture Notes in Electrical Engineering 710,
https://doi.org/10.1007/978-981-15-8815-0_6

generation sources in each area. This motivates the author to consider more than two fuel types of generation units in each area to analyze combined effect.

Moreover, the secondary controllers in LFC and AVR loops are not sufficient to maintain the system stability under large load variations. So, the system needs an additional control strategy along with secondary controller. Therefore, coordinated controlled strategy of TCSC and SMES has been implemented in this work, and no literature is available on testing the TCSC and SMES coordinated control scheme on combined system depicted in Fig. 6.1 to the best of author's knowledge. Several new optimization algorithms are available in literature; their diligence is not encountered in combined effect.

A new DE-AEFA algorithm is fabricated to tune the PID controller gains to demonstrate the combined model of multi-fuel test system with the presence of SMES and TCSC devices. From the above discussion, the objectives of this paper are

(a) To develop a two-area multi-fuel test system for analyzing the combined effect of LFC-AVR model with PID as secondary controller.

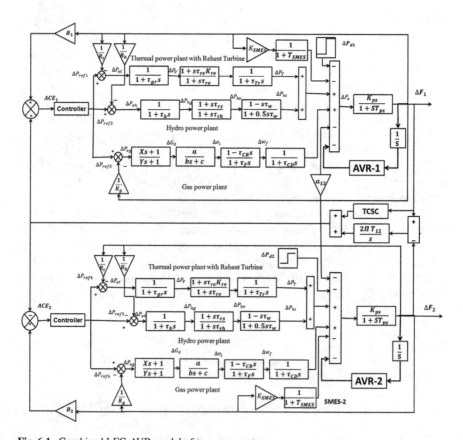

Fig. 6.1 Combined LFC-AVR model of two-area system

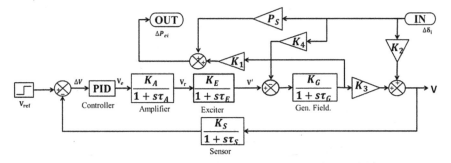

Fig. 6.2 Automatic voltage regulator with cross-coupling coefficients

(b) Controller parameters of PID are tuned with DE-AEFA algorithm.
(c) The impact of SMES and TCSC by integrating them in test system under combined effect is demonstrated.

Considered Test System

The system under investigation comprises of two areas with multi-type generation sources shown in Fig. 6.1 which is considered as test system for investigation. The time and gain constant parameters of the system are considered from [7]. The AVR system is coupled to the LFC loop through power system synchronizer and cross-coupling coefficients. The gain and time constant parameters of the AVR depicted in Fig. 6.2 are considered from [1].

TCSC in Combined System

The power transfer capability of transmission a transmission line may be controlled by using TCSC device. The inherent feature of TCSC enables the alteration of inductive and capacitive reactance compensation. It regulates the line flow without getting more loss in the line subjected to thermal limits.

The benefits of TCSC is utilized in this work to damp out the frequency and voltage fluctuations incurred in the system when subjected to small perturbations. The mathematical modeling of TCSC utilized in this work is considered from [8], and the structure of TCSC is shown in Fig. 6.3.

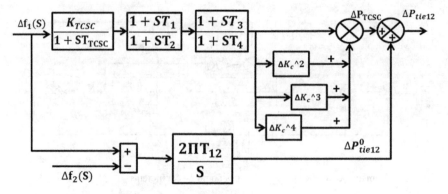

Fig. 6.3 Structure of TCSC as damping controller

Modeling of SMES

SMES devices come under the family of electromagnetic energy storage devices. Among all storage devices, the operational efficiency of SMES is high because of its static operation. SMES consists of a conductor coil and a power electronic conversion device. The conducting coil is enclosed in a container, maintained at cryogenic temperature, and the loss incurred with coil is nearly zero. Under normal conditions, SMES absorbs energy from the grid and stores it in the conductor coil; whenever the sudden small perturbations occur, SMES delivers energy instantly before the action taken up the generators speed governors and turbine. Thus, SMES enhances stability in the system by meeting the sudden loads. In this work, SMES devices are placed in both areas. The input given to the SMES device is frequency deviation (Δf), and output from SMES is change in control vector ($\Delta PSMES$) [3] (Fig. 6.4).

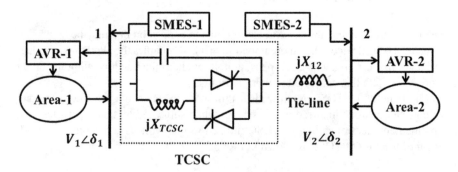

Fig. 6.4 Single-line representation of considered system with coordinated scheme

Objective Function and Optimization Algorithm

Objective Function

In power system LFC-AVR, the generation and load demand mismatch is compensated by minimization of area control error (ACE) [9]. It is dragged down to zero by adjusting the controller parameters while satisfying the constraints. The tuning procedure of the controller is done subjected to error squared over integral (ISE) minimization performance index given in Eq. 6.1.

$$\text{ISE} = \int_0^{T_{\text{sim}}} [\Delta f_1^2 + \Delta f_2^2 + \Delta P_{\text{tie}1,2}^2 + \Delta V_1^2 + \Delta V_2^2] \, dt \quad (1)$$

Differential Evolution—Artificial electric field algorithm (DE-AEFA)

A new DE-AEFA algorithm is presented in this paper and combines the differential evolution (DE) [10] algorithm with artificial electric field algorithm (AEFA) [11]. The problem formulated in this work is analyzing the stability of interconnected power system with considering the effect of AVR coupling. Moreover, the AVR loop will be connected to the LFC loop through some cross-coupling coefficients; with this the total problem becomes more complex. So, optimizing the complex problem needs a novel evolutionary algorithm having stochastic nature with deterministic transition rules. The proposed DE-AEFA algorithm possesses these characteristics. However, AEFA algorithm is adapted to the system and came to know that the problem converges are unseasonable and particles having the tendency to easily drop into the best solution. So, an additional approach and methodology are essential to strengthen AEFA algorithm. DE is one of the best among evolutionary algorithms. It uses mutation, recombination and selection operations in search of global best solutions. Moreover, mutation and recombination execute different roles. Recombination is a convergence operator which aims to pull the population towards local best solutions, and mutation is divergence operator which bankrupts populations and attempts to discover solution in a wide space. These benefits of DE have been applied to the AEFA to make the proposed algorithm more efficient in searching and finding global best solution. The step-by-step procedural of proposed DE-AEFA algorithm is given in [12].

The superiority of proposed algorithm is tested on standard benchmark sphere function. Figure 6.5 depicts the variation of initial and final values of sphere function for 100 trails. From Fig. 6.5, it is revealed that in most of the cases, the initial and

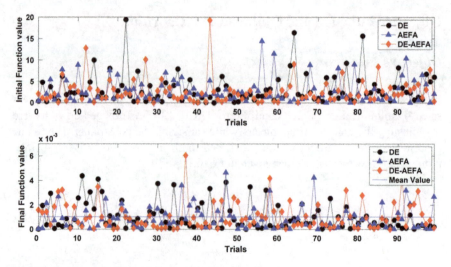

Fig. 6.5 Variation of initial and final values for sphere function

final values of sphere function are below the mean value, and this happens because of the effectiveness of proposed DE-AEFA algorithm (Fig. 6.6).

Simulation Results and Analysis

System Performance with Different Optimization Algorithms

In combined system, PID controller is taken as secondary controller for both LFC and AVR loops. The system analysis is done for applying 1%SLP in area-1. Controller parameters are optimized with PSO, AEFA and DE-AEFA algorithms one at a time. The variations in frequency, voltage and power flow in tie-line due to disturbance impressed in area-1 are shown in Fig. 6.7, and the response corresponding settling time (T_s) is noted in Table 6.1, and it is observed that the objective function of proposed DE-AEFA-based PID controller is improved by 96.73 and 93.86% than that of PSO and AEFA optimizations (Fig. 6.8).

Performance Comparison of System Responses with and Without Considering AVR Coupling

However, a comparison is also made in the system response with and without considering automatic voltage regulator using PID controller optimized by DE-AEFA algorithm. Response in Fig. 6.9 illustrates that there may be more deviation in system

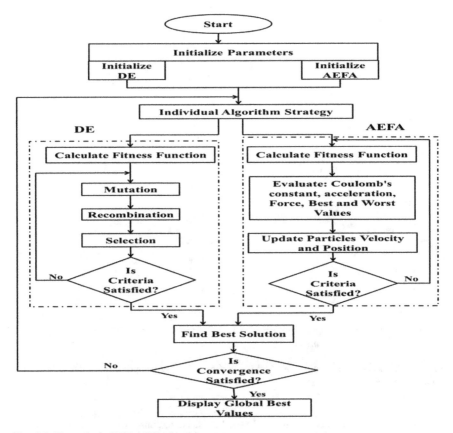

Fig. 6.6 Flow chart of DE-AEFA algorithm

frequency when AVR is incorporated in LFC loops, because the frequency deviation is analogues to real power generation variation and load demand. System voltage had a considerable impact on megawatt power.

Performance of the System with Incorporation of SMES and TCSC

Later, the system is incorporated with SMES and TCSC devices. SMES is connected in both the areas, and TCSC is placed in the tie-line with series. With the incorporation of these devices, both frequency and voltage variations are improved and the deviation in amount of power flow through tie-line is enhanced with TCSC. By considering these two devices, the stability of the system improvised and reaches steady state in less time. The behavior responses of the system with SMES and TCSC for 1%SLP in area-1 are depicted in Fig. 6.10, and the T_S is tabulated in Table 6.2. The optimized

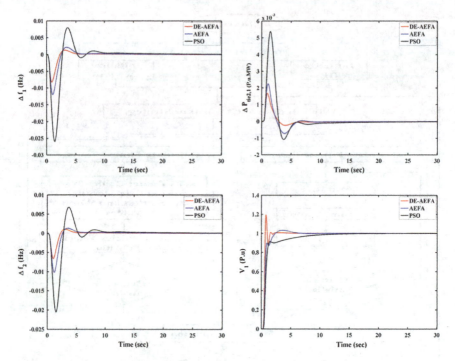

Fig. 6.7 Performance comparison of system responses with different optimization algorithms, $\Delta f1$, ΔP_{tie21}, $\Delta f2$, $V1$

Table 6.1 Numerical results of combined LFC-AVR system responses when 1%SLP in area-1

Optimized controller	Settling time T_S (s)					ISE × 10^{-3}
	$\Delta f1$	$\Delta f2$	ΔP_{tie21}	$V1$	$V2$	
PSO:PID	23.28	22.26	24.96	13.67	13.72	30.73
AEFA:PID	14.64	8.12	9.98	6.96	7.20	16.27
DE-AEFA:PID	10.11	7.78	7.21	4.87	4.90	0.998

gains of the PID controller with studied algorithms and in different cases are shown in Table 6.3.

It is discernible from Table 2 that the DE-AEFA optimized PID controller performance in the system is improved by 90.26% when SMES incorporated in both the areas and by 43.33% when incorporating both SMES and TCSC in the test system. This infers that the impact of SMES and TCSC is predominant in coordination with proposed DE-AEFA optimized PID controller in the considered system analysis.

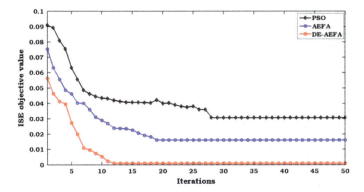

Fig. 6.8 Convergence characteristics of optimization algorithms

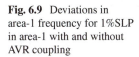

Fig. 6.9 Deviations in area-1 frequency for 1%SLP in area-1 with and without AVR coupling

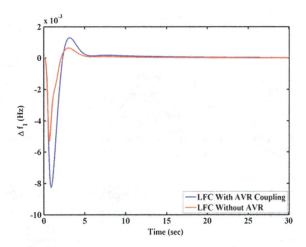

Conclusion

Stabilizing voltage and frequency is essential for maintaining interconnected power system healthy. In this paper, the effect of combined LFC-AVR of two-area multi-type generation sources with coordinated performance of SMES and TCSC with PID controller is analyzed. The analysis is carried out when 1%SLP is applied in area-1. DE-AEFA algorithm is presented to optimize the controller gains in both LFC and AVR loops simultaneously. The system behavioral responses under disturbances are compared with those of PSO, AEFA optimized PID controller. Numerical and graphical results proved that system performance is improved and variations are settled early with DE-AEFA-based controller and judged as the best controller.

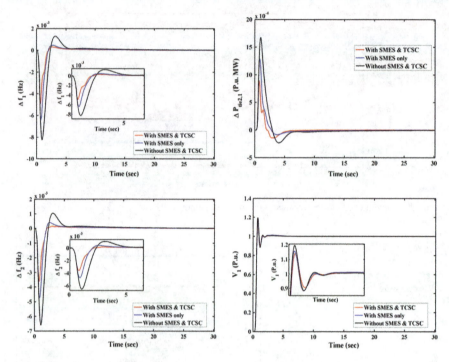

Fig. 6.10 Performance comparison of system responses with and without considering SMES and TCSC, $\Delta f1$, ΔP_{tie21}, $\Delta f2$, $V1$

Table 6.2 Numerical results of combined LFC-AVR with SMES and TCSC

PID: DE-AEFA controller with cases	Settling time T_S (s)					ISE × 10^{-3}
	$\Delta f1$	$\Delta f2$	ΔP_{tie21}	$V1$	$V2$	
Without SMES and TCSC	10.11	7.78	7.21	4.87	4.90	0.998
With SMES Only	9.12	7.08	6.88	4.41	4.39	0.171
With SMES and TCSC	7.93	6.73	6.11	2.25	2.21	0.097

Table 6.3 Optimized gains of PID controller under various cases

Controller	Area-1		Area-2	
	LFC	AVR	LFC	AVR
PID:PSO	$K_P = 2.006$ $K_I = 1.742$ $K_D = 0.259$	$K_P = 0.692$ $K_I = 0.196$ $K_D = 0.294$	$K_P = 1.923$ $K_I = 1.872$ $K_D = 0.260$	$K_P = 0.689$ $K_I = 0.197$ $K_D = 0.303$
PID:AEFA	$K_P = 2.237$ $K_I = 1.926$ $K_D = 0.642$	$K_P = 0.782$ $K_I = 0.199$ $K_D = 0.317$	$K_P = 2.230$ $K_I = 1.892$ $K_D = 0.660$	$K_P = 0.782$ $K_I = 0.199$ $K_D = 0.317$
PID:DE-AEFA	$K_P = 2.651$ $K_I = 2.014$ $K_D = 0.913$	$K_P = 0.812$ $K_I = 0.210$ $K_D = 0.396$	$K_P = 2.457$ $K_I = 2.097$ $K_D = 0.909$	$K_P = 0.809$ $K_I = 0.216$ $K_D = 0.389$
PID:DE-AEFA with SMES only	$K_P = 2.873$ $K_I = 2.183$ $K_D = 1.017$	$K_P = 0.891$ $K_I = 0.210$ $K_D = 0.310$	$K_P = 2.790$ $K_I = 2.162$ $K_D = 1.009$	$K_P = 0.799$ $K_I = 0.312$ $K_D = 0.353$
PID:DE-AEFA with SMES and TCSC	$K_P = 2.901$ $K_I = 2.208$ $K_D = 1.017$	$K_P = 0.903$ $K_I = 0.296$ $K_D = 0.390$	$K_P = 2.893$ $K_I = 2.197$ $K_D = 1.104$	$K_P = 0.914$ $K_I = 0.307$ $K_D = 0.372$

References

1. Saadat H (1999) Power system analysis. McGraw-Hill, USA
2. Gupta A, Chauhan A, Khanna R (2014) Design of AVR and ALFC for single area power system including damping control. In: Recent advances in engineering and computational sciences (RAECS). Chandigarh, pp 1–5
3. Vijaya Chandrakala KRM, Balamurugan S (2016) Simulated annealing based optimal frequency and terminal voltage control of multi-source multi area system. Electr Power Energy Syst 78:823–829
4. Lal DK, Barisal AK (2019) Combined load frequency and terminal voltage control of power systems using moth flame optimization algorithm. J Electr Syst Inf Technol 6(1):8
5. Rajbongshi R, Saikia LC (2018) Combined voltage and frequency control of a multi-area multi-source system incorporating dish-stirling solar thermal and HVDC link. IET Renew Power Gener 12(3):323–334
6. Rajbongshi R, Saikia LC (2017) Combined control of voltage and frequency of multi-area multisource system incorporating solar thermal power plant using LSA optimized classical controllers. IET Gener Trans Distrib 11(10):2489–2498
7. Shankar R, Bhushan R, Kalyan C (2016) Small-signal stability analysis for two-area interconnected power system with load frequency controller in coordination with FACTS and energy storage device. Ain Shams Eng J 7(2):603–612
8. Lal DK, Barisal AK (2017) Comparative performances evaluation of FACTS devices on AGC with diverse sources of energy generation and SMES. Cogent Eng 4:1318466
9. Mukherjee V, Ghoshal SP (2007) Comparison of intelligent fuzzy based AGC coordinated PID controlled and PSS controlled AVR system. Electr Power Energy Syst 29:679–689
10. Stom R, Price K (1997) Differential evolution—a simple and efficient heuristic for global optimization over continuous spaces. J Glob Optim 11(4):341–359
11. Yadav A (2019) Artificial electric field algorithm for global optimization. In: Swarm and evolutionary computation base data. https://doi.org/10.1016/j.swevo.2019.03.013

12. Naga Sai Kalyan Ch, Sambasivarao G (2020) Automatic generation control of multi-area hybrid system considering communication time delays. J Adv Res Dyn Control Syst 12(02):2071–2083

Chapter 7
Optimal Fuzzy-PID Controller Design by Grey Wolf Optimization for Renewable Energy-Hybrid Power System

Himanshu Garg and Jyoti Yadav

Introduction

The power sectors depend most on conventional fossil fuels for its electricity demands which are now becoming very costly and also cause environmental pollution. Moreover, these resources are available in limited quality in nature. Therefore, researchers are exploring alternative hybrid energy [1] technologies like wind, solar, tidal and geothermal renewable resources because of low cost and environment-friendly [2]. The major problem with these energy resources is that the power produced is stochastic in nature which is due to change in climate conditions for particular period of time. This results unbalancing in load demand and power delivered which causes reliability issues [3].

Batteries, flywheels as well as ultracapacitors are some energy storage devices which can store surplus energy produced and supply when more load of demand is more than generated. The efficient control scheme is required to sustain the power quality and minimum grid frequency distortion. Generally, different controllers are required for controlling each energy storing whereas beauty of this hybrid system is that only single central controller is essential to main constant power flow among load and generation side. Therefore, overall hybrid system will be cost-effective, low maintenance and minimum tuning parameters required and also avoid the controller performance deterioration due to interaction of multiple loops.

Various researchers are exploring different control strategies for controlling the proportional, integral and derivative (PID) controller which is furthermost widely used feedback controller in the power application industries. The fuzzy logic control

---

H. Garg (✉) · J. Yadav
Department of Instrumentation and Control Engineering, NSUT, New Delhi, India
e-mail: Himgarg.hg@gmail.com

© The Editor(s) (if applicable) and The Author(s), under exclusive license to Springer Nature Singapore Pte Ltd. 2021
A. K. Singh and M. Tripathy (eds.), *Control Applications in Modern Power System*, Lecture Notes in Electrical Engineering 710,
https://doi.org/10.1007/978-981-15-8815-0_7

(FLC) used widely for the control of nonlinear dynamic systems of many applications. The combination of these two controllers fuzzy-PID [4] will improve the overall capability of controller in handling system's non-linearities and uncertainties in model parameters.

Here, the comparative analysis of PID [5] as well as fuzzy-PID is reported. The energy generated by renewable sources (solar and wind) is stochastic in nature which causes continuous variation in grid frequency. These variations need to be kept within some limit to prevent load from any malfunction. The energy storing elements receive a control signal from designed controller, associated with structure, to release or absorb the power to or from the electricity grid depending on requirement. Also, in case to fulfil short-term load requirement, the diesel engine receives a signal from the controller, associated with system, to release large spurts of power in the grid. The efficacy of fuzzy-PID controller is often limited by non-availability of systematic method to find the rule base. Therefore, effective tuning method is required to find out the optimal controller parameters. Here, grey wolf optimization (GWO) [6] [7] is used for finding fuzzy-PID controller parameters.

The next section gives description of hybrid system and its components. Section 7.3 gives information regarding controller design. Section 7.4 explains GWO together with the objective function. Section 7.5 presents the comparative analysis of various control strategies. Section 7.6 contains the conclusion followed by references in the last section.

Hybrid Power System Components

The system contains different type of systems to generate power and various energy storage elements. Detail is provided in the next subsections. The block diagram of the hybrid power system, consisting of its element, is displayed in Fig. 7.1.

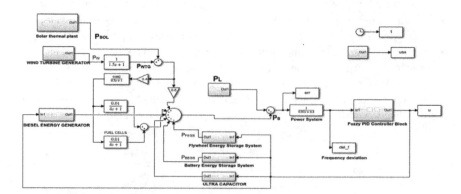

Fig. 7.1 Simulation diagram of complete hybrid power system

Table 7.1 Hybrid power system parameters of various components

Elements	Gain	Time constant
Solar power (SP)	KS = 1.80, KT = 1	TS = 1.80, TT = 0.30
Wind power (WP)	KWP = 1	TWP = 1.50
Diesel engine generator (DEG)	KDEG = 0.0030	TDEG = 2
Flywheel system (FS)	KFS = −0.010	TFS = 0.10
Battery system (BS)	KBS = −0.0030	TBS = 0.10
Ultracapacitor (UC)	KUC = −0.70	TUC = 0.90
Aqua electrolyser (AE)	KAE = 0.0020	TE = 0.50
Fuel cell (FC)	KFC = 0.010	TFC = 4

The transfer functions of the fuel cells (FC), WTG, DEG and that of solar thermal power generator (STPG) are modelled by first-order transfer functions [2]. The values of all parameters are given in Table 7.1.

$$G_{\text{WTG}}(S) = \frac{K_{\text{WTG}}}{1 + (S * T_{\text{WTG}})} = \frac{\Delta P_{\text{WTG}}}{\Delta P_W} \tag{7.1}$$

$$G_{\text{STPG}}(S) = \frac{K_S}{1 + (S * T_S)} * \frac{K_T}{1 + (S * T_T)} = \frac{\Delta P_{\text{STPG}}}{\Delta P_{\text{Sol}}} \tag{7.2}$$

$$G_{\text{FCk}}(S) = \frac{K_{\text{FC}}}{1 + (S * T_{\text{FC}})} = \frac{\Delta P_{\text{FCk}}}{\Delta P_{\text{AE}}}, K = 1, 2 \tag{7.3}$$

$$G_{\text{DEG}}(S) = \frac{K_{\text{DEG}}}{1 + (S * T_{\text{DEG}})} = \frac{\Delta P_{\text{DEG}}}{\Delta u} \tag{7.4}$$

Modelling of Various Energy Storage Components

Fuel cells (FC), ultracapacitor (UC) and battery system (BS) are used as energy storage elements in this power system. The order of FC, BC and UC is a first-order system. The values of time constant and gains provided in Table 7.1.

Modelling of the Aqua Electrolyser

. Making use of the subsystem of the power produced from the energy of renewable source, the aqua electrolyser yields hydrogen, e.g. wind and solar photovoltaic cells. Representation of dynamics of aqua equalizer can be done by the transfer function for small signal analysis. It uses $(1 - Kn)$ portion of net power of wind TG and solar TPG to generate hydrogen, that is being used again by two fuel cells just to generate power and then fed into the grid [1].

$$GA = K_{AE}/1 + sT_{AE}$$
$$= \Delta P_{AE}/((\Delta P_{WP} + \Delta P_{SP})(1 - K_n))$$
$$Kn = P_t/(P_{WP} + P_{SP}), F_n = 0.6 \tag{7.5}$$

Transfer Function Modelling for Power System

$$G_{SYSTEM}(s) = \Delta f/\Delta P_e = 1/(D + I_s) \tag{7.6}$$

where G_{SYSTEM} = T.F. of the system. 'I_s' is constant value of inertia, and 'D' is damping constant. The value of 'I_s' is 0.03 The value of 'D' is 0.4. ΔP_e = error in the power and also in demand load. Δf = the fluctuation in frequency deviation of the system.

Wind, Solar and Demand Load Power Modelling

The equation for modelling the fluctuations on generation of wind, solar power and demand load is as below [1]:

$$P = \xi * \varphi \sqrt{\eta} \left(1 - G(s)\right) + \eta) * \frac{\beta}{\eta} * ¥ \tag{7.7}$$

where P stands for load power. ξ stands for indiscriminate element of power. η stands for the average power. ¥ stands for time-based signal switching with gain that controls the sudden change in mean output of power. (φ and β) are factors to standardize the powers of ξ such that the per-unit level is matched. $G(s)$ is the low pass transfer function.

Parameters of the equation for generation of solar power are [1]:

-

$$\xi \sim U(-1,1); \varphi = 0.7; \beta = 0.1; \eta = 2; ¥ = 1.111 * H(t) - 0.555 * H(t-40);$$

$$G(s) = \frac{1}{10^4 s + 1} \qquad (7.8)$$

Parameters of the equation for generation of demand load are:

-

$$\xi \sim U(-1,1); \varphi = 0.8; \eta = 100, \beta = 0.1, ¥ = H(t) - \left(\frac{0.8}{\xi} * H(t-40)\right);$$

$$G(s) = \frac{300}{(300 * s) + 1} - \frac{1}{(1800 * s) + 1} \qquad (7.9)$$

Equation for the parameters used for the generation of wind power is [1]:

$$\xi \sim U(-1, 1), \varphi = 0.8, \beta = 10, \eta = 2,$$

$$G(s) = \frac{1}{10^4 s + 1} \qquad (7.10)$$

where Heaviside step function denoted by $H(t)$. After modelling various energy storage elements and other components, we observed that there is a need of the continuous flow of power in between generating system and the load. For this, we need a controller that can minimize the frequency deviation of the system. Section 7.3 gives information about designed controllers.

Design of Controller

PID Controller

This controller consists of derivative, proportional as well as integral control actions. The error is linked to the proportional part, integral is the one that integrates the available error and hence improvises the steady state, and the derivative is the one that derivates error and hence improvises transient response [5]. Here, $e(t)$ is error signal which is difference between reference frequency (f_{ref}) and frequency obtained at the output $f(t)$.

$$e(t) = f_{\text{ref}}(t) - f(t) \qquad (7.11)$$

The output of controller is denoted by $y(t)$ in time domain such that K_I is integral gain. K_p = the proportional gain, and K_d = derivative gain.

$$y(t) = K_p * e(t) + K_d * \frac{d_e(t)}{dt} + K_I \int_0^t e(t) * d\tau \qquad (7.12)$$

Fuzzy-PID Controller

Typical fuzzy-PID controller consists of three inputs which are the error, integration as well as the derivative of error. In this form, FLC requires large number of rule base which is required for three inputs [3]. For example, seven membership functions, each containing input, deliver with the $7 \times 7 \times 7$, that is 343 number of rules; therefore, it becomes very tough to maintain this complexity. Therefore, the most commonly used and simplified fuzzy-PID controller configuration is fuzzy PI + fuzzy PD shown in Fig. 7.2 [4]. In this research, we have optimized the parameters using the technique called GWO technique to reduce the frequency deviation [4] (Table 7.2).

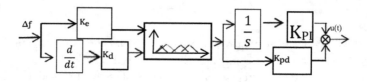

Fig. 7.2 Fuzzy-PID controller block diagram [4]

Table 7.2 Rule base of fuzzy logic controller

$e \frac{de}{dt}$	N_H	N_M	N_L	Z_R	P_L	P_M	P_H
N_H	N_H	N_H	N_H	N_H	N_M	N_L	Z_R
N_M	N_H	N_H	N_H	N_M	N_L	Z_R	P_L
N_L	N_H	N_H	N_M	N_L	Z_R	P_L	P_M
Z_R	N_H	N_M	N_L	Z_R	P_L	P_M	P_H
P_L	N_M	N_L	Z_R	P_L	P_M	P_H	P_H
P_M	N_L	Z_R	P_L	P_M	P_H	P_H	P_H
P_H	Z_R	P_L	P_M	P_H	P_H	P_H	P_H

Grey Wolf Optimization

Over the last two decades, various techniques of optimizing have become very popular. We have used grey wolf optimization (GWO) algorithm.

The GWO algorithm imitates the control order or hierarchy as well as the pattern to hunt by grey wolves in nature. To imitate the leadership hierarchy, four types of grey wolves are employed which are alpha (α), beta (β), omega (ω) and delta (δ) [7].

Social Hierarchy

To design mathematical model, we consider alpha (α)-fittest spot, beta (β)-second best spot and delta (δ)-third best spot.

Encircling Prey

During a hunt, these wolves circle around the prey. Mathematically, it can be expressed as [1]:

$$\vec{D} = \left| \vec{V} \cdot \vec{W_P}(t) - \vec{W}(t) \right| \tag{7.13}$$

$$\vec{W}(t+1) = \left| \vec{W_P}(t) - \vec{P}.\vec{D} \right| \tag{7.14}$$

Here, 't' is the present iteration, $\vec{P}$ and $\vec{V}$ are coeff. vectors, $\vec{W_P}$ is the positional vector of pray, and $\vec{W}$ is positional vector of the GWO. $\vec{P}$ as well as $\vec{V}$ are calculated as given below:

$$\vec{P} = 2\vec{p}.\vec{q_1} - \vec{p} \tag{7.15}$$

$$\vec{V} = 2.\vec{q_2} \tag{7.16}$$

Where components of $\vec{p}$ are decreasing from [2, 0] along the path of iterations. The vectors $\vec{q_1}$, $\vec{q_2}$ are random vectors in range [0, 1].

Hunting

Grey wolves hold the skill to identify the place of the prey and make a circle around them. Generally, the hunt is directed by alpha. The hunting can also be joined by beta and delta now and then [7]. In order to calculate the i hunting actions mathematically of grey wolf, we presume that the 'α', 'β' and 'δ' contains desirable information regarding the position of the preys. Hence, we keep the better solutions acquired till now and update the position of more grey wolves by using 'α' and 'β' position.

$$\vec{D_\alpha} = \left|\vec{V_1} \cdot \vec{W_\alpha} - \vec{W}\right|, \vec{D_\beta} = \left|\vec{V_2} \cdot \vec{W_\beta} - \vec{W}\right|, \vec{D_\delta} = \left|\vec{V_3} \cdot \vec{W_\delta} - \vec{W}\right|$$
$$\vec{W_1} = \left|\vec{W_\alpha}(t) - \vec{P_1}.\vec{D_\alpha}\right|, \vec{W_2} = \left|\vec{W_\beta}(t) - \vec{P_2}.\vec{D_\beta}\right|, \vec{W_3} = \left|\vec{W_\delta}(t) - \vec{P_3}.\vec{D_\delta}\right|$$
$$\vec{W}(t+1) = \frac{\vec{W_1} + \vec{W_2} + \vec{W_3}}{3} \tag{7.17}$$

Attacking the Prey (Exploitation)

Whenever the prey becomes still, the grey wolves complete their hunting by attacking that prey. To design the approach precisely, we reduce the $\vec{p}$ value. The varying array of $\vec{P}$ is reduced by $\vec{p}$, i.e. $\vec{P}$ is an arbitrary value in the range $[-p, p]$ where p is declined from 2 to 0 in the path of iterations. The next spot of a search agent is any spot b/w its present spot and the location of the pray when random values of $\vec{P}$ are in $[-1, 1]$.

For optimization, the objective function (say 'O') has been taken as the performance index for simulation period, by using the sum of fluctuation of controller's signal (u) from its steady state value (uss) and the square of frequency deviation.

$$O = 0.5 * (\Delta f)^2 + 0.5 * (u - \text{uss})^2 \tag{7.18}$$

uss is defined as the steady state control signal:

$$\text{uss} = 0.81 H(t) + 0.17 H(t - 40) + H(t - 80) \tag{7.19}$$

Results and Discussions

In the present work, the hybrid power system is controlled by PID [6] and fuzzy-PID [4] controller to achieve minimum frequency deviation. Later, the gains of fuzzy-PID controller are optimized with the help of GWO algorithm [7]. The simulation is carried out using MATLAB software. To optimize the controller parameters, we have used grey wolf optimization Fig. 7.4 for various controller structure is deviation of control signals that demonstrates controller actions needed in fuzzy-PID tuned GWO is lesser as compared to another controller. Figure 7.5 is 'frequency deviation' of various types of control structures, and it is seen that the oscillation band of fuzzy-PID with GWO is lesser in contrast to other used controllers. Table 7.3 shows the values of parameters for different controllers (Fig. 7.3).

This limited variation in fuzzy-PID tuned GWO is extremely significant as mechanical elements like DEG, BS, FS, etc. are being initialized by using control signal. Because of this, the oscillation reduces.

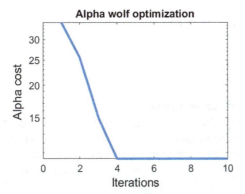

Fig. 7.3 shows the convergence curve of GWO for fuzzy-PID

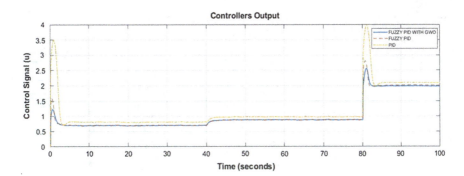

Fig. 7.4 Control signal deviations for various controllers

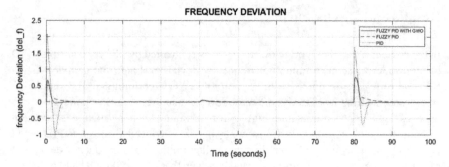

Fig. 7.5 Frequency deviation for various controllers

Table 7.3 Parameters of controllers

Controller	K_P	K_I	K_D	K_{PI}
PID	0.95	1.83	0.02	–
Fuzzy-PID	2.23	0.81	0.05	10.3
GWO tuned fuzzy-PID	1.44	1.33	0.79	0.08

Robustness Test

Variations in Parameters of Ultracapacitor

In this hybrid power system, modification of the variables of ultracapacitor (UC) would interrupt structure much in contrast to another components as UC shares maximum volume of power in contrast to various components linked in response track. Hence, to check the worst condition, testing for robustness is required. 50 and 30% decrease and increase in time constant as well as in gain which is done. Table 7.4 displays the integral square of error (ISE) for all UC.

Table 7.4 Variation of parameter of ultracapacitor for robustness test

Cases	Integral square of error (ISE)		
	PID	Fuzzy-PID	GWO tuned fuzzy-PID
Standard	1.51	1.32	1.15
30% Inc.	1.92	1.35	1.2
30% Inc.	1.19	1.36	1.24
50% Inc.	2.3	1.42	1.21
50% Inc.	1.37	0.90	0.76

Robustness Test After Disconnecting Various Energy Storage Elements

By removing various elements of hybrid system, we have verified the robustness of the simulated values of controller parameter. We have considered three cases in which FS, BS and DEG are disconnected one by one. As we removed these components, it either increases or decreases the system performance or increases as per the conditions. From Table 7.4, it is observed that the controller and the frequency deviation effect is minimum with GWO tuned fuzzy-PID in comparison with fuzzy-PID and PID controllers. It is observed by concluding Tables 7.3, 7.4 and 7.5 that values of integral square of error (ISE) are least for all maximum cases of GWO tuned fuzzy-PID controller as compared to the fuzzy-PID and PID controllers.

The comparison of proposed approach with literature is given in Table 7.6 which shows that frequency and control signal deviation and the ISE value are minimum with GWO tuned fuzzy-PID. Table 7.6 shows that GWO tuned fuzzy-PID controller-based system is robust and effectual under parameter variations as well as component removal.

Table 7.5 Robustness test after disconnecting various energy storage elements

Controller	Component removed	ISE	ISDCO	O
PID	Battery system (BS)	1.71	2.72	4.42
	Diesel energy generator (DEG)	1.74	2.34	4.08
	Flywheel system (FS)	1.45	2.98	4.45
Fuzzy-PID	Battery system (BS)	1.43	2.77	4.20
	Diesel energy generator (DEG)	1.33	2.79	4.12
	Flywheel system (FS)	1.46	2.83	4.31
GWO tuned fuzzy-PID	Battery system (BS)	1.26	2.69	3.95
	Diesel energy generator (DEG)	1.17	2.88	4.05
	Flywheel system (FS)	1.26	2.93	4.2

Table 7.6 Comparison of results with literature

Parameters	Proposed approach	Literature [2]
Frequency deviation	0.6	0.8
Controller signal	2.5	3
ISE	1.15	1.55
ISDOE	2.91	2.96
O	4.06	4.51

Conclusion

The main aim of proposed research work is to keep persistent power flow between generating station and load with minimum grid frequency variations with the use of single controller. To achieve this objective, centralized control scheme is employed for hybrid power system using PID, fuzzy-PID and GWO tuned fuzzy-PID controllers. The robustness test has also been performed under ultracapacitor parameter variations and by removing elements of hybrid power system. It is, hereby, concluded that fuzzy-PID tuned with GWO performs better than the PID as well as standard fuzzy-PID controllers. The result shows that the frequency and control signal deviation and ISE value are minimum in GWO tuned fuzzy-PID in contrast to other designed controllers. The conclusion is that GWO tuned fuzzy-PID controller shows the suitability to effectively maintain persistent power flow between the load and generating stations for hybrid power system.

References

1. Singh J, Singh V, Rani A, Yadav J, Mohan V (2018) Performance analysis of fractional order fuzzy-PID controller for hybrid power system using WOA. In: 2nd International conference on trends in electronics and informatics, pp 976–984
2. Pan I, Das S (2016) Fractional order fuzzy control of hybrid power system with renewable generation using chaotic PSO. ISA Trans 62:19–29
3. Yadav J, Rani A, Singh V (2016) Performance analysis of fuzzy-PID controller for blood glucose regulation in type-1 diabetic patients. J Med Syst 40(12):254
4. Pan I, Das S, Gupta A (2011) Tuning of an optimal fuzzy PID controller with stochastic algorithms for networked control systems with random time delay. ISA Trans 50:28–36
5. Astrom KJ, Hagglund T (1995) PID controllers: theory, design, and tuning, 2nd edn. Instrument Society of America
6. Jun YC, Jin L, Bing GC (2005) Optimization of fractional order PID controllers based on genetic algorithms. In: Proceedings of the fourth international conference on machine learning and cybernetics, Guangzhou
7. Soni V, Parmar G, Kumar M, Panda S (2016) Hybrid grey wolf optimization-pattern search (HGWO-Ps) optimized 2d of PID controllers for load frequency control in interconnected thermal power plants. ICTACT J Soft Comput 06(03):1244–1256

Chapter 8
A New Classical Method of Reduced-Order Modelling and AVR System Control Design

Nafees Ahamad, Afzal Sikander, and Gagan Singh

Introduction

Approximation methods are very important in control systems to make a system simple and cost-effective. The simplified model should retain the essential characteristics of the original higher-order system and map them closely. Some authors have proposed several order reduction techniques in the existing literature such as Pade, Routh, truncation method, Mikhailov criterion, and pole clustering [1–15]. These techniques are mainly categorized as time and frequency domain reduction techniques. In the frequency domain, the first method was introduced by Davison [1] in 1966, and then, after ongoing research in 1974, the Padé approximation method was proposed by Shamash [2]. This method is simple in terms of calculations and matches the initial time moments and minimizes the difference between steady-state values of the reduced and original system. The main problem of this technique is that the simplified system can be unstable, even if the original model is stable. Routh [3] and modified Routh approximation [4] improve the drawbacks of the Pade approximation method. There are several methods for order reduction based on different criteria such as Routh stability criterion [5], truncation method of reduction [6], Cauer continued fraction methods [7], factor division [8], and error minimization [9–16]. There is a mixed method that uses the Mikhailov criterion and factor division algorithm [17] which acquires the stability characteristics of the simplified system if the

---

N. Ahamad (✉) · G. Singh
Department of Electrical and Electronics & Communications Engineering, DIT University, Dehradun, India
e-mail: naf_001@yahoo.com

A. Sikander
Department of Instrumentation and Control Engineering, Dr. B. R. Ambedkar National Institute of Technology, Jalandhar, India
e-mail: afzals@nitj.ac.in

© The Editor(s) (if applicable) and The Author(s), under exclusive license to Springer Nature Singapore Pte Ltd. 2021
A. K. Singh and M. Tripathy (eds.), *Control Applications in Modern Power System*, Lecture Notes in Electrical Engineering 710,
https://doi.org/10.1007/978-981-15-8815-0_8

original is stable. These mixed techniques are developed by utilizing the advantages of two different techniques [18, 19].

A new classical computationally simple and very effective method based on direct coefficients comparison is proposed in this study, where a higher-order system is made equivalent to the desired reduced-order system. After cross-multiplication of the numerators and denominators of both the systems, the coefficient of power of 's' is compared, which gives linear homogeneous equations in terms of coefficient of the given system and desired reduced system. For a given nth-order system and desired reduced kth ($n < k$) order system, the no. of linear equations is $n + k$ of which only $2k + 1$ are independent equations and rest are dependent equations. These independent equations can be put in matrix form which can be converted into row echelon form by using the Gauss-Jordan method [20]. This results in a matrix that has one free variable. The non-trivial solution can be obtained for this homogeneous system by selecting any value for the free variable. Hence, any nth order system can be converted into a kth ($k < n$) order system. To prove the efficiency of the proposed classical approach, one benchmark example is considered from literature which is reduced to second order by this method. Another example is a practical problem of a power system, in which an AVR [21] is reduced by the proposed method and then its PID controller [21–25] is designed using the PSO technique [26]. For a unit step input, the response of the original and reduce system is drawn. It is discovered that the projected technique provides a close estimation to the original system in comparison with the Padé approximation [2] and some of the already existing methods as per the "integral of the square error (ISE)" [12] criterion. The proposed diminished order-based PID controller also works satisfactorily with the full-order AVR system.

Problem Statement

Let us assume a nth-order continuous time SISO system, given by the transfer function

$$G(s) = \frac{N(s)}{D(s)} = \frac{a_{n-1}s^{n-1} + a_2s^2 + a_1s + \cdots a_0}{b_n s^n + b_2 s^2 + b_1 s + \cdots b_0} \tag{8.1}$$

where $a_0, a_1, \ldots a_{n-1}$ and $b_0, b_1 \ldots b_n$ are the scalar constants.

The above system can be reduced to kth ($k < n$) order which is exemplified by the following transfer function

$$G_k(s) = \frac{N_k(s)}{D_k(s)} = \frac{d_{k-1}s^{k-1} + d_2s^2 + d_1s + \cdots d_0}{e_k s^k + e_2 s^2 + e_1 s + \cdots e_0} \tag{8.2}$$

where $d_0, d_1, \ldots d_{k-1}$ and $e_0, e_1 \ldots e_n$ are the scalar constants.

The reduced model should be such as to preserve the substantial features of the original higher-order system and to bring the error between the step response of the two systems as minimum as possible.

Proposed Methodology

Step 1 Comparing Eqs. (8.1) and (8.2) directly and multiplying the corresponding numerator and denominator.

$$(a_{n-1}e_k)s^{n+k-1} + (a_{n-2}e_k + a_{n-1}e_{k-1})s^{n+k-2}$$
$$+ (a_{n-3}e_k + a_{n-2}e_{k-1} + a_{n-1}e_{k-2})s^{n+k-3}$$
$$+ (a_{n-4}e_k + a_{n-3}e_{k-1} + a_{n-2}e_{k-2} + a_{n-1}e_{k-3})s^{n+k-4} \cdots$$
$$+ a_0 e_0 = (b_n d_{k-1})s^{n+k-1} + (b_{n-1}d_{k-1} + b_n d_{k-2})s^{n+k-2}$$
$$+ (b_{n-2}d_{k-1} + b_{n-1}d_{k-2} + b_n d_{k-3})s^{n+k-3}$$
$$+ (b_{n-3}d_{k-1} + b_{n-2}d_{k-2} + b_{n-1}d_{k-3} + b_n d_{k-4})s^{n+k-4} \cdots + b_0 d_0$$
(8.3)

Step 2 Comparing the coefficients of power of 's' and make Table 8.1.
All the equations given in Table 8.1 are not independent. Only $2k + 1$ equations out of $n + k$ are independent because there are $2k + 1$ unknown coefficients (e_k, e_{k-1} … e_0, d_{k-1}, d_k … d_0), and hence, rest $n - k - 1$ equation can be ignored.

Step 3 Write the above equations in matrix form as

Table 8.1 Coefficients corresponding to a different power of 's'

s^{n+k-1}	$a_{n-1}e_k = b_n d_{k-1}$
s^{n+k-2}	$a_{n-2}e_k + a_{n-1}e_{k-1} = b_{n-1}d_{k-1} + b_n d_{k-2}$
s^{n+k-3}	$a_{n-3}e_k + a_{n-2}e_{k-1} + a_{n-1}e_{k-2} = b_{n-2}d_{k-1} + b_{n-1}d_{k-2} + b_n d_{k-3}$
s^{n+k-4}	$a_{n-4}e_k + a_{n-3}e_{k-1} + a_{n-2}e_{k-2} + a_{n-1}e_{k-3} = b_{n-3}d_{k-1} + b_{n-2}d_{k-2} + b_{n-1}d_{k-3} + b_n d_{k-4}$
–	–
–	–
s^0	$a_0 e_0 = b_0 d_0$

$$\begin{bmatrix} a_{n-1} & 0 & 0 & \cdots 0 & -b_n & 0 & \cdots 0 \\ a_{n-2} & a_{n-1} & 0 & \cdots & -b_{n-1} & -b_n & \cdots \\ a_{n-3} & a_{n-2} & a_{n-1} & \cdots & -b_{n-2} & -b_{n-1} & \cdots \\ a_{n-4} & a_{n-3} & a_{n-2} & \cdots 0 & -b_{n-3} & -b_{n-2} & \cdots 0 \\ \cdot & a_{n-4} & a_{n-3} & \cdots a_{n-1} & -b_{n-4} & -b_{n-3} & \cdots -b_n \\ \cdot & \cdot & a_{n-4} & \cdots a_{n-2} \cdot & -b_{n-4} & \cdots -b_{n-1} \\ \cdot & \cdot & \cdot & \cdots a_{n-3} \cdot & \cdot & \cdots -b_{n-3} \\ a_0 & \cdot & \cdot & \cdots a_{n-4} \cdot & \cdot & \cdots -b_{n-4} \\ 0 & a_0 & \cdot & \cdots & -b_0 & \cdot & \cdots \\ \cdot & 0 & a_0 & \cdots & 0 & -b_0 & \cdots \\ \cdot & \cdot & 0 & \cdots & \cdot & 0 & \cdots \\ \cdot & \cdot & \cdot & & \cdot & \cdot & \cdots \\ 0 & 0 & 0 & \cdots & 0 & 0 & \cdots -b_0 \end{bmatrix} \begin{bmatrix} e_k \\ e_{k-1} \\ e_{k-2} \\ \cdot \\ \cdot \\ \cdot \\ e_0 \\ d_{k-1} \\ d_{k-2} \\ d_{k-3} \\ \cdot \\ \cdot \\ \cdot \\ d_0 \end{bmatrix} \quad (8.4)$$

It represents a homogeneous system $[A][x] = [0]$ and always has the solution $[x] = 0$. Any non-zero solution to $[A][x] = [0]$, if they exist, is called non-trivial solution. The size of the above matrices are $(n+k)$ by $(2k+1)$, $(2k+1)$ by 1, and $(2k+1)$ by 1, respectively.

Step 4 To get the solution of the above homogeneous equation, consider $2k+1$ rows only (first k rows and last one row) and interchange the last two rows and then finally convert this coefficient matrix into row echelon form by using Gauss-Jordan method. Now equation becomes

$$\begin{bmatrix} 1 & g_{1,1} & g_{1,2} & \cdots & g_{1,2k} \\ 0 & 1 & g_{21} & \cdots & g_{2,2k-1} \\ 0 & 0 & 1 & \cdots & g_{3,2k-2} \\ 0 & 0 & 0 & \cdots & g_{4,2k-2} \\ \cdot & \cdot & \cdot & & \cdots \\ \cdot & \cdot & \cdot & & \cdots \\ \cdot & \cdot & \cdot & & \cdots \\ 0 & 0 & 0 & \cdots & 1 \end{bmatrix} \begin{bmatrix} e_k \\ e_{k-1} \\ e_{k-2} \\ \cdot \\ \cdot \\ \cdot \\ e_0 \\ d_{k-1} \\ d_{k-2} \\ d_{k-3} \\ \cdot \\ \cdot \\ \cdot \\ 0 \end{bmatrix} = \begin{bmatrix} 0 \\ 0 \\ 0 \\ \cdot \\ \cdot \\ \cdot \\ \cdot \\ \cdot \\ \cdot \\ \cdot \\ \cdot \\ \cdot \\ 0 \end{bmatrix} \quad (8.5)$$

The sizes of the above matrices are $(2k+1)$ by $(2k+1)$, $(2k+1)$ by 1, and $(2k+1)$ by 1, respectively.

Let

$$A = \begin{bmatrix} 1 & g_{1,1} & g_{1,2} & \cdots & g_{1,2k} \\ 0 & 1 & g_{21} & \cdots & g_{2,2k-1} \\ 0 & 0 & 1 & \cdots & g_{3,2k-2} \\ 0 & 0 & 0 & \cdots & g_{4,2k-2} \\ \cdot & \cdot & \cdot & \cdots & \cdot \\ \cdot & \cdot & \cdot & \cdots & \cdot \\ \cdot & \cdot & \cdot & \cdots & \cdot \\ 0 & 0 & 0 & \cdots & 1 \end{bmatrix} \quad x = \begin{bmatrix} e_k \\ e_{k-1} \\ e_{k-2} \\ \cdot \\ e_0 \\ d_{k-1} \\ d_{k-2} \\ d_{k-3} \\ \cdot \\ \cdot \\ \cdot \\ d_0 \end{bmatrix} \& B = \begin{bmatrix} 0 \\ 0 \\ 0 \\ \cdot \\ \cdot \\ \cdot \\ \cdot \\ \cdot \\ \cdot \\ \cdot \\ 0 \end{bmatrix}. \text{ So the system}$$

is represented by the following equation

$$[A][x] = [B] \tag{8.6}$$

Here, the leading variable d_0 is called a free variable. Assuming any value to this free variable gives many solutions to the original homogeneous system. Again, let $d_0 = t$, then B matrix becomes as

$$B = \begin{bmatrix} 0 & 0 & \cdots & 0 & t \end{bmatrix}^T \tag{8.7}$$

Step 5 Solution of Eq. (8.6) is given by the following relation

$$[x] = [A]^{-1}[B] \tag{8.8}$$

Now the reduced kth order system is given by Eq. (8.2) with coefficients calculated as above.

Numerical Examples, Results, and Discussion

Example 1 Let us take an eighth-order benchmark control problem given by the following transfer function [19]

$$\frac{35s^7 + 1086s^6 + 13285s^5 + 82402s^4 + 278376s^3 + 51812s^2 + 482964s^1 + 194480}{s^8 + 21s^7 + 220s^6 + 1558s^5 + 7669s^4 + 24469s^3 + 46350s^2 + 45952s^1 + 17760} \tag{8.9}$$

Let the desired system is of second order, so now Eq. (8.2) becomes

$$G_2(s) = \frac{N_2(s)}{D_2(s)} = \frac{d_0 + d_1 s}{e_0 + e_1 s + e_2 s^2} \qquad (8.10)$$

As explained above, equating Eq. (8.1) with Eq. (8.2) and writing only five equations in matrix form after interchanging last two rows.

$$\begin{bmatrix} 35 & 0 & 0 & -1 & 0 \\ 1086 & 35 & 0 & -21 & -1 \\ 13285 & 1086 & 35 & -220 & -21 \\ 0 & 0 & 194480 & 0 & -17760 \\ 82402 & 13285 & 1086 & -1558 & -220 \end{bmatrix} \begin{bmatrix} e_2 \\ e_1 \\ e_0 \\ d_1 \\ d_0 \end{bmatrix} = \begin{bmatrix} 0 \\ 0 \\ 0 \\ 0 \\ 0 \end{bmatrix} \qquad (8.11)$$

Converting Eq. (8.11) into row echelon form

$$\begin{bmatrix} 1 & 0 & 0 & \frac{-1}{35} & 0 \\ 0 & 1 & 0 & -\frac{35}{1225} & \frac{-1}{35} \\ 0 & 0 & 1 & -\frac{185711}{42875} & \frac{351}{1225} \\ 0 & 0 & 0 & 1 & -\frac{39383085}{451463441} \\ 0 & 0 & 0 & 0 & 1 \end{bmatrix} \begin{bmatrix} e_2 \\ e_1 \\ e_0 \\ d_1 \\ d_0 \end{bmatrix} = \begin{bmatrix} 0 \\ 0 \\ 0 \\ 0 \\ 0 \end{bmatrix} \qquad (8.12)$$

Put $d_0 = 1$ and find other coefficients
$e_2 = 0.0025$, $e_1 = 0.0036$, $e_0 = 0.0913$, $d_1 = 0.0872$, and $d_0 = 1$
So reduced system is given as

$$G_2(s) = \frac{0.08723s + 1}{0.002492s^2 + 0.003576s + 0.09132} \qquad (8.13)$$

The unit step plot of the actual problem and simplified systems is presented in Fig. 8.1, and integral of square error (ISE) is compared in Table 8.2 by different methods. We see that the proposed classical technique offers better results compared to some of the current classical methods.

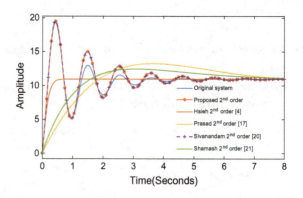

Fig. 8.1 Unit step response for original and simplified systems

Table 8.2 ISE comparison of different simplified systems

Systems	Simplified systems $G_2(s)$	$\text{ISE} = \int_0^\infty (y - y_r)^2 dt$
Proposed method	$\dfrac{0.08723s+1}{0.002492s^2 + 0.003576s + 0.09132}$	2.4960
Pade App [2]	$\dfrac{-100.4s - 76.35}{s^2 - 9.894 - 6.972}$	Unstable
Hsieh et al. [4]	$\dfrac{1.6666s + 61.271473}{0.047619s^2 + s + 5.59533}$	33.6621
Prasad et al. [17]	$\dfrac{8.690832s + 4.498007}{s^2 + 0.836381s + 0.41076}$	115.3929
Sivanandam et al. [20]	$\dfrac{35s + 401.21}{s^2 + 1.436s + 36.63}$	2.4995
Shamash [21]	$\dfrac{13.09095s + 5.271465}{s^2 + 1.245549s + 0.48139}$	90.7035

Example 2 Consider a closed-loop transfer function of an automatic voltage regulator (AVR) without PID controller used in a synchronous generator [21] given as

$$\frac{V_t(s)}{V_{\text{ref}}(s)} = \frac{10 + 0.1s}{11 + 1.51s + 0.555s^2 + 0.045s^3 + 0.0004s^4} \quad (8.14)$$

where $V_t(s)$ is the terminal voltage and $V_{\text{ref}}(s)$ is the reference voltage of the generator. In the above AVR system, the open-loop transfer function has been simplified to second order by the proposed technique which results in the following open-loop transfer function

$$\left(\frac{V_t(s)}{V_{\text{ref}}(s)}\right)_{\text{red}} = \frac{25}{2.5 + 3.75s + 1.35s^2} \quad (8.15)$$

Now, two different proportional–integral–derivative (PID) controllers are designed using particle swarm optimization (PSO) one for full-order and second for reduced-based AVR system. The PSO parameters chosen to design these controllers are Swarm Size $= 50$, Maxi Iterations $= 200$, Kappa $= 1$, Phi1 $=$ Phi2 $= 2.05$, Chi $= 0.72984$, and $C1 = C2 = 1.4962$, and the two objective functions named as "integral time-absolute error (ITAE) and integral time-square error (ITSE)" [12] are taken which are given as $f_{\text{ITAE}} = \int_0^{T_{ss}} t|y(t) - y_r(t)|dt$ and $f_{\text{ITSE}} = \int_0^{T_{ss}} t|y(t) - y_r(t)|^2 dt$. Different step responses are drawn in Fig. 8.2. This figure shows that even a reduced-based PID controller performs well with the original AVR system which confirms the efficacy of the proposed methodology. The values of different controller's coefficients such as K_p, K_i, and K_d are revealed in Table 8.3 with their time response specifications. "Rise time (t_r), settling time (t_s), and maximum overshoot (M_p)" [12] of reduced-based PID controller are comparable with that of full-order controllers.

Fig. 8.2 Unit step response of different systems

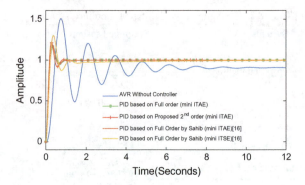

Table 8.3 Controller parameters and time-domain specifications

System	PID controller parameter			Time response specifications		
	K_p	K_i	K_d	M_p (%)	t_r (s)	t_s (s)
Without controller	-	-	-	65.723	0.2607	6.9865
PID/full order	1.23	0.393	0.85	16.474	0.16245	0.85372
PID/reduced order	1.4999	1	0.5399	21.488	0.12916	0.74146
PID/PSO/ITAE [14]	1.3541	0.9266	0.4378	18.818	0.14933	0.81453
PID/PSO/ITSE [14]	1.7774	0.3827	0.3184	30.063	0.16098	3.3994

Conclusion

In this article, the authors proposed a new classical method to decrease the order of the LTI SISO system, and thereafter, a reduced-based PID controller design for the automatic voltage regulator used in electrical power generation. In the suggested technique, the higher-order system (nth order) transfer function is directly compared with the reduced-order (kth order) transfer function ($k < n$). The obtained equations are put in matrix form, and then this matrix is converted into row echelon form using the Gauss-Jordan method. This results in one free variable, the value of which can be assumed depending on the given transfer function. After selecting the free variable, the coefficients of the required kth order system can be computed. Two examples are considered to prove the effectiveness of the techniques. PID controllers are also designed for automatic voltage regulator (AVR) based on reduced-order as well as full-order system using particle swarm optimization (PSO) technique taking ITAE as error criterion. As we see, Pade fails for the eighth-order system whereas the proposed technique gives results with good accuracy. The obtained ISE value is less as compared to other classical techniques. Reduced-order-based PID controller designed for the AVR system works well with full-order controller.

References

1. Davison E (1966) A method for simplifying linear dynamic systems. IEEE Trans Automat Contr 11:93–101
2. Shamash Y (1974) Stable reduced-order models using Padé-type approximations. IEEE Trans Automat Contr 19:615–616. https://doi.org/10.1109/TAC.1974.1100661
3. Hutton M, Friedland B (1975) Routh approximations for reducing order of linear, time-invariant systems. IEEE Trans Automat Contr 20:329–337. https://doi.org/10.1109/TAC.1975.1100953
4. Hsieh CS, Hwang C (1989) Model reduction of continuous-time systems using a modified Routh approximation method. IEE Proc D Control Theory Appl. https://doi.org/10.1049/ip-d.1989.0022
5. Krishnamurthy V, Seshadri V (1978) Model reduction using the Routh stability criterion. IEEE Trans Automat Contr. https://doi.org/10.1109/TAC.1978.1101805
6. Shamash Y (1981) Truncation method of reduction: a viable alternative. Electron Lett. https://doi.org/10.1049/el:19810070
7. Parthasarathy R, John S (1981) Cauer continued fraction methods for model reduction. Electron Lett. https://doi.org/10.1049/el:19810554
8. Lucas TN (1984) Biased model reduction by factor division. Electron Lett. https://doi.org/10.1049/el:19840402
9. Mukherjee S, Mishra RN (1987) Order reduction of linear systems using an error minimization technique. J Franklin Inst. https://doi.org/10.1016/0016-0032(87)9007-8
10. Howitt GD, Luus R (1990) Model reduction by minimization of integral square error performance indices. J Franklin Inst. https://doi.org/10.1016/0016-0032(90)90001-Y
11. Sikander A, Thakur P (2017) Reduced order modelling of linear time-invariant system using modified cuckoo search algorithm. Soft Comput
12. Sikander A, Prasad R (2017) New technique for system simplification using Cuckoo search and ESA. Sadhana
13. Bhatt R, Parmar G, Gupta R, Sikander A (2018) Application of stochastic fractal search in approximation and control of LTI systems. Micro Tech
14. Sikander A, Prasad R (2019) Reduced order modelling based control of two wheeled mobile robot. J. Int Manuf
15. Ahamad N, Sikander A, Singh G (2019) Substructure preservation based approach for discrete time system approximation. Micro Tech
16. Mittal AK, Prasad R, Sharma SP (2004) Reduction of linear dynamic systems using an error minimization technique. J Inst Eng Electr Eng Div
17. Prasad R, Sharma SP, Mittal AK (2003) Linear model reduction using the advantages of Mihailov criterion and factor division. J Inst Eng Electr Eng Div
18. Kumar V, Tiwari JP (2012) Order reducing of linear system using clustering method factor division algorithm. Int J Appl Inf Syst 3
19. G Vasu, P Murari, Kumar RP (2012) Reduction of higher order linear dynamic SISO and MIMO systems using the advantages of improved pole clustering and PSO. In: Proceedings of the international conference on advances in computer, electronics and electrical engineering
20. Gopal M (2002) Control systems: principles and design. Tata McGraw-Hill Education
21. Ekinci S, Hekimoglu B (2019) Improved kidney-inspired algorithm approach for tuning of PID controller in AVR system. IEEE Access. https://doi.org/10.1109/ACCESS.2019.2906980
22. Sikander A, Thakur P, Bansal RC, Rajasekar S (2018) A novel technique to design cuckoo search based FOPID controller for AVR in power systems. Comp Electr Eng
23. Uniyal I, Sikander A (2018) A comparative analysis of PID controller design for AVR based on optimization techniques. Int Comm Ctrl Devices
24. Verma P, Patel N, Nair NKC, Sikander A (2016) Design of PID controller using cuckoo search algorithm for buck-boost converter of LED driver circuit South. Power Electr

25. Goyal R, Parmar G, Sikander A (2019) A new approach for simplification and control of linear time invariant systems. Micro Tech
26. Kennedy J (2011) Particle swarm optimization. In: Encyclopedia of machine learning. Springer, pp 760–766

Chapter 9
Unscented Kalman Filter Based Dynamic State Estimation in Power Systems Using Complex Synchronized PMU Measurements

Shubhrajyoti Kundu, Anil Kumar, Mehebub Alam,
Biman Kumar Saha Roy, and Siddhartha Sankar Thakur

Introduction

The main objective of SE is to obtain the voltage phasor at the different buses with the utilization of existing measurements. The static state estimation (SSE) approach weighted least square estimation (WLSE) has been broadly used due to its fast convergence property. However, state prediction cannot be entertained which is a major advantage of DSE. System observability can be ensured due to prediction capabilities of DSE. The degree of accuracy of the traditional SE technique can be enhanced to a great extent using dynamic-based approach.

A lot of approaches have been suggested to solve the problem of SE. The SE problem has been solved through WLS-based approach in [1, 2]. In [3], tracking SE has been performed where the estimator tracks the changes in measurement vector and corresponding changes in the state vector. Kundu et al. proposed a novel approach of static SE based on linear model of complex PMU measurement [4]. The EKF-based DSE has been implemented in [5] for estimating power system state variables. The authors utilized the Extended fractional Kalman filter (EFK) [6] for power system DSE. UT, i.e., a derivative-free transformation technique, was implemented in [7] to show the shortcoming of derivative-based approach by providing a direct method of transforming mean and co-variance information. In

---

S. Kundu · A. Kumar · M. Alam (✉) · B. K. S. Roy · S. S. Thakur
Electrical Engineering Department, NIT Dugapur, Durgapur, West Bengal 713209, India
e-mail: mehebubjgec1990@gmail.com

A. Kumar
e-mail: anil2k3535@gmail.com

B. K. S. Roy
e-mail: bk.saharoy@ee.nitdgp.ac.in

© The Editor(s) (if applicable) and The Author(s), under exclusive license to Springer Nature Singapore Pte Ltd. 2021
A. K. Singh and M. Tripathy (eds.), *Control Applications in Modern Power System*, Lecture Notes in Electrical Engineering 710,
https://doi.org/10.1007/978-981-15-8815-0_9

[8], authors have implemented UKF-based approach for DSE problem. Chakrabarty et al. [9] applied Cubature Kalman filter-based approach for DSE problem in power systems. In Sreenath et al. [10] introduced Unscented RTS smoother based estimator considering hybrid measurements.

Previously, the SE problem has been solved using WLSE technique in which nonlinear equation of power system network has been transformed into linear equation using Taylor series method which is a major challenge in SE. In order to cope with this problem, UKF, i.e., based on UT can be implemented to solve state estimation problems in power systems. From the existing literature, it can be seen that most of the literature has either discussed SSE problems or solved power system DSE problem utilizing EKF-based approach. Though the authors in [8–10] have used UKF or CKF-based approach, however, the authors haven't exploited the linear relationship existing between complex PMU measurements and complex state variables.

Many researchers have also implemented various methods for optimally allocating PMUs. Taufik et al. [12] implemented binary bat algorithm for optimal allocation of PMUs. In Kundu et al. [13], obtained an optimal allocation of PMUs considering maximum connectivity of buses to lines.

Although SE has emerged as a matured area of work, still further investigation is continued to enrich the SE software and estimate better state at the ECC. In this paper, a new DSE approach has been developed based on UKF which utilizes the complex line currents and complex bus voltage as sets of measurement vector and treats complex bus voltage as state variables.

Problem Formulation

Traditional SE approach

The relation between n dimensional state vector v and measurement vector Zmeas can be illustrated as

$$\text{Zmeas} = h(v) + u \tag{9.1}$$

where $h(v)$ is load flow equation of dimension $m \times n$. Here, u is m dimensional measurement noise vector. Measurement noise vector must satisfy the following statistical properties.

$$E(u) = 0, E(u.u^T) = R \tag{9.2}$$

Here R is measurement error covariance matrix and E is mean. The objective function in WLSE must be minimized to obtain estimated state vector x.

$$J(v) = [\text{Zmeas} - h(v)]^T R^{-1}[\text{Zmeas} - h(v)] \tag{9.3}$$

For the minimization of above objective function $J(v)$ its derivative must be zero. Minimization of $J(v)$ leads to iterative solution as

$$\Delta V = [H^T R^{-1} H]^{-1} H^T R^{-1} \Delta Z \text{meas} \tag{9.4}$$

ΔZ is mismatch vector and H is Jacobian matrix.

The Unscented Transformation

The estimation error depends on the approximation of non-linear function about the operating point, if this approximation is not performed properly then there exists estimation error. The statistics of random variable x of dimension n that goes under non-linear transformation is determined by UT. To observe the statistics of the random variable we generate a set of points of length $2n + 1$ called sigma points. To create $2n + 1$ no. of samples, i.e., sigma points of covariance P and mean m, we introduce a matrix χ. We can define,

$$\chi_o = m \tag{9.5}$$

$$\chi_i = m + \left(\sqrt{(n+\psi)P}\right)_i, \quad i = 1, \ldots, n \tag{9.6}$$

$$\chi_{n+i} = m - \left(\sqrt{(n+\psi)P}\right)_i, \quad i = 1, \ldots, n \tag{9.7}$$

$$\psi = \alpha^2 (n + \text{kappa}) - n, \tag{9.8}$$

In general $10^{-4} \leq \alpha \leq 1$, kappa $= 3 - n$. Cholesky decomposition of covariance matrix P is required to avoid the presence of complex numbers in the sigma point from which lower triangular matrix has been chosen for sigma point calculation. The weight of sigma points corresponding to different column is described below

$$W_0^{(m)} = \frac{\psi}{n + \psi} \tag{9.9}$$

$$W_0^{(c)} \frac{\psi}{n + \psi} + (1 - \alpha^2 + \varphi) \tag{9.10}$$

$$W_i^{(m)} = W_i^{(c)} = \frac{1}{2(n + \psi)} i = 1, \ldots, 2n \tag{9.11}$$

Power System Dynamic Model

The power system dynamics model proposed by Deb & Larson can be given as

$$x(k+1) = x(k) + \vartheta(k) \tag{9.12}$$

According to this model, the Predicted states and corresponding error covariance matrix is expressed as

$$x^p(k+1) = x^e(k) \tag{9.13}$$

$$P^{cov}(k+1) = P^{cov}(k) + Q(k) \tag{9.14}$$

$\vartheta(k)$, and $Q(k)$ are the model uncertainty and diagonal covariance matrix respectively.

UKF-Based DSE Modeling

Calculation of Sigma Points

$2n + 1$ no. of samples, i.e., sigma points are calculated using the state variable x and covariance matrix P at the time instant $k - 1$

$$X_{k-1} = [x_{k-1}, \ldots, x_{k-1}]\sqrt{n+\psi}\left[0\sqrt{P_k^{pri}} - \sqrt{P_k^{pri}}\right] \tag{9.15}$$

State Prediction

The calculated sigma point at time instant $k - 1$ is updated through nonlinear function f as

$$X_{estk}^i = f(X_{k-1}^i, k-1) \tag{9.16}$$

The size of matrix X_{estk}^i is n × (2n + 1).

$$x_k^{pri} = \sum_{i=0}^{2n} W_i^m X_{estk}^i \tag{9.17}$$

$$P_k^{\text{pri}} = \sum_{i=0}^{2n} W_i^c \left[(X_{estk}^i - x_k^{\text{pri}})(X_{estk}^i - x_k^{\text{pri}})^T \right] + Q_{k-1} \qquad (9.18)$$

Where x_k^{pri} and P_k^{pri} represent the mean vector of the predicted state variables and covariance matrix, respectively and Q_k is uncorrelated covariance matrix.

State Filtering

Predicted sigma points are calculated as shown in (9.23) utilizing the mathematical equation expressed in (9.21) and (9.22).

$$X_k^{\text{pri}} = [x_k^{\text{pri}}, \ldots, x_k^{\text{pri}}] + \sqrt{n + \psi} \left[0 \sqrt{P_k^{\text{pri}}} - \sqrt{P_k^{\text{pri}}} \right] \qquad (9.19)$$

The predicted sigma points are evaluated through function of complex PMU measurements as given in (9.24)

$$Y_k^{\text{pri}} = h(X_k^{\text{pri}}, k) \qquad (9.20)$$

Mean corresponding to the propagated points are calculated as

$$\mu_k = \sum_{i=0}^{2n} W_i^m Y_k^{\text{pri}i} \qquad (9.21)$$

$$D_k = \sum_{i=0}^{2n} W_i^c \left[(Y_k^{\text{pri}i} - \mu_k)(Y_k^{\text{pri}i} - \mu_k)^T \right] + R_k \qquad (9.22)$$

$$F_k = \sum_{i=0}^{2n} W_i^c \left[X_k^{\text{pri}i} - x_k^{\text{pri}})(Y_k^{\text{pri}i} - \mu_k)^T \right] \qquad (9.23)$$

D_k and F_k are measurement covariance and cross-covariance matrix, respectively. Kalman gain K_k, is obtained as

$$K_k = D_k F_k^{-1} \qquad (9.24)$$

Estimated state variables and corresponding error covariance matrix can be calculated as

$$x_k = x_k^{\text{pri}} + K_k [y_k - \mu_k] \qquad (9.25)$$

$$P_k = P_k^{-1} - K_k D_k K_k^T \qquad (9.26)$$

Proposed Algorithm

The UKF-based proposed DSE algorithm is described below:

Step 1 Read the measurement, i.e., complex line currents, and complex bus voltage.
Step 2 Initialize the state and covariance (X_o, P_o)
Step 3 Compute the sigma points.
Step 4 Obtain the mean vector of the predicted states and covariance matrix using sigma points.
Step 5 Calculate state mean, measurement and cross-covariance matrix utilising Measurements
Step 6 Calculate Kalman gain using measurement and cross covariance matrix.
Step 7 Calculate estimated state mean vector and error covariance matrix.
Step 8 Update the error covariance matrix
Step 9 Display the results.

Simulation Details

The load change at each bus in IEEE test system has been considered at an interval of an hour. So we have 24 different time instant by considering the practical situation as mentioned in IEEE Reliability test system [11]. Change in load is shared by each generator based on their corresponding participation factors. Load flow analysis has been performed for 24-time samples to obtain the true value of power system states. The proposed method has been simulated utilizing complex bus voltage and complex branch current as set of measurement vector. A Gaussian noise of mean value zero and standard deviation (SD) of 0.5% has been introduced in true measurement vector for the practical scenario of actual power system. Deb and Larsen method is used for prediction of states in which predicted states at $(k + 1)$th time instant remains unchanged to states at kth time instant. The programming and simulation for IEEE test system are carried out using MATLAB software of version 8.1 (R2013a).

Indices for Performance Parameter

Different performance parameter has been evaluated for the proposed UKF-based approach. The performance parameter of conventional WLSE is compared with proposed method UKF.

The average absolute state error (AE) is computed as

$$AE(k) = \frac{1}{2*(n\text{bus}-1)} \sum_{i=1}^{2*(n\text{bus}-1)} \left[x_i^{est}(k) - x_i^{tr}(k)\right] \quad (9.27)$$

x(k) represents voltage vector for time step of k. $x_i^{est}(k)$ is the estimated voltage vector and $x_i^{tr}(k)$ denotes the true value. J index, i.e., the performance evaluation index can be computed as

$$J(k) = \frac{\sum_{i=1}^{nm} \left|z_i^{est}(k) - z_i^{tr}(k)\right|}{\sum_{i=1}^{nm} \left|z_i^{meas}(k) - z_i^{tr}(k)\right|} \quad (9.28)$$

Where nm is entire measurement set for each time sample. $z_i^{est}(k)$ is estimated complex bus voltages and complex line currents at kth time step. similarly, $z_i^{tr}(k)$ and $z_i^{meas}(k)$ are true and measured value for kth time step. The maximum absolute state (ME) error is found as

$$m(k) = \max\left[\left|x_i^{est}(k)\right| - \left|x_i^{tr}(k)\right|\right] \quad (9.29)$$

Statistical Parameter

To show the supremacy of the proposed algorithm, statistical parameter is also calculated. The maximum of mean square error (ω)

$$\omega = \max\left[\frac{1}{t_s} \sum_{k=1}^{t_s} (x_i^{est}(k) - x_i^{tr}(k))^2\right] \quad (9.30)$$

t_s is total number of time steps, $x_i^{est}(k)$ and $x_i^{tr}(k)$ are estimated and actual value of voltage vector. The maximum standard deviation error (λ) is calculated as

$$\frac{1}{t_s - 1}\left[\sum_{k=1}^{t_s} (x_i^{est}(k) - x_i^{tr}(k)^2)\right]^{0.5} \quad (9.31)$$

t_s is total number of time step, $x_i^{est}(k)$ and $x_i^{tr}(k)$ are estimated and actual value of voltage vector. The maximum of sum squared error ($\jmath$) is calculated as

$$\jmath = \max\left[\sum_{k=1}^{t_s}(x_i^{est}(k) - x_i^{tr}(k))^2\right] \quad (9.32)$$

The average of absolute error (δ). The mathematical formulation for δ is expressed as

$$\delta = \frac{1}{t_s} \sum_{k=1}^{t_s} \frac{1}{(2*n_{bus}-1)} \sum_{k=1}^{(2*n_{bus}-1)} |x_i^{est}(k) - x_i^{tr}(k)| \qquad (9.33)$$

Results and Discussions

The proposed approach is implemented on various IEEE systems, i.e., IEEE 6, 14, 30, 57, and 118 bus test systems. Due to limitation of space, some useful results are demonstrated here. Figures 9.1 and 9.2 show the comparison of active power at bus 29 and 40 of IEEE 57 and IEEE 118 bus system respectively. From Figs.9.1 and 9.2, it can be clearly visualized that the proposed approach provides much closer to true value compared to WLS-based approach. Further, Fig. 9.3 shows the comparison of reactive power at bus 22 of IEEE 118 bus system. It can also be noted that the

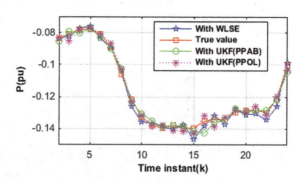

Fig. 9.1 Estimated bus active power at bus 29 of IEEE 57 bus system

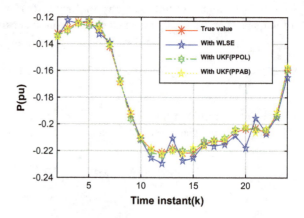

Fig. 9.2 Estimated bus active power at bus 40 of IEEE 118 bus system

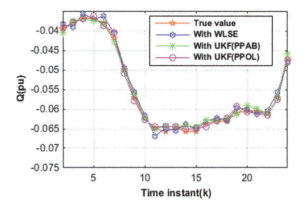

Fig. 9.3 Estimated bus reactive power at bus 22 of IEEE 118 bus system

estimation obtained through proposed UKF considering optimally allocated PMUs (PPOL) is almost similar to that of estimation obtained considering PMUs allocated at all the buses (PPAB). PMUs are allocated optimally based on the approach as mentioned in [13].

J index is an important performance parameter that shows overall performance of estimated power system states. For IEEE 57 bus test system, comparison of J index for WLSE and UKF-based algorithm has been shown in Fig. 9.4 It is observed that for most of the time instant J index is less in UKF considering both PPAB and PPOL. In fact, when PMUs are placed at all the buses (PPAB), the J index obtained through proposed approach lies much below as shown in blue color for IEEE 57 bus.

To validate the estimated state vector, AE and ME are also calculated. Both the error has been reduced to a great extent in UKF-based approach than WLSE approach. AE and ME are shown in Figs. 9.5 and 9.6 for IEEE 57 and IEEE 30 bus system, respectively. It is also observed that AE obtained considering PPAB and PPOL is almost the same. This is also true for ME calculation.

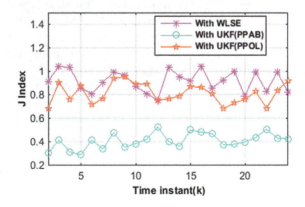

Fig. 9.4 Comparison of J index using UKF and WLSE-based approach (IEEE 57 bus)

Fig. 9.5 Comparison of AE index using UKF and WLSE-based approach (IEEE 57 bus)

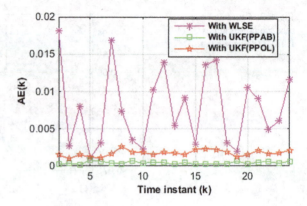

Fig. 9.6 Comparison of ME using UKF and WLSE-based approach ((IEEE 30 bus)

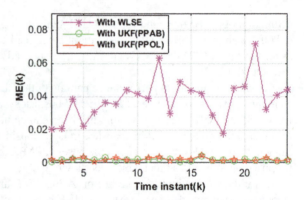

Statistical parameters are also calculated to show the effectiveness of proposed UKF model. For the proposed model, statistical parameters are calculated by considering both PPAB and PPOL. For both the cases in UKF, statistical parameter is very less as compare to WLSE technique. This shows the supremacy of proposed UKF model with respect to the traditional WLSE algorithm. The statistical parameters obtained through WLS approach and proposed approach are listed in Table 9.1.

In this study, all the PMU measurements are considered to be healthy. Although, this assumption is not true in real power systems. Some of the measurements may contain the erroneous value which leads to bad data. This bad data is to be detected and to be removed in order to have a better and reliable estimation of the states. Therefore, detection of bad data in conjunction with SE problem is left for the future scope of this study.

Table 1. Comparison of statistical parameter for different systems

Test system		WLSE	UKF considering PPAB	UKF considering PPOL
IEEE 6 Bus	ω	0.0002157	0.000007303	0.000015585
	λ	0.0150000	0.002800000	0.004000000
	$\jmath\imath$	0.0052000	0.000167980	0.000384700
	δ	0.0060000	0.001200000	0.001600000
IEEE 14 Bus	ω	0.0011000	0.000012230	0.000016890
	λ	0.0345000	0.003600000	0.004200000
	$\jmath\imath$	0.0273000	0.000281310	0.000388580
	δ	0.0094000	0.000792610	0.000014147
IEEE 30 Bus	ω	0.0042000	0.000003123	0.000014147
	λ	0.0665000	0.001800000	0.003800000
	$\jmath\imath$	0.10160000	0.000071830	0.000325380
	δ	0.01140000	0.000501890	0.001000000
IEEE 57 Bus	ω	0.0028000	0.000006910	0.000150170
	λ	0.0544000	0.002700000	0.012500000
	$\jmath\imath$	0.0681000	0.000158970	0.003500000
	δ	0.0096000	0.000440420	0.001600000
IEEE 118 Bus	ω	0.0002250	0.00000361	0.00082711
	λ	0.0153000	0.0019	0.02940000
	$\jmath\imath$	0.0054000	0.000083219	0.01900000
	δ	0.0020000	0.00029884	0.00120000

Conclusions

In this study, UKF-based DSE has been proposed utilizing synchronized complex PMU measurements. The proposed UKF-based DSE approach utilizes linear relationship between complex PMU measurements and complex state variables and is derivative-free approach. In the proposed model, the estimated variables are calculated considering PMUs placed at all buses (PPAB) as well as considering optimal location of PMUs (PPOL). The proposed UKF-based DSE approach has been applied to various IEEE systems. Furthermore, the various statistical parameters obtained through proposed approach (with both PPAB and PPOL) are also compared with WLS approach. The obtained results establish the efficacy of the proposed UKF-based DSE approach over WLS-based approach. From the results, it can be also concluded that the results obtained through UKF considering PPOL are almost similar to that of UKF results obtained considering PPAB. In view of the test results, the proposed UKF-based DSE approach proves to be more suitable for SE analysis compared to the traditional WLS-based approach.

References

1. Schweppe FC, Wildes J (1971) Power system static state estimation, part 1: Exact model. IEEE Trans Power Apparatus Syst 89(1):120–125
2. Meriem M, Bouchra C, Abdelaziz B, Jamal SOB, Faissal EM, Nazha C (2016) Study of state estimation using weighted-least-squares method (WLS). In: 2016 International conference on electrical sciences and technologies in Maghreb (CISTEM). Marrakech, pp 1–5
3. Falcao DM, Cooke PA, Brameller A (1982) Power system tracking state estimation and bad data processing. IEEE Trans Power Apparatus Syst 101(2):325–333
4. Kundu S, Alam M, Thakur SS (2018) State estimation with optimal PMU placement considering various contingencies. In: 2018 IEEE 8th power international Conference (PIICON). Kurukshetra, India, pp 1–6
5. Durga Prasad G, Thakur SS (1998) A new approach to dynamic state estimation of power system. Electric Power Syst Res 45(3):173–180
6. Lu Z, Yang S, Sun Y (2016) Application of extended fractional Kalman filter to power system dynamic state estimation. In: 2016 IEEE PES Asia-Pacific power and energy engineering conference (APPEEC). Xi'an, pp 1923–1927
7. Uhlmann JK (1994) Simultaneous map building and localization for real time application. Transfer thesis Univ. Oxford Oxford, U.K.
8. Valverde G, Terzija V (2011) Unscented Kalman filter for power system dynamic state estimation. IET Gener Transm Distrib 5(1):29–37
9. Sharma A, Srivastava SC, Chakrabarti S (2017) A cubature kalman filter based power system dynamic state estimator. IEEE Trans Instrum Meas 66(8):2036–2045
10. Geetha SJ, Sharma A, Chakrabarti S (2019) Unscented Rauch–Tung–Streibel smoother-based power system forecasting-aided state estimator using hybrid measurements. IET Gener Transm Distrib 13(16):3583–3590
11. Grigg C et al (1999) The IEEE reliability test system 1996. A report prepared by the reliability test system task force of the application of probability methods subcommittee. IEEE Trans Power Syst 14(3):1010–1020
12. Tawfik AS, Abdallah EN, Youssef KH (2017) Optimal placement of phasor measurement units using binary bat algorithm. In: 2017 nineteenth international middle east power system conference (MEPCON). Cairo, pp 559–564
13. Kundu S, Thakur SS (2019) Optimal PMU placement and state estimation in indian practical systems considering multiple bad data and sudden change of load. In: 2019 IEEE international conference on sustainable energy technologies and systems (ICSETS). Bhubaneswar, India, pp 279–284

Chapter 10
Real-Time Electric Vehicle Collision Avoidance System Under Foggy Environment Using Raspberry Pi Controller and Image Processing Algorithm

Arvind R. Yadav, Jayendra Kumar, Roshan Kumar, Shivam Kumar, Priyanshi Singh, and Rishabh Soni

Introduction

Fog is composed of visible cloud water droplets in air near the troposphere. Fog can be contemplated as a type of low-lying cloud, and it is heavily influenced by nearby bodies of water, topography, pollution, and wind conditions. In turn, fog has disturbed many human activities, such as shipping, travel, and warfare because it reduces visibility. Fog affects vision on the road ahead, reducing the time to react to potential danger on the road. In general, fog is classified in three ways: (1) aviation fog (visibility range from 1,000 m or less), (2) thick fog (visibility range drops to

---

A. R. Yadav (✉) · S. Kumar · P. Singh · R. Soni
Department of ECE, PIET, FET, Parul University, Vadodara, India
e-mail: arvind.yadav.me@gmail.com

S. Kumar
e-mail: shivamkumarchs@gmail.com

P. Singh
e-mail: singh.priyanshi95@gmail.com

R. Soni
e-mail: sonirishabh09@gmail.com

J. Kumar
Department of ECE, National Institute of Technology Jamshedpur, Jamshedpur, India
e-mail: jkumar.ece@nitjsr.ac.in

R. Kumar
Department of Electronic and IT, Miami College, Henan University, Kaifeng, China
e-mail: roshan.iit123@gmail.com

© The Editor(s) (if applicable) and The Author(s), under exclusive license to Springer Nature Singapore Pte Ltd. 2021
A. K. Singh and M. Tripathy (eds.), *Control Applications in Modern Power System*, Lecture Notes in Electrical Engineering 710,
https://doi.org/10.1007/978-981-15-8815-0_10

Fig. 10.1 Car accidents due to fog on highways

200 m or less), and (3) dense fog (visibility is only at 50 m or less). Fog is a common phenomenon occurring in winter months; there are also high chances of rain and frost, which could affect braking distance and the control in vehicle (electric as well as other types) as illustrated in Fig. 10.1 accidents due to fog on highways.

In winter season during December–January, most of the northern India experiences fog condition, causing not only loss of economy but also life because of accidents [1]. Under foggy and hazy weather condition, images acquired by visual systems produce severely degraded images not useful for computer vision-based applications. Thus, improving the quality of the degraded image to its clear form is of enormous importance. Therefore, three widely used algorithms: fusion-based strategy, dark channel prior, and white balance, gamma correction and tone mapping had been investigated in [2]. In [3], a satellite-based bispectral brightness temperature difference (BTD) technique was employed for identifying nighttime fog. In 2017, a visibility improvement scheme for foggy images was proposed by Anwar et al. [4], by making use of post-processing technique for dark channel prior (DCP) concept to preserve sharp details and maintain the color quality of the defogged image. Gupta et al. [5] have proposed a collision detection system for vehicles in hilly and dense fog-affected area. They have used GPS to actively send vehicle location coordinates to the server, which processes the vehicle data and predicts potential collision on smart phone devices [5].

Singh et al. [6] proposed a system which makes use of RF transmitter–receiver pair, temperature, humidity and ultrasonic long range sensor, and a controller, to prevent collision of vehicles under fog and blind curve situation. When the vehicle approaches toward each other at blind curves and under fog condition, an alarm is produced to alert the driver for collision avoidance [6]. A fast bilateral filter and Canny edge detector-based image processing application have been implemented using Raspberry Pi for telemedicine applications [7]. A Raspberry Pi (model 2B), RF, ultrasonic sensor, and GPS-based vehicle-to-vehicle communication system for crash avoidance were implemented that ensure blind spot detection crucial while changing the lanes [8]. Raspberry Pi-based driver drowsiness detection system using image processing was proposed by Kulkarni et al. [9]. The webcam was employed to capture the image, and eye blink detection was performed using Haar features to identify the driver's drowsiness [9].

Methodology

The schematic diagram for vehicle/obstruction detection and collision avoidance system under foggy environment using Raspberry Pi controller is illustrated in Fig. 10.2. Though this system is useful for any type of vehicle, in the present work, it has been targeted for electric vehicle because most of the vehicle manufactures across the globe are coming up with variants of electric vehicle segments in near future. The overall approach is divided into two parts namely hardware and software.

Hardware

The hardware used in the proposed system for electric vehicle collision avoidance under foggy environment is briefly described hereunder:
- **Raspberry Pi controller**: Raspberry Pi has been widely used in many applications [10, 12]. It is a unique hardware with 40 nm technology, lightweight (50 g), 1.5–6.7 W power consumption under load along with VideoCore IV GPU, and economically viable [13]. The Raspberry Pi (model 3B) has been chosen for this work as it has Quad Core 1.2 GHz Broadcom BCM2837 64-bit CPU and on-chip graphics processing unit. That is sufficient to meet the requirement of real-time object detection and collision avoidance system as the most of the vehicles driven during foggy environment usually prefer to have speed limit of 40–60 km or less than that on National Highways. It also has on-board Wi-Fi, Bluetooth, USB boot capabilities, 1 GB RAM, CSI camera port for connecting a Raspberry Pi

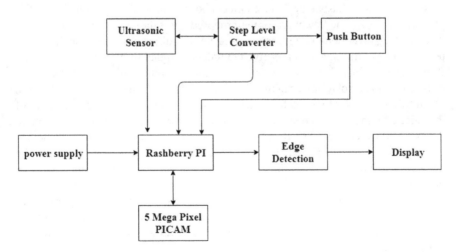

Fig. 10.2 Block diagram of electric vehicle collision avoidance system under foggy environment condition

camera, and DSI display port for connecting a Raspberry Pi touch screen display. Other features of Raspberry Pi which make it more appealing for this application are that, they are low cost and low power device and have immense amount of resources sufficient to carry out real-time processing task.

- **Ultrasonic sensor**: The ultrasonic sensor is used to measure the distance between object by using sound waves. It measures distance by transmitting out a sound wave at a precise frequency and receiving the reflected wave. The HC-SR04 module ultrasonic sensor measures distance from 2 cm (0.065 feet) to 400 cm (13.1234 feet).
- **Step-level converter**: Connecting a 3.3 V device to a 5 V system is a difficult task; hence, a 4-channel 5–3 V bidirectional logic converter module that safely steps down 5 V signals to 3.3 V and steps up 3.3–5 V at the same time is employed.
- **Pi camera**: A 5 MP fixed-focus camera supports 1080p30, 720p60, and VGA90 video modes as well as stills capture. This camera is interfaced to CSI port of Raspberry Pi using a 15 cm ribbon cable. The advantage with the Pi camera is it can be accessed through V4L APIs, and several third-party libraries are available for use as open access library.
- **Power supply**: The power required for the processing task has been supported by a 5 V, 2.4 A power supply micro-USB AC adapter charger.
- **Display purpose**: Raspberry Pi 7″ touch screen display is used for displaying the processed images and parameters.

Software

Raspberry Pi controller without programming is an idiot box. This research work has been developed on Raspberry Pi development board with Python 2.7.3 and OpenCV platform [10, 14]. The flowchart for the proposed algorithm is shown in Fig. 10.3.

Once the stand-alone system is powered up, the ultrasonic sensor shall continuously transmit the signal and receive the reflected signal to measure the distance. If the obstacle/object is within the sensor's range, it shall calculate the distance and display it on the display device of the system installed on electric vehicles dashboard. The ultrasonic sensor for object/obstruction detection only indicates the presence of the object, but the shape and size of the object/obstruction are difficult to get from it. As soon as the other vehicle/any obstruction shall come in the range of the vehicle in which system is installed, the Pi camera connected to Raspberry Pi must capture the image. The captured image shall be applied to edge detection algorithm to enhance the fog image.

The edge detection image shall be displayed on display unit to show presence of objects, shape, and size of the obstruction, which shall be useful in taking decision whether brake need to be applied or the driver can overtake the object from left- or right-hand side.

Fig. 10.3 Flowchart of the proposed approach

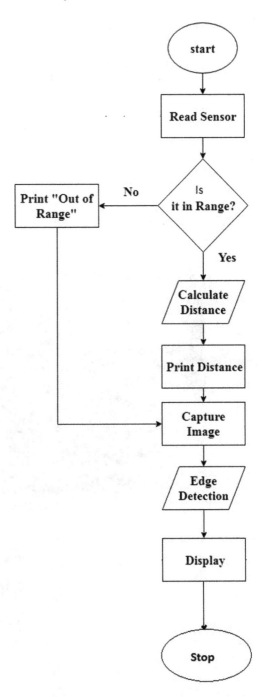

Results and Discussion

The complete setup can be observed in Fig. 10.4 given below. During experimentation phase, computer display unit has been interfaced with the Raspberry Pi using HDMI. The image has been processed by Prewitt, Sobel, and Canny edge detection-based algorithms in Raspberry Pi controller. Figure 10.5 illustrates the original fog image and the results obtained by Sobel operator. It is evident from the processed image that the edge detection-based images provide enhanced image compared to the fog image. Though the images were also processed by Prewitt and Canny edge detection algorithms, the computational requirement of the canny edge detection is more compared to the Sobel edge detection algorithm. Therefore, though the results of Canny edge detection algorithms were much better compared to Sobel, but because of processing requirement trade-off, Sobel edge detection has been considered in this work.

Fig. 10.4 Raspberry Pi setup

Fig. 10.5 a Original fog image, **b** sobel operator-based edge detection Image

Conclusion

Object detection in foggy weather is very difficult task, due to which many accidents take place on the road. The problem has been resolved by developing a system which provides safety feature to the people traveling in foggy ambient condition. The proposed system used an ultrasonic sensor for object detection and distance measurement. As soon as the object has reached within the specified distance range, the foggy image has been captured by Pi camera and processed with edge detection (Sobel operator) technique. The output obtained after processing has been displayed to the user, to give the exact idea about the object in front of it. This system ensures that there will not be any collision of electric vehicles as well as other vehicles due to poor visibility and helps in avoiding road accidents in foggy weather.

References

1. Dey S (2018) On the theoretical aspects of improved fog detection and prediction in India. Atmos Res 202:77–80
2. Pal T, Bhowmik MK, Bhattacharjee D, Ghosh AK (2016) Visibility enhancement techniques for fog degraded images: a comparative analysis with performance evaluation. In: IEEE Region 10 conference (TENCON). IEEE, Singapore, Singapore, pp 2583–2588
3. Ahmed R, Dey S, Mohan M (2015) A study to improve night time fog detection in the Indo-Gangetic Basin using satellite data and to investigate the connection to aerosols. Meteorol Appl 22(4):689–693
4. Anwar MI, Khosla A, Singh G (2017) Visibility enhancement with single image fog removal scheme using a post-processing technique. In: 4th International conference on signal processing and integrated networks (SPIN). IEEE, Noida, India, pp 280–285
5. Gupta AK, Wable G, Batra T (2014) Collision detection system for vehicles in hilly and dense fog affected area to generate collision alerts. In: International conference on issues and challenges in intelligent computing techniques (ICICT). IEEE, Ghaziabad, India, pp 38–40
6. Singh A, Ratnakar R, Rajak A, Gurale N, Pacharaney US (2019) Accident avoidance system in blind curves and fog using integration of technologies. In: International conference on sustainable communication networks and application, LNDECT, vol 39. Springer, pp 119–134
7. Manikandan L, Selvakumar R, Nair SAH, Kumar KS (2020) Hardware implementation of fast bilateral filter and canny edge detector using Raspberry Pi for telemedicine applications. J Ambient Intell Human Comput: 1–7
8. Ghatwai N, Harpale V, Kale M (2016) Vehicle to vehicle communication for crash avoidance system. In: International conference on computing communication control and automation (ICCUBEA). IEEE, Pune, India, pp 1–3
9. Kulkarni S, Harale A, Thakur A (2017) Image processing for driver's safety and vehicle control using raspberry Pi and webcam. In: International conference on power control, signals and instrumentation engineering (ICPCSI). IEEE, Chennai, India, pp 1288–2129
10. Purohit M, Yadav AR (2018) Comparison of feature extraction techniques to recognize traffic rule violations using low processing embedded system. In: 5th international conference on signal processing and integrated networks (SPIN). IEEE, Noida, India, pp 154–158
11. Kumar J, Kumar S, Kumar A, Behera B (2019) Real-time monitoring security system integrated with Raspberry Pi and e-mail communication link. In: 9th international conference on cloud computing data science and engineering (confluence). IEEE, Noida, India, pp 79–84

12. Kumar J, Ramesh PR (2018) Low cost energy efficient smart security system with information stamping for IoT networks. In: 3rd international conference on internet of things: smart innovation and usages (IoT-SIU). IEEE, Bhimtal, India, pp 1–5
13. Mittal S (2019) A Survey on optimized implementation of deep learning models on the NVIDIA Jetson platform. J Syst Architect 97:428–442
14. Sahani M, Mohanty MN (2015) Realization of different algorithms using Raspberry Pi for real time image processing application. In: Intelligent computing, communication and devices. advances in intelligent systems and computing, vol 309. Springer, pp 473–479

Chapter 11
Modified Sine Cosine Algorithm Optimized Fractional-Order PD Type SSSC Controller Design

Preeti Ranjan Sahu, Rajesh Kumar Lenka, and Satyajit Panigrahy

Introduction

Nowadays, flexible AC transmission systems (FACTS) have found an extensive application in the power transmission system. With the advancement in power electronics-based controller technology, the FACTS controller damped system oscillation by improving stability [1, 2]. Among the FACTS devices static VAR compensator (SVC) [3, 4] has popular for its instantaneous response to voltage change. The thyristor controlled series capacitor (TCSC) [5, 6] of the FACTS family damped electromechanical oscillations, regulate transmission voltage, limit short circuit currents and mitigate the power system oscillation. The second-generation FACTS device static synchronous compensator (STATCOM) has quicker response compare to conventional SVC to system abnormality. The STATCOM controls the reactive power flow in the transmission line [7–9]. The SSSC is a second-order FACTS device that has several advantages among other FACTS devices due to its storage element [10, 11]. Most of these devices utilize IGBT and GTOs based voltage source converter (VSC). The SSSC controller directly controls the current flow in the transmission line by regulating the reactive power requirement of the power system. This controller does not change the transmission line impedance like TCSC controller, hence does not suffer from resonance issue [12–15].

In previous literature, several population based optimization algorithm (POA) have been adopted to estimate the controller parameter for the optimal operation of the controller. In [16] a comparative study is carried out between GA and PSO based SSSC controller parameter design. Even though GA can effectively find the global

---

P. R. Sahu (✉)
NIST, Berhampur, Odisha 761008, India
e-mail: preetiranjan.sahu@gmail.com

R. K. Lenka · S. Panigrahy
NIT Rourkela, Rourkela, Odisha 769008, India

© The Editor(s) (if applicable) and The Author(s), under exclusive license to Springer Nature Singapore Pte Ltd. 2021
A. K. Singh and M. Tripathy (eds.), *Control Applications in Modern Power System*, Lecture Notes in Electrical Engineering 710,
https://doi.org/10.1007/978-981-15-8815-0_11

optimal result but takes a very long run time. The PSO suffers from slow convergence with weak local search ability [17]. In [18], the bacteria forging (BF) algorithm has been addressed for power system stability controller design. The main disadvantage in the BF algorithm that it searches the global solution in random directions which may cause long run time. Various new algorithms have been adopted recently for the design of an SSSC controller in recent times like whale optimization algorithm (WOA) [19, 20], modified WOA [21] and hybrid DE-PSO [22] etc.

SCA is a newly developed optimization algorithm which uses sine/cosine functions during algorithm formulation [23]. This approach efficiently transfers from the exploration phase to the exploitation phase by using sine/cosine functions. However, the optimization algorithm suffers from slow convergences and the consequences due to unbalanced distribution among the exploration and exploitation phase. So, the conventional SCA algorithm fails to achieve the solution with better convergence. In this paper FO PD structured SSSC controller is designed using modified SCA (MSCA) algorithm which can overcome the discussed issues of conventional SCA.

The following objectives are carried out in this present study.

1. This chapter primarily focuses on the modeling of an SSSC controller with FO PD damping.
2. The proposed controller provides extra freedom in controller tuning.
3. Finally, for optimum tuning of the FO PD parameters of the SSSC controller, MSCA is applied.

SCA and Its Modification

Sine Cosine Algorithm is a POA first proposed in [23], which uses sine/cosine function in the algorithm to update the position. Generally, POA starts with a randomly selected solution set. Then the iterative process evaluates the selected solution set in formulated objective function with certain system constraints to obtain the global solution. Hence a sufficient number of the solution set with more number of iteration increase the probability of getting an optimal solution. The optimization process in the SCA algorithm progresses its search for an optimal solution in exploitation and exploration phases. The SCA algorithm search for a promising region in the search space in the exploration phase and gradually changes the value of random solutions in exploitation phase to get an optimal solution. During both phases, the position updating formulation is presented below.

$$X_j^{n+1} = \begin{cases} X_j^n + r_1 \times \sin(r_2) \times |(r_3 \times pos_j^n) - X_j^n| & r_4 < 0.5 \\ X_j^n + r_1 \times \sin(r_2) \times |(r_3 \times pos_j^n) - X_{jj}^n| & r_4 \geq 0.5 \end{cases} \quad (1)$$

In (1) X_j^n represents jth-dimension current solution after nth-iteration. The term pos_j defines the destination point position in jth dimension. The random variables in (1) are the main parameters of the SCA algorithm whose values ranges in between [0, 1]. The random parameter r_1 decides the direction of the next position either inside the solution and destination or away from it. r_2 decides the distance travel during movement in the direction of the destination or far away from the destination. As shown in (1), r_3 add weight to the position pos_j to emphasize or de-emphasize impact of the destination in order to define the distance. Whenever $r_3 > 1$ SCA gives emphasis to the effect of destination and for $r_3 < 1$ SCA does not give emphasis to the effect of destination. Finally, r_4 value decides the sine or cosine function be selected to update the position as given in (1).

To converge towards the global solution, the optimization algorithm should able to find a promising region in the search space. To achieve this objective, the optimization algorithm should maintain a proper equilibrium among the phase of exploration and exploitation. To maintain equilibrium between these two phases, the SCA estimate the range of sine and cosine in (1) by changing the value of adaptively as given below:

$$r_1 = a - n\left(\frac{a}{N}\right) \qquad (2)$$

where n define the current state of iteration with N being the maximum number of iteration. The variable a is a constant maintain equilibrium between two phases. By linearly increasing the r_1 with respect to the iteration count, the number of iteration needed is high which increases the algorithm's computation time. Furthermore, the convergence rate also decreases. This paper presents modified SCA (MSCA) by varying r_1 as given in (3) to boost the convergence and get the optimal global solution.

$$r_1 = a - n^{1.5}\left(\frac{a}{N^{1.5}}\right) \qquad (3)$$

The constant a in (3) is set to be 1.6 and the value r_1 is varying non-linearly with the iteration count as in (3) to attain the optimal solution with high convergence. Here, the iteration n is changed to its fractional power of 1.5, the maximum number of iteration N is changed to its fractional power value of 1.5. The comprehensive process flow chart MSCA presented in Fig. 11.1.

System Modeling and Controller Design

SSSC Based System Modeling

It is necessary to evaluate the performance of any damping controller based upon SSSC along with their proper design. Therefore, a SMIB system is being considered as shown in Fig. 11.2, comprising of a synchronous generator that is linked to an infinite-bus with the help of a transformer and an SSSC.

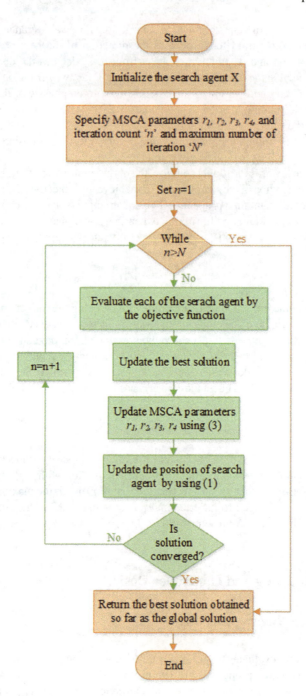

Fig. 11.1 Process flow chart of the MSCA

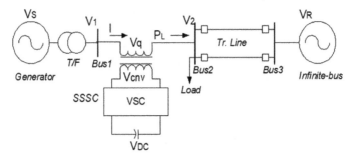

Fig. 11.2 SMIB system with FACTS controller

The Proposed Controller

Recently, the key emphasis is to update the traditional PID controllers using the Fractional Calculus initiative. A thorough explanation of the FO PD type controller is elaborated in [21]. The designed controller in this chapter consisted of FO PD structured lead-lag components as given in Fig. 11.3. The FO PD type design consists of a fractional integrator, proportional gain, derivative gain and a filter. The lead-lag structure components are discussed in [21].

Problem Formulation

In this chapter, First order derivative filter $K = 100$ [21] and $t_{WS} = 10$ s is used. K_{PS}, K_{DS} are the gains of the controller, λ_S is the fractional integrator and the time constants to be measured. For the damping of the power system oscillations, the voltage of the injected series is balanced, and the effective V_q' is given by:

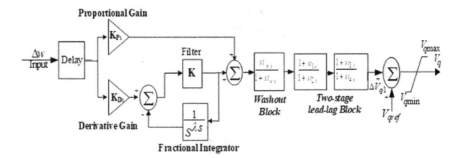

Fig. 11.3 Controller structure of FO PD type SSSC

$$V_q = V_{qref} + \Delta V_{q1} \tag{4}$$

To achieve improved device performance, the MSCA is used to adjust the controller parameters. Objective function description is a first stage prerequisite for controller purposes using the latest heuristic optimizations. The objective function for the speed deviation for the SMIB system is expressed in (5).

$$J = \int_0^t |\Delta\omega| \, t \, dt \tag{5}$$

where, t is the range of simulation time is the change is speed. The performance indices named ITAE are selected as an objective function to be minimized. However, Minimizing objective function J is subject to controller parameter restriction.

$$\text{Minimize } J \tag{6}$$

Subject to

$$\begin{aligned} K_{PS}^{\text{Min}} &\leq K_{PS} \leq K_{PS}^{\text{Max}} & K_{DS}^{\text{Min}} &\leq K_{DS} \leq K_{DS}^{\text{Max}} \\ \lambda_S^{\text{Min}} &\leq \lambda_S \leq \lambda_S^{\text{Max}} & t_{1S}^{\text{Min}} &\leq t_{1S} \leq t_{1S}^{\text{Max}} \\ t_{2S}^{\text{Min}} &\leq t_{2S} \leq t_{2S}^{\text{Max}} & t_{3S}^{\text{Min}} &\leq t_{3S} \leq t_{3S}^{\text{Max}} \\ & & t_{4S}^{\text{Min}} &\leq t_{4S} \leq t_{4S}^{\text{Max}} \end{aligned} \tag{7}$$

It is to be noted that two gains, one fractional integrator and four time constant parameters are needed to be optimized for a SMIB system.

Result and Analysis

A toolbox called Sim Power Systems (SPS) has been used exclusively for all the simulations and designing of the damping controller. Engineers are capable of simulating these Power Systems using the MATLAB-based design tool SPS that helps them design and build models with ease.

Application to SMIB System

For the implementation and efficient performance of the MSCA algorithm, the selection of various parameter values has to be done carefully. The various cases selected are as follows.

Case A: Nominal Loading Condition

In this case, the suggested controller performance is demonstrated at $P_e = 0.8$ p.u loading conditions for nominal loading with respect to the occurrence of a severe disturbance in the system. At time $t = 1s$, a 3-cycles, 3-phase fault is applied at the mid-section of the transmission line. Figures 11.4, 11.5, 11.6 and 11.7 depicts the various responses of the system which leads to the conclusion that the MSCA optimized proposed controller gives improved dynamic response as compared with PSO, GA and original SCA optimized proposed controller.

Case B: Light Loading Condition

To look at the superiority of the planned SSSC controller, its performance is evaluated under light load condition. In this case, $P_e = 0.55$ p.u is set for the generator load and a 100 ms 3-phase fault is applied close to bus-3 at $t = 1$ s. The system response under this possibility which explains the efficacy of the suggested SSSC controller under different working conditions and sort of disruption has appeared in Fig. 11.8. Also, the proposed MSCA method provides enhanced transient response with PSO, GA and original SCA optimized proposed controller.

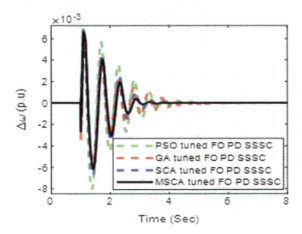

Fig. 11.4 Nominal loading speed deviation response

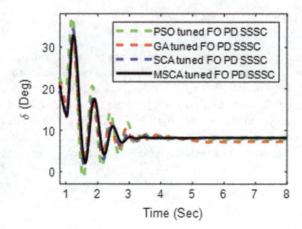

Fig. 11.5 Power angle response

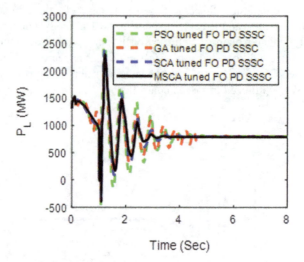

Fig. 11.6 Nominal loading tie-line power

Case C: Heavy Loading Condition

The performance of the suggested SSSC controller is also tested by generator heavy loading i.e., $P_e = 1.0$ p.u. In this case at $t = 1$ s, at the midpoint of the transmission line near bus 4, a 100 ms 3-phase fault is applied. The speed deviation responses are tested in heavy loading condition is displayed in Fig. 11.9. Figure 11.9 shows the MSCA optimized SSSC controller gives more stable performance contrast with PSO, GA and SCA optimized controller.

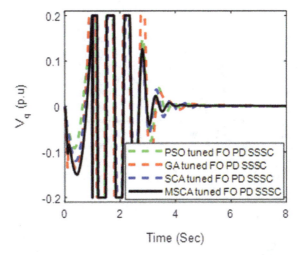

Fig. 11.7 SSSC injected voltage

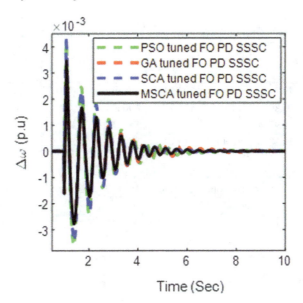

Fig. 11.8 Light loading speed deviation response

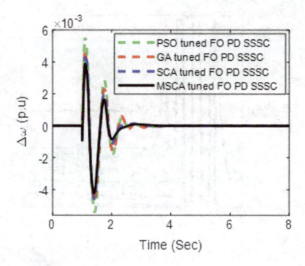

Fig. 11.9 Heavy loading speed deviation response

Table 11.1 PSO, GA, SCA, and MSCA optimized FO PD type SSSC controller parameters

Optimization techniques	Parameters						
	K_{PS}	K_{DS}	λ_S	t_{1S}	t_{2S}	t_{3S}	t_{4S}
PSO	29.5838	3.2471	29.8523	0.7146	1.3028	1.2191	1.5148
GA	101.8034	101.8034	134.4432	1.9009	1.4219	1.1747	0.3071
SCA	9.9774	1.0000	0.4625	0.8613	0.4119	1.7175	0.4375
MSCA	10.0000	6.4371	0.1000	2.0000	0.9717	0.7309	0.1078

Table 11.2 SMIB system ITAE values considering PSO, GA, SCA and MSCA techniques

Cases	Techniques			
	PSO	GA	SCA	MSCA
Case-a ($\times 10^{-4}$)	13.900	10.622	9.802	8.133
Case-b ($\times 10^{-4}$)	9.985	8.733	8.222	7.011
Case-c ($\times 10^{-4}$)	6.010	5.322	4.821	4.422

The optimized parameters of the suggested controller are presented in Table 11.1 for the SMIB system. Table 11.2 shows that the least ITAE values are observed with MSCA methods that take into account different instances compared to PSO, GA and SCA methods.

To improve stability, the existing work can be applied to an integrated power system by means of a wind farm [24].

Conclusions

In this present work, a modified SCA tuned FO PD based SSSC controller is designed to improve power system stability. For the suggested controller design issue, a time-domain objective function is used to reduce the oscillations of the power system. In addition, modified SCA is used to tune the FO PD type FACTS controller. The efficacy of the controller configuration is determined by the use of the SMIB system under different serious disturbances. To reveal the effectiveness of the proposed system, the results of the proposed modified SCA tuned FO PD type SSSC structure are compared with the PSO, GA and original SCA.

References

1. Kundur P (1994) Power system stability and control. McGraw-Hill
2. Hingorani NG, Gyugyi L (2000) Understanding FACTS. IEEE Press, New York
3. Mansour I, Abdesalm DO, Wira P, Merckle J (2009) Fuzzy logic control of a SVC to improve the transient stability of AC power systems. In: 35th annual conference of IEEE industrial electronics, pp 3240–3245
4. Imran Azim M, Fayzur Rahman M (2014) Genetic algorithm based reactive power management by SVC. Int J Electr Comput Eng (IJECE) 4(2):200–206
5. Hadi SP, Wiennetou HI, Mochamad RF (2013) TCSC power oscillation damping and PSS design using genetic algorithm modal optimal control. Int J Eng Comput Sci 13(1):23–30
6. Shahgholian G, Movahedi A, Faiz J (2015) Coordinated design of TCSC and PSS controllers using VURPSO and genetic algorithms for multi-machine power system stability. Int J Control Autom Syst 13(2):398–409
7. Lee YS, Sun SY (2002) STATCOM controller design for power system stabilization with suboptimal control and strip pole assignment. Int J Electr Power Energy Syst 24(9):771–779
8. Mandour ME, Abd-Elazeem SM (2009) Hopf Bifurcation control in a power system with static synchronous compensator. Ain Shams J Electr Eng 44(1):245–255
9. Abd-Elazim SM, Ali ES (2016) Imperialist competitive algorithm for optimal STATCOM design in a multimachine power system. Int J Electr Power Energy Syst 76:136–146
10. Sen KK (1998) SSSC: theory, modelling, and applications. IEEE Trans Power Delivery 13(1):241–246
11. Kazemi A, Ladjevardi M, Masoum MAS (2005) Optimal selection of SSSC based damping controller parameters for improving power system dynamic stability using genetic algorithm. Iran J Sci Tech Trans Eng 29(B1):1–10
12. Panda S, Padhy NP, Patel RN (2008) Power system stability improvement by PSO optimized SSSC-based damping controller. Electr Power Compon Syst 36:468–490
13. Panda S (2011) Differential evolution algorithm for SSSC based damping controller design considering time delay. J Franklin Instit 348(8):1903–1926
14. Panda S (2009) Multi-objective evolutionary algorithm for SSSC-based controller design. Electr Power Syst Res 79(6):937–944
15. Ali ES, Abd-Elazim SM (2014) Hybrid BFOA-PSO approach for optimal design of SSSC based controller. Int J WSEAS Trans Power Syst 9(1):54–66
16. Kennedy J, Eberhart R (1995) Particle swarm optimization. In: Proceedings of IEEE international conference on neural networks, pp 1942–1948
17. Rini D, Shamsuddin S, Yuhaniz S (2011) Particle swarm optimization: technique, system and challenges. Int J Comput Appl 14(1):19–27

18. Ali ES, Abd-Elazim SM (2013) Power system stability enhancement via bacteria foraging optimization algorithm. Int Arab J Sci Eng (AJSE) 38(3):599–611
19. Sahu PR, Hota PK, Panda S (2018) Power system stability enhancement by fractional order multi input SSSC based controller employing whale optimization algorithm. J Electric Syst Inf Technol 5(3):326–336
20. Sahu PR, Hota PK, Panda S (2017) Whale optimization algorithm for fuzzy lead-lag structure SSSC damping controller design. In: 2017 14th IEEE India council international conference INDICON, pp 1–5
21. Sahu PR, Hota PK, Panda S (2018) Modified whale optimization algorithm for fractional-order multi-input SSSC-based controller design. Optim Control Appl Methods 39(5):1802–1817
22. Behera S, Barisal AK, Dhal N, Lal DK (2019) Mitigation of power oscillations using hybrid DE-PSO optimization-based SSSC damping controller. J Electric Syst Inf Technol 6(1):5
23. Mirjalili S (2016) SCA: a sine cosine algorithm for solving optimization problems. Knowl Syst 96:120–133
24. Bhukya J, Mahajan V (2020) Optimization of controllers parameters for damping local area oscillation to enhance the stability of an interconnected system with wind farm. Int J Electric Power Energy Syst 119:105877

Chapter 12
Performance Enhancement of Optimally Tuned PI Controller for Harmonic Minimization

Anish Pratap Vishwakarma and Ksh. Milan Singh

Introduction

In recent times, indices of power quality such as voltage imbalances, flicker and harmonics. are considered to be the most indispensable quantities for measuring and analyzing power quality in uninterrupted process plants. Harmonics are created by the existence of nonlinear load when solid-state devices are used to transform AC power to feed to the electrical loads [1]. Active power filters [2], unified power flow controller (*UPFC*), unified power quality conditioner (*UPQC*), passive inductance capacitance (*LC*) filters [3], dynamic voltage restorer (*DVR*), etc. are some of the available methodologies to reduce harmonics in electrical parameters. Multiple functions are occurred in *APF* such as load balancing, flicker reduction, voltage and current harmonic filtering and reactive power control [1]. Component drafting of filters and control strategies related to the parameters are analyzed under comprehensive review [4]. An adaptive artificial neural network (Adaline) based *PI controller* for accelerated measurement of the compensating current which also improved the operating performance of *SAP* filter [5]. Semiconductor controller-based inverter or rectifier loads are drawn a distorted supply current, but their effect in the substantially lower down using power filter results in overall power quality improvement [4]. For highly nonlinear load like electric furnace, the quality of distorted source current can be improved using p-q theory which usually decreases the lower order harmonics in the distribution system. The invariably chosen tuned parameters of a *PI controller* control the DC link capacitive voltage. Hence, the optimization of *PI* parameter can

---

A. P. Vishwakarma (✉) · Ksh. Milan Singh
National Institute of Technology Meghalaya, Shillong, Meghalaya, India
e-mail: anishvishwakarma9766@gmail.com

Ksh. Milan Singh
e-mail: milan.singh@nitm.ac.in

© The Editor(s) (if applicable) and The Author(s), under exclusive license to Springer Nature Singapore Pte Ltd. 2021
A. K. Singh and M. Tripathy (eds.), *Control Applications in Modern Power System*, Lecture Notes in Electrical Engineering 710,
https://doi.org/10.1007/978-981-15-8815-0_12

be appropriate achieved using harmony search algorithm, cuckoo search algorithm and GA and PSO optimized controller [6]. The *ITAE* criteria have been chosen as the best error extraction criteria in tuned *PID controller* to determine voltage stability [7]. The *EP* algorithm works in the four phases, first phase of the *EP* optimization works to prepare the background for their better performance, and in second phase, this background will be categorized in unimodal and multimodal function. In third and fourth phases, the algorithm is utilized for controlling its variable parameter, and then it finds the global best constraints for minimum value of fitness function, respectively [8, 9]. For highly nonlinear furnace type load, hybrid power filters are also considered with tuned *PI controller* using adaptive optimization techniques such as *HSA* which is also proved as a better solution [10]. In some random distribution system, the harmony search optimization which is based on selecting the best tune or harmony for a particular composition is analyzed successfully in some studies [11, 12]. In the proposed system, the applicability and at the same time the performance enhancement of metaheuristic optimization techniques are being analyzed and tabulated to meet the real-time nonlinearity problems. Also the performance of *HSA* is simulated with *ISE,IAE* and *ITAE* criteria for the very first time. Moreover, the reduced percentage of *THD* is obtained with *ITAE* criteria.

Model Designing

In the proposed filter, the equal and opposite harmonic current is injected in the system using the voltage source converter (*VSC*) based *SAPF* in which the reference current is generated by using *SRF* theory. A gate driver circuit which receives a controlled signal to trigger the *IGBT*-based *VSC* and the current harmonics generated by the filter is injected using an inductor known as interface reactor.

Figure 12.1 shows the proposed schematic compensation system substantially minimizes current harmonics which arises due to nonlinear load. Various optimization techniques have been implemented to control *PI* parameters of a *SRF*-based *PI controller* in conjunction with *VSC*-based *SAPF*.

Control Circuit

Figure 12.2 shows the operation of reference current generation using *SRF* theory for the proposed current sensitive filter to obtain gate triggering current.

First, the control circuit is designed to get reference current through *SRF* theory. Here, the d-q-0 revolving coordinates are obtained after the conversion of a-b-c system which includes the following two stages:

Stage 1: Conversion of three-phase system into two-phase system.
Stage 2: Conversion of two-phase to the d-q revolving coordinate.

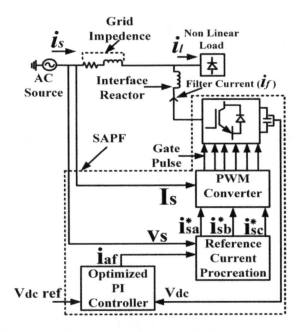

Fig. 12.1 Nonlinear load grid configuration with shunt active filter

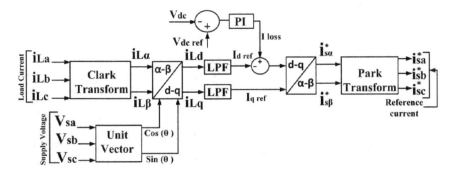

Fig. 12.2 Control strategies of shunt active filter

Second, the *PI controller* comprises of a proportional gain that works to assist for an output which is error proportional and also convergence this output to demolish error present in the signal. Output of controller is given as Eq. (12.1).

$$I_{af} = K_p \Delta + K_i \int_0^t \Delta dt \quad (12.1)$$

$$\Delta = SP - PV \quad (12.2)$$

where Δ is deviating from set point (SP) to calculated values (PV). Usually, Δ is the difference between $V_{dc\,ref}$ and measured DC voltage.

Optimization Technique

The *PI controller* parameter tuning or determining the suitable magnitudes of *PI* parameters to obtain the minimum *THD* is known as optimization. This optimization is usually applied through an objective function.

Objective Function

The goal is to reduce the addition of averages of THD data which is given by Eq. (12.3).

$$\text{ObjectiveFunction} = \text{Min}\left(\sum_{i=0}^{n} \text{Mean(THD)}\right) \quad (12.3)$$

The upper and lower limits of the gain parameters are defined as

$$P_1 \leq K_p \leq P_2 \quad (12.4)$$

$$I_1 \leq K_i \leq I_2 \quad (12.5)$$

Eagle Perching Optimization

The eagle perching optimizer (*EPO*) begins by initializing a random value of "*x*" for "*n*" number of samples of vision about "*K*" number of eagles. Now the best peak valley point can be identified after "*ts*" (number of iterations) by "*K*" number of eagles, with "*L*" scope index or glancing angle per unit area. The optimal values of "*x*" are set to the lowest *THD* values, i.e., the objective function "*F*", while an upper and lower bound are predefined for the *PI controller* K_p and K_i parameter. Table 12.1 gives the novel eagle perching optimization parameters with the resolution range eta as 0.05 which gives the minimum *THD*.

Table 12.1 Eagle perching optimization parameters

E.P parameter	ISE	IAE	ITAE
No. of iteration	50	50	50
Dimension	04	04	04
Resolution range	0.05	0.05	0.05
P	550.4	519.4	595.9
I	526.9	559.4	630.4
K_p	498.4	493.7	952.7
K_i	510.7	523.4	851.4
Current THD (%)	6.51	6.67	5.43
Execution time (s)	52,165.3	52,175.8	51,902.9

Harmony Search Algorithm

Freshly developed *harmony search algorithm (HSA)* contrasted with an application for music optimization. This process is based on guitar tuning, in which music encoder improvises the pitches of their guitars to establish superior harmony. In conjunction with the *HSA* error obtaining criteria such as *IAE, ISE* and *ITAE*, the result of simulation is tabulated for a range of *pitch adjustment ratio (PAR)* from 0.1 to 0.9 and *harmony memory consideration rate (HMCR)* from 0.15 to 0.95 in Table 12.2 (Fig. 12.3).

Table 12.2 Harmony search algorithm parameters

HSA Parameter	ISE			IAE			ITAE		
HMCR	0.15	0.55	0.95	0.15	0.55	0.95	0.15	0.55	0.95
PAR	0.1	0.5	0.9	0.1	0.5	0.9	0.1	0.5	0.9
P	1269	867	1282	903	969	734	757.6	1240	1166
I	667	1059.5	912	877.4	1163	1208	758	1240	880
K_p	1362	1207	1129	1108	1008	971	1239	1245	1044
K_i	1490	1274	1164	1017	945	1193	868	1062	1216
Current THD (%)	4.38	4.73	4.44	4.72	4.81	5.12	4.08	3.81	3.85
Execution time (S)	30,256	31,025	30,256	29,352	29,343	29,340	25,362	25,251	25,263

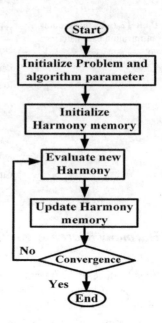

Fig. 12.3 Flow chart of harmony search algorithm

Design Criteria

There are different types of criteria used for designing of optimum controller. These criteria are used to diminish the overshoot time, setting time of response and error in steady-state. Initially, the weighting time (t) is large, thus to reduce that time these criteria are helpful. Some of them are listed below.

ISE Criteria—To suppress large errors, *ISE* is better than *IAE* because the errors are squared and thus contribute more to the value of integral

$$ISE = \int_0^\infty e^2(t)\mathrm{d}t \qquad (12.6)$$

IAE Criteria—*IAE* is better than *ISE* because when we eliminating tiny errors as they get smaller even though when we square small numbers

$$IAE = 1\int_0^\infty |e(t)|\mathrm{d}t \qquad (12.7)$$

ITAE Criteria—The *ITAE* criteria have managed to improve the parameter of the *PI controller* in order to minimize errors that may persist in the system for a long period, since the presence of large time (t) amplifies the effect of even little errors in the integrated value.

$$ITAE = \int_0^\infty t|e(t)|\mathrm{d}t \tag{12.8}$$

Among the above three mentioned criteria, *ITAE* criteria are most widely used because it concludes best values.

Results and Discussion

The system is developed at simulated under the operating condition of nonlinear load for which efficiently satisfy the applicability of optimized *PI controller*-based *SAPF*. The aim is to minimize odd harmonics using conventional *PI* and metaheuristic techniques like *EPO* and *HSA*. The above-mentioned simulation is being performed on the three criteria for obtaining solute errors, i.e., *IAE, ISE* and *ITAE* criteria. Among these three criteria, *ITAE* gives minimum *THD* for both of the optimization techniques as observed and tabulated in Tables 12.1 and 12.2. The *conventional PI controller* gives a 11.63% current *THD*. Further, when optimization techniques are implemented in *conventional PI controller*, *eagle perching* and *harmony search algorithm* give a current *THD* of 5.43% (*ITAE*) and 4.08% (*ITAE—0.1 PAR*), 3.81% (*ITAE—0.5 PAR*) and 3.85% (*ITAE—0.9 PAR*), respectively.

Performance of SAPF with and Without Optimized PI Controller

After simulation, the obtained waveform and the corresponding FFT analysis are as follows. Figure 12.4 shows the source current waveform without SAPF, and nonlinear load is in operating mode. The corresponding FFT analysis is shown in Fig. 12.5. The percentage of THD observed in this condition is 22.74%.

Figures 12.6 and 12.7 show the waveform of source current and its corresponding *FFT* analysis with conventional *PI controller*, respectively. Here, from the waveform, it is clear that after 0.1 s, the distortion is compensated and the obtained *THD* is 11.63%.

Figure 12.8 shows the waveform of distorted source current up to 0.1 s and compensated after 0.1 s with the help of optimized *PI controller* employed for harmonic mitigation. Figure 12.9 shows the corresponding *FFT* analysis in which the *THD* obtained as 5.43%. The above results are although simulated for *IAE* and *ISE* criteria, and also here only *ITAE* criteria are shown as it gives less *THD* as compared two others.

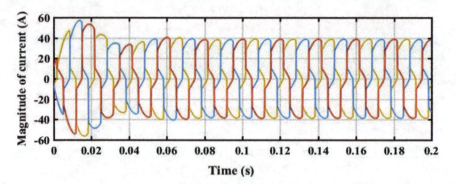

Fig. 12.4 Waveform (distorted) of source current without filter

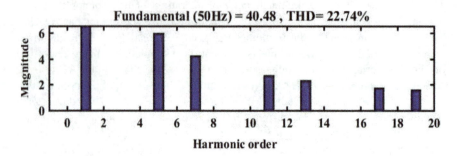

Fig. 12.5 FFT analysis of source current without filter

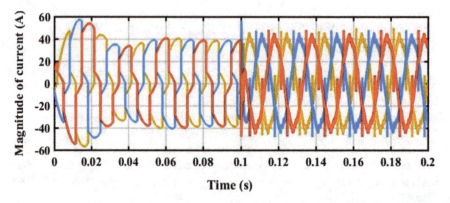

Fig. 12.6 Waveform of source current with filter (compensated with conventional PI controller)

Now, Figs. 12.10 and 12.11 show the simulation results of *HSA* applied in the system for optimization purpose with *ITAE* criteria for both waveform and its corresponding *FFT* analysis, respectively. Figure 12.12 shows the comparative analysis of

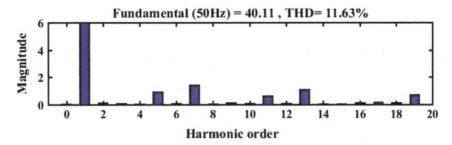

Fig. 12.7 FFT waveform of source current with conventional PI controller

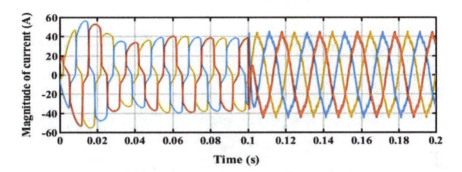

Fig. 12.8 Waveform of source current with optimization technique (EP ITAE)

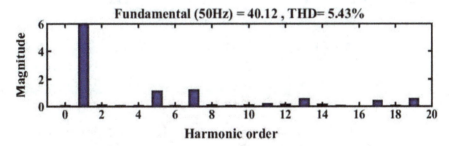

Fig. 12.9 FFT waveform of source current with optimization technique (EP ITAE)

various optimization techniques with the conclusion that the *HSA* provides minimum *THD* (under the range of recommended practice by *IEEE*).

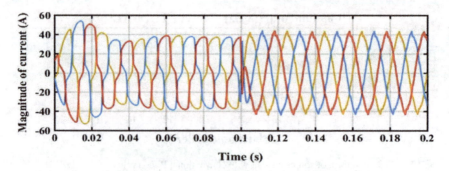

Fig. 12.10 Waveform of source current with optimization technique (HS ITAE MED)

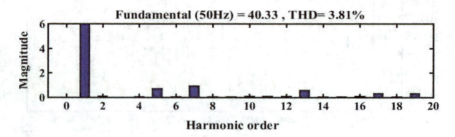

Fig. 12.11 FFT waveform of source current with optimization technique (HS ITAE MED)

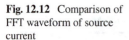

Fig. 12.12 Comparison of FFT waveform of source current

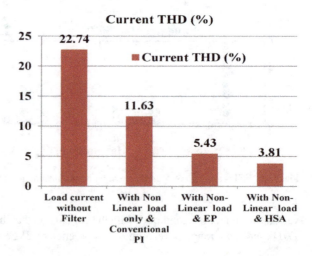

Conclusion

The proposed model allows the harmonic mitigation within the *IEEE* range with the help of optimized *PI controller*. The designed algorithm is based on *SAPF*. The novel optimization techniques such as *eagle perching* and *harmony search* are implemented to obtain minimum harmonic distortion for the criteria (*IAE, ISE* and *ITAE*). The proposed *HSA* provides better optimization as compared to *EP* for highly nonlinear operating system.

References

1. Akagi H (2005) Active harmonic filters. Proce IEEE 93(12):2128–2141
2. Chaoui A (2007) PI controlled three-phase shunt active power filter for power quality improvement. Electric Power Comp Syst 35(12):1331–1344
3. Rajeshwari AB (2017) Voltage harmonic reduction using passive filter shunt passive-active filters for non-linear load. In: 2017 7th international conference on communication systems and network technologies (CSNT), vol 12. IEEE, pp 131–36
4. Singh B, Verma V (2005) Hybrid filters for power quality improvement. IEE Proc Gener Trans Distrib 152(3):365
5. Bhattacharya A, Chandan C (2011) A shunt active power filter with enhanced performance using ANN-based predictive and adaptive controllers. IEEE Trans Industr Electron **58**(2):421–428
6. Jhapte R (2017) Design of optimized hybrid active power filter for electric arc furnace load. In: 2017 recent developments in control, automation and power engineering (RDCAPE). IEEE, pp 105–110
7. Singh M (2016) Performance comparison of optimized controller tuning techniques for voltage stability. In: 2016 IEEE first international conference on control, measurement and instrumentation (CMI). IEEE, pp 11–15
8. Khan AT (2018) Model-free optimization using eagle perching optimizer arXiv:1807.02754 [Cs]
9. Betts KS (2000) The wrong place to perch. Environ Sci Technol 34(13):292A–293A
10. Jhapte R (2018) Design and analysis of optimized hybrid active power filter for electric arc furnace load. In: Bhattacharyya S et al (ed) Advanced computational and communication paradigms, vol 475. Springer, Singapore, pp 250–258
11. Al-Omoush AA (2019) Comprehensive review of the development of the harmony search algorithm and its applications. IEEE Access 7:14233–14245
12. Gu J, Wu D (2018) A random distribution harmony search algorithm. In: 2018 tenth international conference on advanced computational intelligence (ICACI), IEEE, pp 432–437.
13. Singh B (1999) A review of active filters for power quality improvement. IEEE Trans Industr Electron 46(5):960–971
14. Vishwakarma AP, Milan Singh K (2020) Comparative analysis of adaptive PI controller for current harmonic mitigation. In: 2020 international conference on computational performance evaluation (ComPE), Shillong, India, pp. 643–648

Chapter 13
Design and Performance Analysis of Second-Order Process Using Various MRAC Technique

Saibal Manna and Ashok Kumar Akella

Introduction

Aerospace engineering in the 1950s was searching solution for the design of autopilots. The aim was to discover a controller that could operate well under flight conditions. There has been developed an advanced controller such as the adaptive controller that could guarantee high performance. On that basis several adaptive flight control schemes were suggested, the one suggested by Whitaker being the most well-known in many research paper [1–3]. MRAC is the system that supports the Whitaker Scheme that is used even today and can solve an automatic pilot problem. This approach was originally based on the gradient approach and was called the "MIT law." In this rule, the controller adjusts its parameter so that the system follows the given RM [2, 3]. In addition, this method was progressed by park in 1960s, by using Lyapunov stability [1–3].

It is difficult to control processes with various complexities and recurrent disorders. In such situations, traditional feedback controllers [4] typically fail because they are primarily equipped for certain operating conditions. Therefore, adaptive controllers [5] with large operating areas are preferred to deal with fluctuations in process dynamics and unwanted disruptions. If the model of the process is recognized, controllers can be built for their linearized counterparts using model-based approaches [6] under different working situations. However, most manufacturing techniques are much complex to be fully understood. In these situations, it is a safe choice to use adaptive controllers.

Recursive computation is an online process on which controller parameters (CP) are modified in real-time based on the predefined algorithm and predicted values

---

S. Manna (✉) · A. K. Akella
NIT Jamshedpur, Jamshedpur, Jharkhand 831014, India
e-mail: mannasaibal1994@gmail.com

[7]. MRAC is another renowned adaptation tool [8, 9], in which the required process performance is based on the RM. The adjustment mechanism based on MIT rule [10] modifies the CP with a specified adaptation gain and the difference between closing loop response and the performance of an RM.

Here, an updated MRAC with PD feedback for second-order underdamped system is developed and this updated controller is called MRAC-PD. Enhanced responses with lower swings are observed during set-point monitoring compared to traditional controller and faster process recovery for MRAC-PD is observed in the presence of load changes and this is also verified with the measured like integral absolute error (IAE), integral time absolute error (ITAE) and integral square error (ISE) values.

MRAC

The simple block diagram of conventional MRAC is illustrated in Fig. 13.1. Basically, it has two loops: internal loop includes plant and controller, and outside loop adjust the CP. These loops have two specific objectives: system stability and adopt a pattern of RM so that the error between the process and reference system outcome is towards zero as shown by Eq. (13.1) [1].

$$e_1 = y - y_m \tag{1}$$

MIT Rule

This rule is primarily focused on seeking a criterion of minimization which updates the CP to eliminate the error towards zero.

$$J(\theta) = \frac{e_1^2}{2} \tag{2}$$

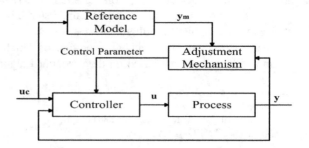

Fig. 13.1 MRAC block diagram

Parameters in negative gradient direction must be adjusted. In such circumstances, J becomes small:

$$\frac{d\theta}{dt} = -\gamma \frac{\partial J}{\partial \theta} = -\gamma e_1 \frac{\partial e_1}{\partial \theta} \tag{3}$$

where $\frac{\partial e_1}{\partial \theta}$ is the sensitivity.

The RM and plant model has been selected in the following type [11]:

$$\frac{d^2 y_m}{dt^2} = -a_{m1}\frac{dy}{dt} - a_{m2}y + b_m u_c \tag{4}$$

$$\frac{d^2 y}{dt^2} = -a_1\frac{dy}{dt} - a_2 y + bu \tag{5}$$

In Eqs. (13.4) and (13.5), a, b, a_m, b_m are the system parameters, u_c and u are the command and control signals. The output on conventional controller is indicated as

$$u = \theta_1 u_c - \theta_2 y - \theta_3 \dot{y} \tag{6}$$

where the vector parameter (θ) is $= [\theta_1 \; \theta_2 \; \theta_3]$. The adaptation method for standard MIT rule is [12]:

$$\frac{d\theta_1}{dt} = -\gamma \left(\frac{1}{p^2 + a_{m1}p + a_{m2}} u_c \right) e_1 \tag{7}$$

$$\frac{d\theta_2}{dt} = \gamma \left(\frac{1}{p^2 + a_{m1}p + a_{m2}} y \right) e_1 \tag{8}$$

$$\frac{d\theta_3}{dt} = -\gamma \left(\frac{1}{p^2 + a_{m1}p + a_{m2}} \dot{y} \right) \dot{e}_1 \tag{9}$$

where $p = \frac{d}{dt}$.

Proposed MRAC

The performance of updated controller is explored here. Figure 13.2 displays the simple block diagram of this controller. Proportional-derivative (PD) feedback is connected to the standard model and the ultimate modified controller is referred to as the MRAC-PD. In comparison to the conventional MRAC controller, the output of this proposed controller is tested on second-order underdamped and with dead time system. A Second-order system as given in Eq. (13.10) is considered as RM.

The control law for this controller is expressed by,

Fig. 13.2 MRAC-PD block diagram

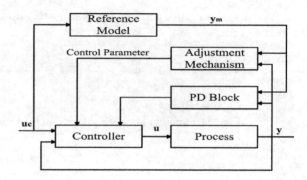

$$u = \theta_1 u_c - \theta_2 y - \theta_3 \dot{y} - \left(k_p e_1 + k_d \frac{de_1}{dt}\right)$$

where k_d and k_p is derivative and proportional gain. The transfer function (TF) of the RM is expressed by

$$G_m(s) = \frac{9}{s^2 + 4.2s + 9} \tag{10}$$

The TF of the Second-order and with the dead time system is given by

$$G(s) = \frac{2}{s^2 + 1.6s + 4} \tag{11}$$

$$G(s) = \frac{2}{s^2 + 1.6s + 4} e^{-0.10s} \tag{12}$$

The control action for the proposed MRAC-PD is already suggested by integrating, based on a fixed γ, the resulting outcome of a PD action with the adaptive control action of the MRAC controller. θ_1, θ_2 and θ_3 are derived from the MIT rules as shown by. $\frac{d\theta_1(t)}{dt} = -\gamma\left(\frac{2}{s^2+1.6s+4} u_c(t)\right) e_1(t) \frac{d\theta_2(t)}{dt} = \gamma\left(\frac{2}{s^2+1.6s+4} y(t)\right) e_1(t)$ and

$$\frac{d\theta_3(t)}{dt} = -\gamma\left(\frac{2}{s^2 + 1.6s + 4} \dot{y}(t)\right) \dot{e}_1(t) \tag{13}$$

MRAC's combined effect alongside PD feedback, that is, MRAC-PD, increases process behavior during transient and stable responses in secondary processes as compared to typical MRACs performance.

Simulation and Result

For the simulation analysis, a second-order and a dead time model are taken. During simulation study, a step input is taken having magnitude 5. In both cases, γ is considered as 0.05 and the PD parameter considered as $k_p = 5.2$ and $k_d = 1.2$.

It was found that, when using second-order system simulation, oscillations are present in the case of traditional MRAC, but there is nearly no oscillation in the case of the proposed MRAC-PD. The output of the method hits the value set smoothly. Therefore, in the response of traditional MRAC, a high peak is observed, while the MRAC-PD provides very small oscillation in process output. Figures 13.3 and 13.4 shows the process response with MRAC and MRAC-PD.

The comparison of these two methods is given in Fig. 13.5. The error plot of the MRAC and MRAC-PD is shown in Fig. 13.6a, b. The control parameters $\theta_1, \theta_2, \theta_3$ for both methods are also shown in Fig. 13.7a, b.

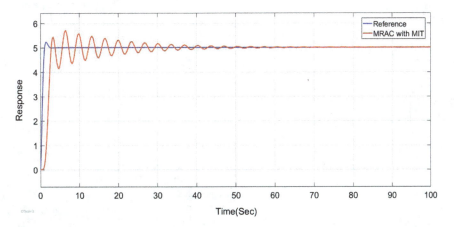

Fig. 13.3 Second-order process response with MRAC

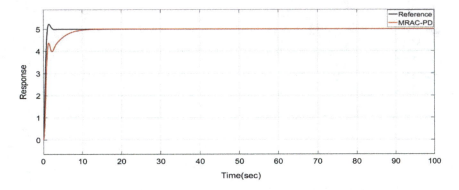

Fig. 13.4 Second-order process response with MRAC-PD

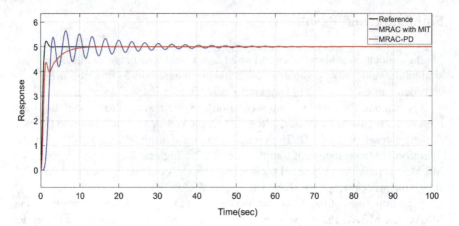

Fig. 13.5 Comparison of two method response for second-order process

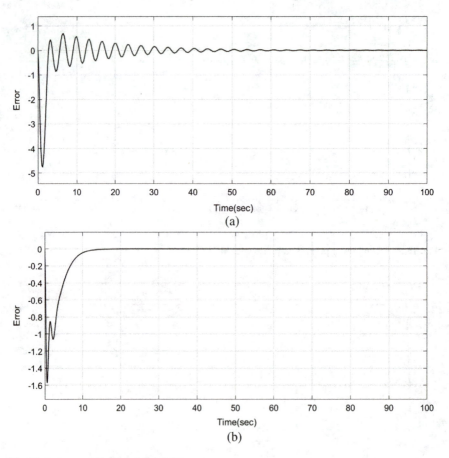

Fig. 13.6 Error **a** MRAC, **b** MRAC-PD

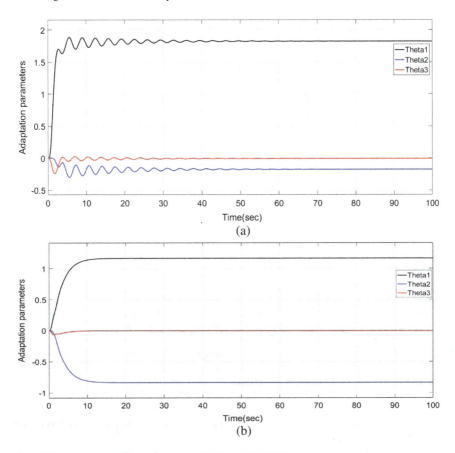

Fig. 13.7 Control parameter response **a** MRAC, **b** MRAC-PD

The increase of the MRAC-PD in process efficiency compared to MRAC is ensured for the second-order method by quantifying the ITAE, IAE, and ISE indexes as described in Table 13.1.

The system parameters keep same for second-order system with dead time. The results for this method are listed below and the performances indexes are shown in

Table 13.1 Performance evaluation of two controllers for second-order process

Type of controller	Set point tracking			Performance indices		
	Settling time (s)	Overshoot (%)	Rise time (s)	ISE	IAE	ITAE
MRAC	46	13.07	1.275	1.5×10^{-06}	0.001225	0.1225
MRAC-PD	11		3.772	1.1×10^{-11}	3.4×10^{-06}	0.0003397

Table 13.2 Performance evaluation of two controllers for second-order process with dead time

Type of controller	Set Point tracking			Performance indices		
	Settling time (s)	Overshoot (%)	Rise time	ISE	IAE	ITAE
MRAC	51	15.40	1.225 s	0.00208	0.0456	9.124
MRAC-PD	11	–	813 ms	1.6×10^{-11}	3.4×10^{-06}	0.000397

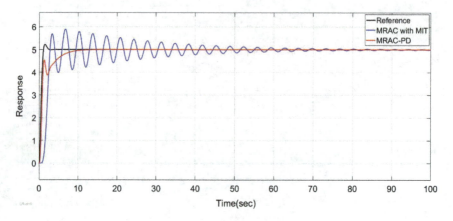

Fig. 13.8 Comparison of two method response (with dead time)

Table 13.2. Figure 13.8 shows the comparison response between two methods and Fig. 13.9 shows the error of these two methods.

The simulation outcomes are taken from a second-order under-damped with the dead time and MRAC-PD gives an overall improved process response during steady and transient operation. During set-point tracking, the peak overshoot and settling time are decreased in both cases. Performance indices values are less related to MRAC.

Conclusion

Adaptive control strategies are highly recommended for process controls of time-varying in nature and applications where frequent disturbances occur. MRAC is popular among the reported systems for adaptive control technology for their straight-forward design and remarkable performance. Here, we added a new PD feedback with traditional MRAC. It is called the MRAC-PD. A simulation experiment in a second-order underdamped process and with dead time is carried out to determine the superiority of this proposed controller. In load disturbance and set point tracking

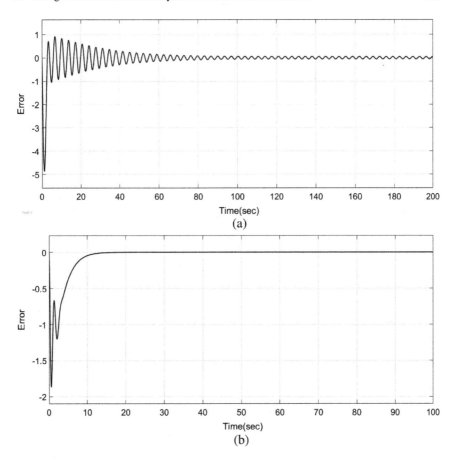

Fig. 13.9 Error **a** MRAC, **b** MRAC-PD (with dead time)

processes, improved output is noticed. In the future, we will be looking at ways of checking the output on higher level process models of the suggested controller.

References

1. Astrom K, Wittenmark B (2008) Adaptive control. Courier-Dover, USA
2. Astrom K, Kumar P-R (2014) Control: a perspective. J IFAC Int Feder Autom Control Elsevier 50(1):1–43
3. Ioannou P, Fidan B (2006) Adaptive control tutorial. Society for Industrial and Applied Mathematics, USA
4. Seborg DE, Edgar TF, Mellichamp DA (2004) Process dynamic and control, 2nd edn. Wiley, New York
5. Astrom KJ, Wittenmark B Adaptive control, 2nd edn, Dover Publications

6. Datta S, Nath UM, Dey C (2015) Design and implementation of decentralized IMC-PI controllers for real time coupled tank process. In: Michael Faraday IET international summit: MFIIS-2015, pp 93–98
7. Wang X, Zhao J, Tang Y (2012) State tracking model reference adaptive control for switched nonlinear systems with linear uncertain parameters. J Control Theory Appl 10(3):354–358
8. Swarnkar P, Jain S, Nema RK (2011) Effect of adaptation gain in model reference adaptive controlled second order system. ETASR Eng Technol Appl Sci Res 1(3):70–75
9. Benjelloun K, Mechlih H, Boukas EK (1993) A modified model reference adaptive control algorithm for dc servomotor. In: Second IEEE conference on control applications, Vancouver, B.C., pp 941–946
10. Parks PC (1966) Liapunov redesign of model reference adaptive control systems. IEEE Trans Autom Control 11(6)
11. Dorf R, Bishop R (2011) Modern control systems. Prentice Hall, USA
12. Oltean S, Abrudean M, Dulau M (2005) Model reference adaptive control for inverted pendulum. In: Inter disciplinarily in engineering scientific conference, pp 431–436

Chapter 14
Investigation Analysis of Dual Loop Controller for Grid Integrated Solar Photovoltaic Generation Systems

Aditi Chatterjee and Kishor Thakre

Introduction

Amalgamation of energy sources based on renewable energy (RE) and conventional energy for power production has intensified the use of power converters [1]. The power output of solar photovoltaic generation (SPVG) systems is not persistent all the times, so an input side controller has to be integrated with the system to excerpt all-out power from the source regardless of the weather conditions. A grid side control system has to be incorporated to enhance the power quality of the output current of power converter interfacing the RE based distributed generation (DG) plant with the distribution grid. To address these control issues various control strategies and topologies have been proposed in literature.

In [2] diverse single phase inverter topologies for interfacing PV system to distribution grid has been focused on. In [3] transformerless topologies are described.

Some of the significant control stratagems i.e. the hysteresis current control (HCC), proportional integral (PI) control, dead beat current (DBC) control, proportional resonant (PR) control and model predictive control (MPC) has been investigated for single-phase grid integrated DG plants [4]. In [5] fuzzy logic based current controller is introduced for photovoltaic system. MPC-based current control technique is proposed for PV based DG plant operating in grid-connected manner [6]. In [10] MPC voltage control system is presented for 1-phase inverter for standalone renewable energy system.

A. Chatterjee (✉)
IGIT, Sarang, Dhenkanal, Odisha, India
e-mail: contactaditi247@gmail.com

K. Thakre
GEC, Jhalawar, Rajasthan, India
e-mail: thakrekishor26@gmail.com

© The Editor(s) (if applicable) and The Author(s), under exclusive license to Springer Nature Singapore Pte Ltd. 2021
A. K. Singh and M. Tripathy (eds.), *Control Applications in Modern Power System*, Lecture Notes in Electrical Engineering 710,
https://doi.org/10.1007/978-981-15-8815-0_14

In this work, a comparative study of sole stage and double stage SPVG system is presented thereafter a constrained delay compensated model predictive current controller (CDC-MPCC) is introduced for 1-phase grid tied SPVG system. The MPC-based current controller is designed with a switching frequency constrained cost function and the delay effect is considered while designing the controller.

Comparative Analysis of Sole and Double Stage Grid Tied SPVG Systems

In case of sole stage topology, as illustrated in Fig. 14.1a. The maximum power extraction and the grid side control action are executed by the inverter. The reliability of this structure is higher because of reduced number of components but the output voltage of the solar array is stepped down by the H-bridge converter. The high power PV systems adapt to this topology.

Power progression transpires at two consecutive stages in double stage structure. There is a boost converter that performs MPPT cascaded by an inverter as shown in Fig. 14.1b. The DC-DC converter excerpts the maximum possible power from the source and boosts the output voltage level to desired value. The low power SPVG plant adapts this structure [2]. The DC-AC converter regulates the power exchange between the SPVG and the distribution grid and also alleviates the output current harmonics. During the power flow between the two converters there can be a power inequity between the two power conversion stages. To balance the inequity a capacitance is placed in between the two converters. The DC and AC power variance is a sinusoidal component at twice the grid frequency delivered to the grid. The DC-link capacitor current constituent a DC part and a double grid frequency AC part [13]. Discarding the DC part, the capacitor current can be simplified as (14.1).

$$i_{dc}(t) = I_{dc} \cos(2 \cdot \omega_g \cdot t) \tag{14.1}$$

$$i_{dc}(t) = C_{dc} \frac{dv_{dc}(t)}{dt} \tag{14.2}$$

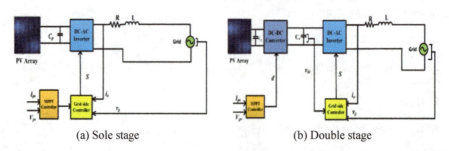

(a) Sole stage   (b) Double stage

Fig. 14.1 Converter topologies for grid interfaced SPVG systems

The AC constituent of capacitor voltage is expressed by (14.3), where C_{dc} is the DC-link capacitance.

$$\hat{v}_{dc}(t) = \frac{I_{dc}}{C_{dc}} \int_T \cos(2 \cdot \omega_g \cdot t) dt \qquad (14.3)$$

The above Eq. (14.3) can be simplified as (14.4):

$$\hat{v}_{dc}(t) = \frac{I_{dc}}{2 \cdot \omega_g \cdot C_{dc}} \sin(2 \cdot \omega_g \cdot t) \qquad (14.4)$$

The peak to peak amplitude of capacitor voltage oscillations is:

$$\Delta v_{dc} = \frac{P_{dc}}{2 \cdot \omega_g \cdot C_{dc} \cdot v_{dc}} \qquad (14.5)$$

where P_{dc} is the DC power and it is equivalent to PV source power output. The capacitor voltage oscillations can transmit back through the boost converter up to the photovoltaic generator, depreciating the performance of MPPT controller. The inverter output waveform quality is also deteriorated due to the increased voltage ripples. The voltage ripples can be condensed by using a larger value of capacitance as observed from (14.5). The foremost shortcoming of double stage system is usage of bulk capacitance. Hence, incorporation of an effective current control strategy is essential to improve the output power quality.

The non-linear I-V curve of solar array can be used to obtain a relationship between the voltage amplitude (14.5) and MPPT efficiency. By neglecting the parallel and series resistance of solar array model the array output current at MPP and its 1st and 2nd derivatives are given by (14.6)–(14.8).

$$I_{mpp} = I_{ph} - I_o (e^{\frac{q \cdot V_{mpp}}{A \cdot K \cdot T \cdot N_s}} - 1) \qquad (14.6)$$

$$\frac{\partial I_{mpp}}{\partial V_{mpp}} = \frac{-I_o}{A \cdot K \cdot T \cdot q^{-1} N_s} e^{\frac{q \cdot V_{mpp}}{A \cdot K \cdot T \cdot N_s}} \qquad (14.7)$$

$$\frac{\partial^2 I_{mpp}}{\partial V_{mpp}^2} = \frac{-I_o}{(A \cdot K \cdot T \cdot q^{-1} N_s)^2} e^{\frac{q \cdot V_{mpp}}{A \cdot K \cdot T \cdot N_s}} \qquad (14.8)$$

where, I_{ph}, I_0, I_{mpp} are photo-generated current, module saturation current, and maximum current output at MPP, respectively. The number of cells in series is N_s, v_{pv} is the PV array output voltage, V_{mpp} is the voltage at maximum power point and A is diode ideality factor. The electron charge and the Boltzman constant are denoted by q and K, respectively and T is the operating temperature of the cell in Kelvin. The current–voltage curve of solar array across the MPP can be approximated by means of second-order Taylor series (14.9).

$$i_{pv} = I_{mpp} + \frac{\partial I_{mpp}}{\partial V_{mpp}}(v_{pv} - V_{mpp}) + \frac{1}{2} \cdot \frac{\partial^2 I_{mpp}}{\partial V_{mpp}^2}(v_{pv} - V_{mpp})^2 \qquad (14.9)$$

Equation (14.9) can be written as (14.10)

$$i_{pv} = \gamma + \beta v_{pv} + \alpha v_{pv}^2 \qquad (14.10)$$

where, γ, β and α are given by (14.11), (14.12) and (14.13) respectively.

$$\gamma = I_{mpp} - V_{mpp}\frac{\partial I_{mpp}}{\partial V_{mpp}} + \frac{1}{2}\frac{\partial^2 I_{mpp}}{\partial V_{mpp}^2} V_{mpp}^2 \qquad (14.11)$$

$$\beta = \frac{\partial I_{mpp}}{\partial V_{mpp}} - V_{mpp}\frac{\partial^2 I_{mpp}}{\partial V_{mpp}^2} \qquad (14.12)$$

$$\alpha = \frac{1}{2}\frac{\partial^2 I_{mpp}}{\partial V_{mpp}^2} \qquad (14.13)$$

The voltage transformation ratio of the boost converter is:

$$M(D) = \frac{v_{dc}}{v_{pv}} \qquad (14.14)$$

The oscillations of voltage across, the DC-link causes voltage oscillations across the PV panel as given by the following equation:

$$\Delta v_{pv} = \frac{\Delta v_{dc}}{M(D)} \qquad (14.15)$$

The voltage output of PV panel can be articulated as sum of MPP voltage and the voltage component at twice the grid frequency as in (14.16).

$$v_{pv}(t) = V_{mpp} + \hat{v}(t) = V_{mpp} + \Delta v_{pv} \cdot \sin(2 \cdot \omega_g \cdot t) \qquad (14.16)$$

The instantaneous power can be calculated by (14.17), (14.18).

$$p_{pv}(t) = v_{pv}(t).i_{pv}(t) \qquad (14.17)$$

$$p_{pv}(t) = (V_{mpp} + \hat{v}(t)) \cdot \gamma + (V_{mpp} + \hat{v}(t))^2 \cdot \beta + (V_{mpp} + \hat{v}(t))^3 \cdot \alpha \qquad (14.18)$$

The average power over one fundamental period of grid voltage is given by (14.19).

$$p = \frac{1}{T} \int_T p_{pv}(t) \cdot dt \tag{14.19}$$

$$p = \gamma \cdot V_{mpp} + \beta \cdot V_{mpp}^2 + \alpha \cdot V_{mpp}^3 + \left(\frac{3 \cdot \alpha \cdot V_{mpp} + \beta}{2}\right) \Delta v_{pv}^2 \tag{14.20}$$

$$p = p_{mpp} + \frac{3 \cdot \alpha \cdot V_{mpp} + \beta}{2} \Delta v_{pv}^2 \tag{14.21}$$

Substituting (14.15) in (14.21), the average power can be expressed as (14.22):

$$p = p_{mpp} + \frac{3 \cdot \alpha \cdot V_{mpp+\beta}}{2} \left(\frac{\Delta v_{dc}}{M(D)}\right)^2 \tag{14.22}$$

The MPPT efficiency as a function of voltage oscillations can be evaluated as (14.23).

$$\eta_{mppt} = \frac{p}{p_{mpp}} \tag{14.23}$$

Substituting (14.22) in (14.23) the MPPT efficiency can be written as in (14.24).

$$\eta_{mppt} = 1 + \left(\frac{3 \cdot \alpha \cdot V_{mpp} + \beta}{2 \cdot p_{mpp} \cdot M^2(D)}\right) \Delta v_{dc}^2 \tag{14.24}$$

From (14.24), it can be concluded that the MPPT efficiency will be higher for lower value of voltage oscillation amplitude as α and β are negative terms. Hence, the amplitude of voltage oscillation for desired MPPT efficiency is given by (14.25).

$$\Delta v_{dc} = M(D) \sqrt{\frac{2 \cdot P_{mpp}(\eta_{mppt} - 1)}{3 \cdot \alpha \cdot V_{mpp} + \beta}} \tag{14.25}$$

By substituting a desired value of MPPT efficiency in (14.25) a constraint can be put on the magnitude of voltage oscillation magnitude. Equation (14.25) depicts that a boost DC–DC converter, with $M(D)$ value greater than 1, will make possible a smaller value of DC-link capacitance for a given MPPT efficiency. From the above explanation, it can be concluded that a DC–DC converter with $M(D) > 1$, aids in mitigating the detrimental effects of oscillations at DC-link, for all frequency ranges. On the other hand, sole-stage topology has no mechanism for reduction of voltage oscillation amplitude, as it includes only one large electrolytic capacitor at the PV terminal. This mechanism improves the reliability of the system in terms of reduced number of component usage but reduces the MPPT efficiency and also deteriorates the quality of inverter output power.

Control System Structure for Grid-Tied Solar Photovoltaic Generator System

Figure 14.2 illustrates the control architecture of double stage grid interfaced SPVG system. It incorporates three controllers, i.e. the fuzzy logic maximum power point tracking (FLC-MPPT) controller, voltage controller, and current controller.

Modelling of SPVG System and Design of FLC-MPPT Controller

In this simulation work, Kyocera makes KC200GT solar panel is used. The details of modelling of PV array is not mentioned in this chapter as it is reported in literature [11]. The fundamental concept of perturb and observe (P and O) MPPT algorithm is that at highest power point the variation in solar array power output is zero. A FLC-based P and O MPPT controller is introduced in [12]. The two inputs to the FLC are given by (14.26) and (14.27) and variation in duty cycle of boost converter is the output (14.28).

$$e(k) = \frac{dP(k)}{dV(k)} \qquad (14.26)$$

$$\Delta er(k) = er(k) - er(k-1) \qquad (14.27)$$

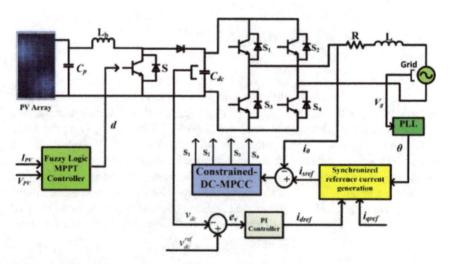

Fig. 14.2 Control system structure for grid-tied SPVG systems

Table 14.1 FLC rule base

	er				
Δer	NeL	NeS	Z	PoS	PoL
NeL	PoS	PoL	PoL	NeL	NeS
NeS	Z	PoS	PoS	NeS	Z
Z	Z	Z	Z	Z	Z
PoS	Z	NeS	NeS	PoS	Z
PoL	NeS	NeL	NeL	PoL	PoS

$$\Delta dC = dC(k) - dC(k-1) \tag{14.28}$$

There are five membership functions (MF) positive small (PoS), positive large (PoL), zero (Z), negative small (NeS), negative large (NeL). Triangular MF is used and the rule based is formulated based on the principle of P and O algorithm as in Table 14.1.

Grid Side Controller

The proposed control system consists of two nested control loops as illustrated in Fig. 14.2.

Voltage Controller. A conventional proportional integral (PI) controller is designed for DC voltage control as depicted in Fig. 14.3.

The current reference generated by the PI controller is given by (14.29)

$$i_{dref}(s) = e_v(s) * \left(k_p + \frac{k_i}{s}\right) \tag{14.29}$$

Current Controller. The reference current regulating the active power is generated by the PI control loop. The reactive reference current is set as per the load demand. The reference current is obtained by (14.30).

$$i_{sref} = i_{dref} \sin \omega t - i_{qref} \cos \omega t \tag{14.30}$$

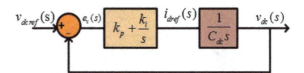

Fig. 14.3 PI-DC-link voltage controller

Constrained Delay Compensated Model Predictive Controller

The MPC stratagem with several variations has been implemented for control of power converters for drives and renewable energy technologies [6–10].

In this work, a MPC with switching frequency reduction is designed for dual-stage SPVG system. This assists in reducing the switching frequency along with minimizing the reference current tracking error. The scheme is shown in Fig. 14.4.

Analysis of Discretized System Model Dynamics

The converter switching states tabulation is specified in Table 14.2. There are four output voltage levels denoted by v_o. The switching statuses of the 1-phase inverter are given by S_a and S_b and depends on the DC-link voltage.

The discretized dynamic equation of the system model is expressed as (14.31):

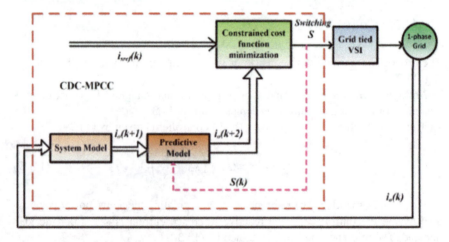

Fig. 14.4 Constrained model predictive current controller block diagram

Table 14.2 Switching table for 1-phase inverter

S_1	S_2	S_3	S_4	S_a	S_b	v_o
0	0	1	1	0	0	0
1	0	0	1	1	0	v_{dc}
0	1	1	0	0	1	$-v_{dc}$
1	1	0	0	1	1	0

$$v_o(k) = Ri_o(k) + L\frac{i_o(k+1)}{T_{sp}} - L\frac{i_o(k)}{T_{sp}} + v_g(k) \tag{14.31}$$

From (14.31) in view of the voltage level that is applying during the present sampling interval, estimated value of current at $(k + 1)$ instant is evaluated as:

$$\hat{i}_o(k+1) = \left(1 - \frac{RT_{sp}}{L}\right)i_o(k) + \frac{T_{sp}}{L}(v_o(k) - v_g(k)) \tag{14.32}$$

The estimated value from (14.32) is used to calculate the current at the next time instant for all four switching statuses of the inverter:

$$i_o^P(k+2) = \left(1 - \frac{RT_{sp}}{L}\right)\hat{i}_o(k+1) + \frac{T_{sp}}{L}(v_{on}(k+1) - v_g(k)) \tag{14.33}$$

A cost function is weighed for each of four predictions.

Cost Function Assessment

The cost function optimizes the error between the magnitude of reference and forecasted current and is given by (14.34).

$$g_i = (e_\alpha)^2 + (e_\beta)^2 + \lambda n_f \tag{14.34}$$

where, e_α and e_β are error components calculated by (14.35).

$$\begin{bmatrix} e_\alpha \\ e_\beta \end{bmatrix} = \begin{bmatrix} i_{sref\alpha} - i_{o\alpha}^P(k+2) \\ i_{sref\beta} - i_{o\beta}^P(k+2) \end{bmatrix} \tag{14.35}$$

The number of switches that change when the switching signal is applied is n_f and λ is the weighting factor. For single-phase inverter, there are four switches and four voltage vectors as tabulated in Table 14.2. The switching state vector that describes the switching of each inverter leg is $S = (S_a, S_b)$. The number of switches changing from time (k) to $(k + 1)$ can be written as (14.36):

$$n_f = |S_a(k+1) - S_a(k)| + |S_b(k+1) - S_b(k)| \tag{14.36}$$

Since the cost functions consist of two terms which are physically different i.e. one is current and the other is change in number of switches a weighting factor is multiplied with the second term. The primary aim of the cost function is to reduce the current error and the secondary term is a constraint which minimizes the switching frequency. The output voltage level which yields the minimum value of cost function is opted for switching.

Hardware Implementation

The model is first simulated in Matlab/Simulink software and the simulation parameters are taken from [6]. For experimental implementation, a Semikron makes inverter is used. dSPACE 1104 is used to execute the digital control algorithm as shown in the experimental set up in Fig. 14.5. The sensors LA 55-P and LV 25-P are deployed for output current and grid voltage sensing respectively. The total harmonic distortion (THD) of output current signal is measured by a power analyzer.

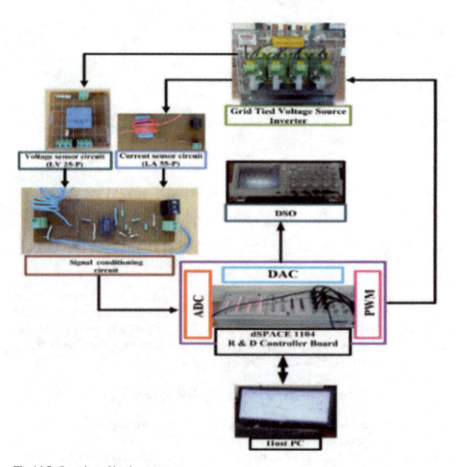

Fig. 14.5 Snapshot of hardware set up

Hardware Results

Figure 14.6a depicts the output current signal shown by channel 1 is aligned with the grid voltage signal shown by channel 2. This demonstrates that the power factor is almost unity. To analyze the power quality of output current waveform a THD plot is obtained. The experimental THD magnitude is found to be 2.23% as shown in Fig. 14.6b and as per the IEEE 1547 standards, it should be less than 5%.

To study the dynamics of the proposed current control scheme. The reactive current reference is varied. It is shown in Fig. 14.7a, that the output current waveform is lagging the grid voltage when reactive current reference is made positive and it is leading when it is negative as depicted by Fig. 14.7b.

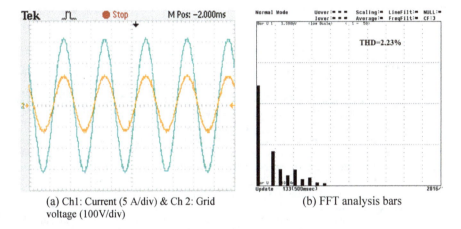

(a) Ch1: Current (5 A/div) & Ch 2: Grid voltage (100V/div)

(b) FFT analysis bars

Fig. 14.6 Steady-state experimental waveforms

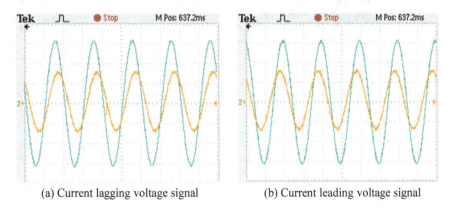

(a) Current lagging voltage signal

(b) Current leading voltage signal

Fig. 14.7 Transient state experimental waveforms Ch 1: current (4 A/div) and Ch 2: grid voltage (100 V/div)

Conclusion

This research work focuses on the development of an effective constrained delay compensated model predictive current control system for grid-connected solar photovoltaics. An analytical comparison between the sole stage and double stage grid integrated SPVG is illustrated from which it can be concluded that the double stage topology is a more resounding topology for small and medium power SPVG systems. The control system structure for grid-tied SPVG system is discussed in details. The fuzzy logic based perturb and observe MPPT controller extracts the all-out power from the solar array and the proposed current control systems mitigate the current harmonics. During transient state, the controller executes very fast dynamics. The performance of the control strategy is authenticated by developing a hardware prototype in the laboratory using DS1104.

References

1. Carrasco JM, Franquelo LG, Bialasiewicz JT, Galvan E, Guisado RCP, Prats MAM, Leon JI, Moreno-Alfonso N (2006) Power-electronic systems for the grid integration of renewable energy sources: a survey. IEEE Trans Ind Electron 53(4):1002–1016
2. Kjaer SB, Pedersen JK, Blaabjerg F (2005) A review of single-phase grid-connected inverters for photovoltaic modules. IEEE Trans Ind Appl 41(5):1292–1306
3. Patrao I, Figueres E, Espin FG, Garcera G (2011) Transformerless topologies for grid-connected single-phase photovoltaic inverters. Renew Sustain Energy Rev 15:3423–3431
4. Chatterjee A, Mohanty KB (2018) Current control strategies for single phase grid integrated inverters for photovoltaic applications-a review. Renew Sustain Energy Rev 92:554–569
5. Ali A et al (2017) Harmonic distortion performance in PV systems using fuzzy logic controller. In: Chadli M et al (ed) ICEECA 2017, LNEE, vol 522. Springer, pp 328–337
6. Chatterjee A, Mohanty KB, Kommukuri VS, Thakre K (2017) Design and experimental investigation of digital model predictive current controller for single phase grid integrated photovoltaic systems. Renew Energy 108:438–448
7. Rodriguez J, Pontt J, Silvam CA, Correa P, Lezana P, Cortes P, Ammann U (2007) Predictive current control of a voltage source inverter. IEEE Trans Industr Electron 54(1):495–503
8. Pavlou KG, Vasiladiotis M, Manias SN (2012) Constrained model predictive control strategy for single-phase switch-mode rectifiers. IET Power Electron 5(1):31–40
9. Errouissi R, Muyeen SM, Al-Durra A, Leng S (2016) Experimental validation of a robust continuous nonlinear model predictive control based grid-interlinked photovoltaic inverter. IEEE Trans Industr Electron 63(7):4495–4505
10. Talbi B, Krim F, Laib A et al (2020) Model predictive voltage control of a single-phase inverter with output LC filter for stand-alone renewable energy systems. Electric Eng
11. Chatterjee A, Mohanty KB (2014) Development of stationary frame PR current controller for performance improvement of grid tied PV inverters. In: 9th International Conference on Industrial and Information Systems (ICIIS). IEEE, Gwalior, pp 1–6
12. Boukezata B, Chaoui A, Gaubert JP, Hachemi M (2016) An improved fuzzy logic control MPPT based P&O method to solve fast irradiation change problem. J Renew Sustain Energy 8:043505
13. Chen YM, Wu HC, Chen YC, Lee KY, Shyu SS (2010) The AC line current regulation strategy for the grid-connected PV system. IEEE Trans Power Electron 25(1):209–218

Chapter 15
Damping Enhancement of DFIG Integrated Power System by Coordinated Controllers Tuning Using Marine Predators Algorithm

Akanksha Shukla and Abhilash Kumar Gupta

Introduction

In present-day power grids, small-signal stability refers to the damping of low-frequency oscillation (LFO) modes in the system [1]. The LFO modes usually in the frequency range of 0.2–2 Hz [2]. The LFOs are inherent to the power system and advent of unforeseen events or contingencies may amplify the oscillations' amplitude. This may lead to synchronism loss in the system and even blackout in some cases [3].

The electricity generation scenario is gradually shifting its focus to renewable sources to reduce the carbon footprint and environmental degradation. Among several renewables gaining ground, wind seems to be most advantageous overall [4]. The variable speed wind turbine generators (WTG), having doubly-fed induction generators (DFIG) are usually preferred [5]. The wind-based power plants are located in high wind potential areas, usually remote places, with long and relatively weak transmission lines which affects the power flow and system damping [6]. The DFIGs being asynchronous machines do not directly impact the system damping. However, due to reduced inertia and converter interaction with nearby generators, it may have negative impact on system small signal stability [7]. The system LFO modes should have a acceptable damping, which is usually 5–10%, for the system to remain oscillatory stable.

Conventionally power system stabilizers (PSSs) are employed for improving the small-signal stability of the system [8]. In current scenario with increasing renewables

---

A. Shukla (✉) · A. K. Gupta
Department of Electrical Engineering, GLA University, Mathura, India
e-mail: akanksha.shukla@gla.ac.in

A. K. Gupta
e-mail: abhilash.gupta@gla.ac.in

© The Editor(s) (if applicable) and The Author(s), under exclusive license to Springer Nature Singapore Pte Ltd. 2021
A. K. Singh and M. Tripathy (eds.), *Control Applications in Modern Power System*, Lecture Notes in Electrical Engineering 710, https://doi.org/10.1007/978-981-15-8815-0_15

share in the grid, it becomes difficult to damp LFOs with PSSs alone, even after proper tuning. There is a need to support the PSS by adding additional controllers to augment the modes damping in such scenarios. The literature suggests that with increasing renewable integration the LFOs damping can be improved by using PSS along with supplementary controllers. The authors in [9] shows that wind power plants can be used for damping improvement of interarea modes. In [10], a PSS has been employed along with DFIG to improve the modes damping. A coordinated AVR-PSS design is proposed in [11] to achieve small signal stability. Many authors suggested the use of power oscillation dampers (PODs) with wind units to damp the oscillations [12–16]. The authors used the PMU data for wide area LFO damping in presence of wind in [17]. However, most of the papers tune the controllers for small wind penetrations. This might not guarantee optimal operation at higher DFIG penetrations. Also, the literature suggests that the actual impact of wind penetration on system damping is evident with more than 15% penetration of wind. Thus, there is a need to develop an efficient method for augmenting small signal stability for high wind penetration in the system.

The selection of wide area input signals for damping controllers is equally important. In [18], author proposes adaptive signal selection for wide area controllers employed for damping improvement. The proper input signal is one which minimizes interaction among controllers and improve the observability of critical modes [19]. The tuning of controller's parameters is usually obtained using analytical, numerical, meta-heuristic techniques, etc. [20]. However, out of them, metaheuristic techniques are preferred as they do not require past knowledge of the problem and can be employed for complex problems [21]. Thus, there is a need to carry out the controllers tuning effectively to damp out the critical LFO modes.

In this work, a robust coordinated damping improvement strategy is proposed to improve the system's small signal stability using PODs and PSSs. The strategy aims to provide robust results without using any complex formulation. The geometric modal observability is utilized for wide area PODs signal selection. A new meta-heuristic technique Marine Predators Algorithm (MPA) [22] has been utilized to tune the parameters of controllers simultaneously and results are compared with particle swarm optimization and grey wolf optimization to show its robustness over other techniques. Further, eigenvalue analysis and time-domain simulations are executed to demonstrate the capability of suggested control approach.

Proposed Methodology

The work carried out in this paper aims to develop a robust coordinated tuning strategy for damping controllers employed by synchronous generators (SGs) and DFIGs. For improving the small-signal stability of high wind integrated system, PODs are employed with DFIGs and they are tuned in coordination with PSSs already installed with SGs. The DFIG modelling employed in this work can be referred from [23].

Input Signals for Controllers

The input signals are chosen based on the observability of critical modes in those signals. For this, the geometric measure of modal observability has been put to use for finding out the suitable input signal. For jth mode it is given by

$$MO_m = \frac{|C_m \phi_j|}{\|C_m\| \|\phi_j\|} \tag{15.1}$$

where C_m is the mth row of C matrix. The active power flow of all the transmission lines are checked for the observability values using Eq. (15.1) and the one having the highest value is chosen as the input signal. The wide area signal is measured with the help of phasor measurement units (PMUs) placed in the system optimally [24].

Fitness Function Formulation

As discussed in the introduction section, the aim is to develop a simple formulation for fitness function so as to have less complexity for large systems application. The proposed coordinated control problem is formulated as an optimization problem which finds the optimal values of controllers' parameters. The objective of improving small signal stability of the system can be achieved by improving the critical LFO modes damping ratios and also shifting the critical modes to more stable left-hand region of s-plane.

These two objectives are combined into a single fitness function which is to be minimized to shift the critical modes into the stable region and is given as

$$f = \min \left(\sum_{n=1}^{oc} \sum_{\xi_{j,n} \leq \xi_{des}} (\frac{\xi_{des} - \xi_{j,n}}{\xi_{des}})^2 + \sum_{n=1}^{oc} \sum_{\sigma_{j,n} \geq \sigma_{des}} (\frac{\sigma_{des} - \sigma_{j,n}}{\sigma_{des}})^2 \right)$$

subject to: $K^{LB} \leq K \leq K^{UB}$

$T_{1,3}^{LB} \leq T_{1,3} \leq T_{1,3}^{UB}$

$T_{2,4}^{LB} \leq T_{2,4} \leq T_{2,4}^{UB}$

(15.2)

In Eq. (15.2), the oc represents the operating conditions for which the parameters are optimized, $\sigma_{j,n}$ and $\xi_{j,n}$ are the real part and the damping ratio of the jth eigenvalue corresponding to the nth operating condition. The desired values, σ_{des} and ξ_{des} are set at -1.0 and $0.1(10\%)$, respectively. The lower and upper bound values for gain K is set as 0.1 and 50, respectively, and for time constants the lower and upper bounds are 0.01 and 0.1, respectively ($T_1 = T_3, T_2 = T_4$).

The proposed tuning of controllers is a highly non-linear problem in nature and thus meta-heuristic techniques are preferred for such optimization. In this work, an advanced and recently introduced nature-inspired meta-heuristic technique MPA has been employed for optimizing the controller parameters. The details about MPA can be referred from [22].

Simulations and Results

The proposed control approach is tested on benchmark IEEE 39-bus New England test system. It has 10 SGs with total generation of 6139 MW. The complete system data used in this work can be referred from [25]. The PSSs are installed at all SGs except the slack. The system is modified to integrate three wind farms of 512 MW each to achieve a wind penetration of 25%. The single line diagram of modified system is shown in Fig. 15.1. The wind farms are located at bus number 1, 9 and 12.

In this modified system, when PSSs are untuned, eigen analysis has been carried out to find out the LFO modes in the system. The results obtained are shown in Table 15.1. There are nine modes present in the system. Out of them, eight are local and one is interarea mode as shown in the Table 15.1. As discussed in the previous section, the desired damping for every LFO mode is 10%. From the Table 15.1, it

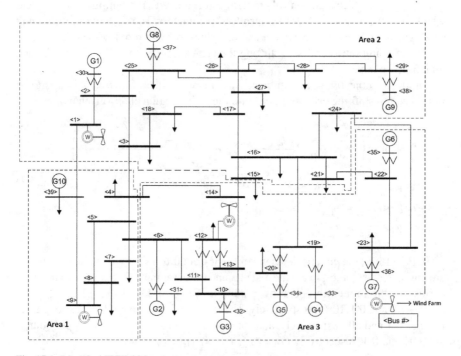

Fig. 15.1 Modified IEEE 39 bus test system

Table 15.1 LFO modes in base case (PSSs untuned)

S. No.	Frequency (Hz)	Damping ratio (%)	Mode type
1	1.6459	19.35	Local
2	1.6148	21.81	Local
3	*1.4587*	*08.45*	*Local*
4	1.4183	28.97	Local
5	1.3629	22.30	Local
6	1.1592	20.64	Local
7	1.1046	29.71	Local
8	0.9459	19.22	Local
9	*0.6015*	*05.04*	*Interarea*

can be observed that in base case there are two modes with damping less than 10%, shown in bold. One is local mode and the other is interarea mode. So, these are the critical LFO modes in the base case for this system. The aim is to develop a robust control to improve damping of these critical modes above 10% such that the system remains small signal stable even in critical operating conditions.

PODs Input Signals

In order to improve the modes damping, PODs are employed with DFIGs. Then, they are coordinately tuned with PSSs. As discussed in previous section, the wide area input signal for PODs has been selected on the basis of observability results. The active power flow of the transmission lines has been checked for the observability values using Eq. (15.1) and the one having the highest value is chosen as the input signal. Figure 15.2 shows that the line 16 has the highest value of observability of critical interarea mode. Thus, the active power flow between bus 9 and 39 has been fed as input to PODs, measured using PMUs.

Coordinated Tuning of Controllers

The robust tuning of the PSSs and PODs has been carried out using MPA optimization technique. The effectiveness of this algorithm has been evaluated by comparing it with well-established particle swarm optimization (PSO) and grey wolf optimizer (GWO). The comparison has been done on the basis of statistical results and convergence characteristics. Same search parameters (population = 50, maximum iterations = 100, runs = 10) are taken for all the three techniques. The parameters taken for different optimization techniques are provided in Table 15.2. The comparison has

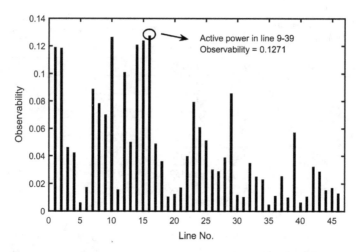

Fig. 15.2 Observability Results

Table 15.2 Parameter settings for the three algorithms

Algorithm	Parameter	Value
MPA	Constant number (P)	0.5
	Fish aggregating devices (FADs)	0.2
	Power law index (beta)	1.5
GWO	Coefficient vector (a)	Linearly decreased from 2 to 0
PSO	Cognitive and social constant (C_1, C_2)	2, 2
	Inertia weight (w_{min}, w_{max})	0.4, 0.9

been done for the highest problem dimension, i.e., when there are 9 PSSs and three PODs in the system which gives a dimension of 36 (each contributing three variables).

The comparison on the basis of statistical results is shown in Table 15.3. The statistical analysis confirms the superiority of MPA over other algorithms. The standard

Table 15.3 Comparison of different optimization statistical results

	Algorithms		
	MPA	GWO	PSO
Average	0.1732	2.9762	4.7278
Standard deviation	0.3711	1.8459	2.1965
Median	0	2.6352	5.7431
Best	0	0	0.9358
Worst	1.1425	6.5437	7.3781

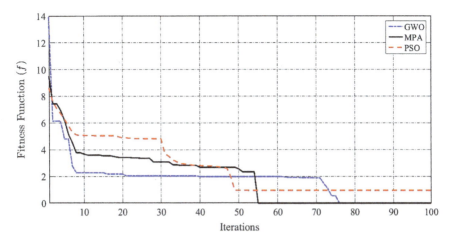

Fig. 15.3 Convergence characteristics comparison

deviation closer to zero for MPA indicate that this algorithm can effectively reach optimal solution irrespective of the initial population. Thus, have exploration capability and does not get stuck in local optima. Further, this is supported by lower best and worst fitness values obtained in MPA as compared to GWO and PSO algorithms. The median value 0 obtained in case of MPA signifies the exploitation capability of MPA algorithm. Hence, it can be said that MPA has better exploration and exploitation capability. Further, the consideration of FAD effect has avoided stagnation in local optima.

The superiority of MPA algorithm is further validated in convergence characteristic as shown in Fig. 15.3. It shows that MPA algorithm converges faster than the other two. Also, it is observed that global optimal solution is not obtained in case of PSO algorithm. Therefore, MPA is employed in this work for parameter optimization.

The MPA is used for this coordinated control of damping controllers with search parameters taken as discussed above. In order to achieve a fairly robust damping performance the controllers are simultaneously tuned for multiple operating scenarios. These are base case scenario, 20% decrement in load, 20% increment in load and outage of tie-line between buses 14–15. The performance of tuning has been evaluated by analyzing outputs for three scenarios:

(1) Case 1: Base case scenario having untuned controllers, PODs not employed.
(2) Case 2: System has tuned power system stabilizers, PODs not employed.
(3) Case 3: System has coordinately tuned damping controllers (both PSS and POD).

The system performance is compared for these three cases using eigenvalue analysis and time-domain analysis. In addition, some critical scenarios are simulated to assess the controllers robustness to varying system conditions.

Table 15.4 Eigen-analysis results

Mode	Case 1			Case 2			Case 3		
	Real part	f (Hz)	DR (%)	Real part	f (Hz)	DR (%)	Real part	f (Hz)	DR (%)
1	−1.76	1.6459	19.35	−2.15	1.6425	23.91	−3.45	1.7037	33.75
2	−1.96	1.6148	21.81	−2.06	1.6360	22.67	−2.14	1.6296	23.44
3	−0.87	1.4587	08.45	−1.08	1.3735	11.42	−2.54	1.4255	29.11
4	−2.45	1.4183	28.97	−2.83	1.4192	31.68	−2.45	1.1931	28.69
5	−2.02	1.3629	22.30	−1.80	1.3590	20.41	−3.70	1.1132	35.32
6	−1.82	1.1592	20.64	−1.77	1.1237	19.97	−3.14	1.0174	31.90
7	−2.56	1.1046	29.71	−2.77	1.0237	30.14	−3.96	0.8270	39.40
8	−1.73	0.9459	19.22	−2.10	0.9825	23.42	−4.01	0.6582	40.91
9	−0.41	0.6015	05.04	−0.79	0.5450	07.98	–	–	–

Eigen-Analysis Results. The eigenvalue analysis has been carried out for the above three cases and the results are shown in Table 15.4. The results show that there are two critical modes in Case 1 with damping less than 10% when PSSs are untuned. When PSSs are tuned in Case 2, the local mode damping improves to become greater than 10% but the interarea mode still has low damping ratio. However, with application of coordinated tuning in Case 3, all modes have damping greater than the desired limit. Also, the number of LFO modes reduces to 8 in Case 3 with application of coordinated tuning. The real part of eigen values of all LFO modes also becomes smaller than the desired value of −1.0. The improved system damping in Case 3 clearly shows the efficacy of proposed approach of tuning using MPA.

Time-Domain Analysis Results. To further prove the effectiveness of proposed tuning approach and to verify the eigen-analysis results, time-domain analysis has been performed. At 1.0 s, a 3-phase fault is carried out at bus 16 and the output of SGs are observed to check the oscillations damping in the system. The result is shown in Figs. 15.4 and 15.5.

It can be observed from the figures (Figs. 15.4 and 15.5) that in the base case the oscillations grow with time making system small signal unstable. However, the oscillations got damped out in Case 2 (only PSS) and Case 3 (PSS + POD) within time and thus system remains oscillatory stable. The oscillations are slightly higher in Case 2 as compared to Case 3, as evident from the figures. Thus, the time domain simulation result verifies the eigen-analysis results.

However, in case of critical contingencies the PSS tuning (Case 2) might fail as there are no supplementary controllers employed to assist in damping in such case. The proposed control improves system damping in those scenarios also. Therefore, to verify the robustness of the proposed coordinated tuning of controllers' various critical contingencies have been simulated and the results are compared for the three cases. Two such scenarios are discussed here:

15 Damping Enhancement of DFIG Integrated Power System … 173

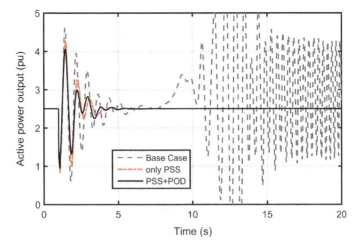

Fig. 15.4 SG1 active power output

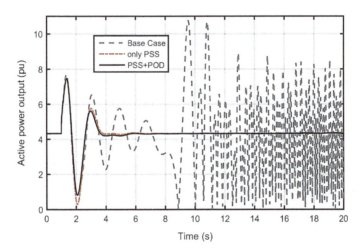

Fig. 15.5 SG10 active power output

1. The load is increased by 40%, along with a 3-phase fault of 6 cycles at bus 29 and simultaneous outages of tie-lines between bus 3–4 and 21–22.
2. A 100 MW reduction in outputs of wind farms keeping same fault and line outages as scenario 1.

The active power output of SG1 is shown in Figs. 15.6 and 15.7 for these two scenarios, respectively. In scenario 1, the system has large undamped oscillations in base case (Case 1) as shown in Fig. 15.6. With tuned PSS (Case 2), they die out slowly, whereas with application of proposed control (Case 3) the oscillations damp out quickly within 6 s. Similarly, in scenario 2, as shown in Fig. 15.7, the

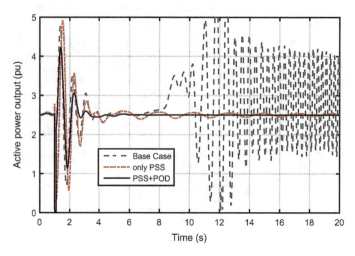

Fig. 15.6 For critical scenario 1

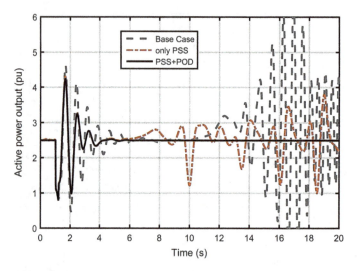

Fig. 15.7 For critical scenario 2

oscillations die out with application of coordinated tuning whereas the oscillations remain in base case and tuned PSS case. These two scenarios highlight the efficacy of the robust coordinated tuning approach employed in this work for damping the LFO modes in the system. The results verify that the controllers coordinated very well with each other when tuned with MPA and improves system damping in presence of high penetration of wind. Thus, the proposed work paves the way for stability improvement in future power systems with high renewable integration.

Conclusions

The paper proposes a robust damping improvement strategy for modern grids having high DFIG integration. It involves coordinated tuning of PSSs and PODs parameters using an advanced meta-heuristic algorithm MPA. The MPA prove to be superior to GWO and PSO in optimizing the controller parameters. The eigen-analysis and time-domain simulations proves the efficacy of proposed approach in improving the small signal stability of the wind integrated power system. The proposed approach proves to be robust to changing system conditions and provides excellent damping even on occurrence of critical contingencies. The proposed work paves the way for small signal stability improvement in future power systems with high wind penetration.

References

1. Gurung S et al (2020) Comparative analysis of probabilistic and deterministic approach to tune the power system stabilizers using the directional bat algorithm to improve system small-signal stability. Elec Power Sys Res 181:106176
2. Pal B, Chaudhuri B (2006) Robust control in power systems. Springer Science & Business Media
3. Gupta AK et al (2017) Dynamic impact analysis of DFIG-based wind turbine generators on low-frequency oscillations in power system. IET GTD 11(18):4500–4510
4. Jia Y et al (2020) Parameter setting strategy for the controller of the DFIG wind turbine considering the small-signal stability of power grids. IEEE Access 8:31287–31294
5. Quintero J et al (2014) The impact of increased penetration of converter control-based generators on power system modes of oscillation. IEEE Trans Power Syst 29(5):2248–2256
6. Tan A et al (2016) A novel DFIG damping control for power system with high wind power penetration. Energies 9(7):521
7. Gupta AK, Verma K, Niazi KR (2019) Robust coordinated control for damping low frequency oscillations in high wind penetration power system. Int Trans Electric Energy Syst 29(5):e12006
8. Kundur P et al (1994) Power system stability and control, vol 7. McGraw-hill, New York
9. Singh M et al (2014) Interarea oscillation damping controls for wind power plants. IEEE Trans Sustain Energy 6(3):967–975
10. Edrah M, Lo KL, Anaya-Lara O (2016) Reactive power control of DFIG wind turbines for power oscillation damping under a wide range of operating conditions. IET Gener Trans Distrib 10(15):3777–3785
11. Golpira H, Bevrani H, Naghshbandy AH (2012) An approach for coordinated automatic voltage regulator–power system stabiliser design in large-scale interconnected power systems considering wind power penetration. IET Gener Transm Distrib 6(1):39–49
12. Elkington K, Ghandhari M (2013) Non-linear power oscillation damping controllers for doubly fed induction generators in wind farms. IET Renew Power Gener 7(2):172–179
13. Surinkaew T, Ngamroo I (2014) Robust power oscillation damper design for DFIG-based wind turbine based on specified structure mixed H2/H∞ control. Renew Energy 66:15–24
14. Saadatmand M, Mozafari B, Gevork B, Soleymani S (2020) Optimal PID controller of large-scale PV farms for power systems LFO damping. Int Trans Electric Energy Syst e12372
15. Surinkaew T, Ngamroo I (2016) Hierarchical co-ordinated wide area and local controls of DFIG wind turbine and PSS for robust power oscillation damping. IEEE Trans Sustain Energy 7(3):943–955

16. Zhang C et al (2017) Coordinated supplementary damping control of DFIG and PSS to suppress inter-area oscillations with optimally controlled plant dynamics. IEEE Trans Sustain Energy 9(2):780–791
17. Khosravi-Charmi M, Turaj A (2018) Wide area damping of electromechanical low frequency oscillations using phasor measurement data. Int J Electr Power Energy Syst 99:183–191
18. Surinkaew T, Ngamroo I (2017) Adaptive signal selection of wide-area damping controllers under various operating conditions. IEEE Trans Industr Inf 14(2):639–651
19. Kunjumuhammed LP, Pal BC (2014) Selection of feedback signals for controlling dynamics in future power transmission networks. IEEE Trans Smart Grid 6(3):1493–1501
20. Del V, Kumar G, Mohagheghi S, Hernandez JC, Harley RG (2008) Particle swarm optimization: basic concepts, variants and applications in power systems. IEEE Trans Evol Comput 12(2):171–195
21. Mirjalili S, Mirjalili SM, Lewis A (2014) Grey wolf optimizer. Adv Eng Softw 69:46–61
22. Faramarzi A, Heidarinejad M, Mirjalili S, Gandomi AH (2020) Marine predators algorithm: a nature-inspired metaheuristic. Exp Syst Appl 113377
23. Milano F (2010) Power system modelling and scripting. Springer Science & Business Media
24. Gupta AK, Verma K, Niazi KR (2017) Contingency constrained optimal placement of PMUs for wide area low frequency oscillation monitoring. In: Proceedings of IEEE 7th international conference on power systems. Pune, India, pp 1–6
25. Canizares C, Fernandes T et al (2015) Benchmark systems for small signal stability analysis and control. https://resourcecenter.ieee-pes.org/pes/product/technical-reports/PESTR18PES-TR

Chapter 16
IoT-Integrated Voltage Monitoring System

Himanshu Narendra Sen, Ashish Srivastava, Mucha Vijay Reddy, and Varsha Singh

Introduction

IoT is acquiring control and monitoring across the globe. IoT or Internet of Things connects the devices (mechanical or electrical) with the internet world. Most of the devices now come with IoT enabled features that allows the users to interface one IOT enabled device with other IOT enabled devices for data interchange. IoT has made tremendous progress in the field of electrical engineering. The Readings about load at different substations are made available in remote devices like smartphone or laptop using IoT. All the power system stations from generation to transmission and distribution are being IoT enabled. In the field of Power electronics, IoT is being utilized to make the circuits more compact and efficient. Power electronics is used for controlling devices such as air conditioners, home appliances, and electric vehicles that are enabled for IoT-based communication with the user. Data from these devices are sent to the cloud server and is then received at the user end. IoT makes the data readily available to the user anytime. This allows better monitoring, analyzing and reduces the time spent on taking any corrective action if the desired output from a device is not generated.

For wireless communication, ESP32S is used in the proposed work. It is the latest and cheapest IoT Wi-Fi and Bluetooth ready module. The ESP32, unlike its predecessor ESP8266 has a dual-core. A comparative analysis has been done between ESP32, ESP8266, CC32, Xbee [1]. Where a single-core of ESP8266 was utilized in handling Wi-Fi and Bluetooth operations. Additional microcontroller/Arduino is required for data processing and interfacing sensors and digital input–output. The aforesaid issues are resolved in ESP32S as it contains dual-core named Protocol

---

H. N. Sen · A. Srivastava · M. V. Reddy · V. Singh (✉)
National Institute of Technology, Raipur, Raipur, Chhattisgarh 492010, India
e-mail: vsingh.ele@nitrr.ac.in

© The Editor(s) (if applicable) and The Author(s), under exclusive license to Springer Nature Singapore Pte Ltd. 2021
A. K. Singh and M. Tripathy (eds.), *Control Applications in Modern Power System*, Lecture Notes in Electrical Engineering 710,
https://doi.org/10.1007/978-981-15-8815-0_16

CPU and Application CPU [2]. The ESP32S also has multiple built-in peripherals which help in sensing the voltage as described in this chapter. One port of the device is used for Wi-Fi/Bluetooth and internal peripherals and other for application code. Due to all these features and compactness, ESP32S is well suited for designing and implementing a web server for real-time photovoltaic system monitoring [3], a solar water pumping system by using a smartphone [4], developing electronic nose system for monitoring LPG leakage [5]. It has also been used for monitoring home temperature, gas leakage, water level and controlling the fan, motor, gas knob, etc. [6]. In the agriculture field it is popularly used for maintaining good crop health by monitoring soil temperature, pressure, wind velocity, humidity, etc. and feeding them to ESP32 [7]. In paper [8] IoT has been used with power electronic devices for educational purposes to help bachelor's students. In [9] authors discuss how power electronics circuit when combined with the Internet of Things (IoT) devices and software creates a fully autonomous system which helps to control street lighting and reduce the consumption. A survey of various IoT cloud Platforms has been done in [10]. Likewise, an investigation on IoT middleware platform for smart application development is reported in [11]. ThingSpeak is an IoT analytics platform service providing cloud services for storing, analyzing, and visualizing data [12]. ThingSpeak can be handled for different applications just like in the case of a smart irrigation system [13]. DC to DC converters are popularly used in power systems, electric vehicles and small electronic devices where it's not feasible to use large batteries for supply. A DC–DC boost converter is responsible for stepping up the voltage from Input supply to load side output. A dc-dc boost converter module XL6009 is used for testing. The features which it has are a wide input range of 5–32 V and a wide output voltage of 5–35 V. The built-in 4A MOSFET switches provide efficiency up to 94% in the conversion process [14]. In this chapter the DC-DC converter is connected to a programmed ESP32 ensuring continuous monitoring of the data. Data storing and accessing is done through ThingSpeak in real-time. The data in ThingSpeak can be made available to a single client or multiple or can be made public as per the user's need.

This chapter discusses one such application of IoT which is the monitoring of the voltage of the boost converter. This helps in getting real-time data anywhere around the world. The proposed work aims an IoT-based voltage monitoring system that can be used in any industry placed anywhere in the world and the voltages observed can be monitored by any person across the globe when configured with appropriate settings. The main objective of the work is to provide and insight as to how the voltage data that is accessible by anyone and can be used for safety and monitoring purposes. The chapter is organized in the following sections: Sect. 16.2 discusses the operation of the proposed system through the block diagram and Sect. 16.3 demonstrates the flow char for better insight. Implementation through hardware and result is shown in Sect. 16.4 followed by conclusion in Sect. 16.5.

Fig. 16.1 Block diagram of the proposed system

Proposed Methodology

Figure 16.1 shows the block diagram for the working of the prototype. The block diagram has several intermittent implementation stages. The first stage has a regulated 9 V DC supply that is connected to a boost converter. The output of the boost converter is fed to a potential divider circuit. This is done to limit the current and voltage within the permissible limit of ESP32S. The ESP32S senses the real-time change in the voltage and current that is obtained from the variable output boost converter. ESP32S is configured with an open-source Thing Speak cloud platform to monitor the result using Wi-Fi as well as server protocols.

A regulated DC supply is provided at the input of the boost converter. The converter steps up the input voltages to the desired value and the output voltages obtained are supplied to the ESP32S. The output voltage of the converter cannot be directly fed into ESP32S since the maximum voltage limit for the ESP32S module is 3.3 V. A simple potential divider consisting of two resistors steps down the voltage to a level below 3.3 V, which is then suitable to be fed to the ESP32S. The program is written accordingly in Arduino IDE and the data about the voltage is being sent to the ThingSpeak server. This data then can be accessed by the person at the remote system using his ThingSpeak ID. Data can be made public or private as per the need.

System Flowchart

The flowchart in Fig. 16.2 illustrates the entire methodology adopted and the entire sequence which has been followed for the experiment setup. A 9 V regulated DC supply was connected to an XL6009 Boost Converter which steps up the voltages and the output of the boost converter was then fed to the potential divider circuit whose output was connected to the ESP32S and using Arduino IDE the data was successfully uploaded to ThingSpeak server and anyone having an account on ThingSpeak can

Fig. 16.2 System flowchart

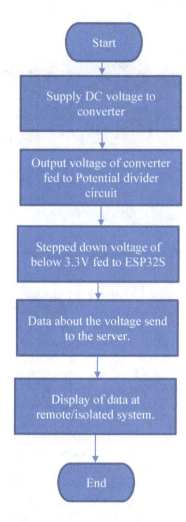

access it as it was made public in ThingSpeak. The main objective of the entire project is the voltage data that is accessible by anyone and can be used for safety and monitoring purposes.

Hardware Setup and Results

To present the feasibility of the working principle discussed in Sects. 16.2 and 16.3 a prototype is made in the laboratory. Table 16.1 shows the list of equipment used in the experimental setup.

Table 16.1 List of equipment

S. No.	Device type	Name of the device
1	Boost converter	XL6009
2	Microcontroller	ESP32S
3	Software	Arduino IDE

A regulated 9 V DC source is connected to the boost converter XL6009 IC as shown in Fig. 16.3. The potential divider circuit is designed to produce a maximum of 3.3 V across its lower voltage arm which is then connected to ESP32S IC as shown in Fig. 16.4. The component ICs and the entire setup are shown in Figs. 16.5.

XL6009 Boost Converter depicted in Fig. 16.3 contains screw to change the boost level and provide options to check for multiple values of output voltages. ESP32S (Fig. 16.4) is the Wi-Fi module that has been used in the proposed work, as it offers a large number of connectivity features like Bluetooth and Wi-Fi through which data can be sent to various devices.

Figure 16.5 exemplifies the prototype of prosed work. A 9 V DC supply is connected to Boost Converter whose output is in turn connected to the voltage divider circuit and the output voltages across lower voltage arm of the voltage divider is connected to ESP32S it senses the output voltage and sends the data to cloud server which is then accessed through ThingSpeak.

The things speak channel on which the voltages will be displayed can be accessed by anyone across the world. This is represented in Fig. 16.6.

Table 16.2 shows an observation table tabulating the input voltage, which is kept constant at 9 V and the output voltages of the boost converter and the potential divider circuit for different positions of the screw present in the XL6009 Boost Converter. The input voltage to ESP32S is scaled-down within 3.3 V that can be seen in Table 16.2.

The performance of real-time voltage-dependent application is quite accurate with fewer computations and time lag of 10–15 s for monitoring and recording of data on remote server. The significance of IoT-based system designed is economical, simple compact, accurate for both potential divider and ESP32S.

Figure 16.7 depicts the variation of output voltages monitored at various time intervals during the entire day. It also states the details about a certain point which include the output voltage, day, date, and time.

Conclusion

A voltage monitoring system is designed using IoT for remote servers. ESP32S is used for the implementation of cloud-based monitoring system. IoT-based voltage monitoring system is used in plants and various other factories for keeping a check on voltage levels. The fact that the voltages observed are readily accessible from

Fig. 16.3 XL6009 (boost converter)

anywhere across the globe makes this model one of its kind. The experimental prototype developed is economical, compact, and accurate.

Fig. 16.4 ESP32S

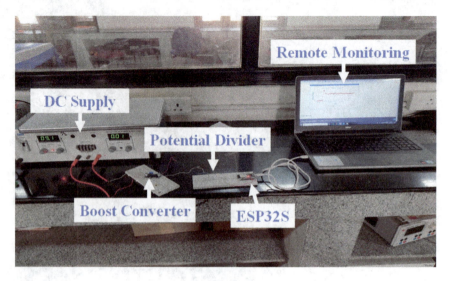

Fig. 16.5 Experimental prototype

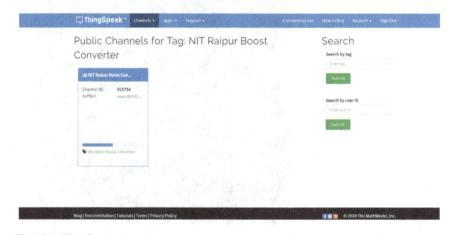

Fig. 16.6 ThingSpeak channel

Table 16.2 Variation in input and output voltages at every stage of the proposed circuit

S. No.	Converter		Potential divider		ESP32S
	V_{in} (V)	V_{out} (V)	V_{in} (V)	V_{out} (V)	V_{in} (V)
1	9	13.5	13.5	1.227	1.227
2	9	18	18	1.636	1.636
3	9	22.5	22.5	2.045	2.045
4	9	27	27	2.454	2.454
5	9	31.5	31.5	2.863	2.863
6	5	7.5	7.5	0.681	0.681
7	5	10	10	0.909	0.909
8	5	12.05	12.5	1.136	1.136
9	5	15	15	1.363	1.363
10	5	17.5	17.5	1.590	1.590

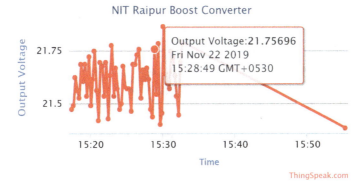

Fig. 16.7 Voltage versus time graph indicating the voltage measured

References

1. Maier A, Sharp A, Vagapov Y (2017) Comparative analysis and practical implementation of ESP32 microcontroller module for IoT. In: 7th IEEE international conference on internet technologies and applications ITA-17, pp 143–148
2. Folty´nek P, Babiuch M, Sura´nek P (2019) Measurement and data processing from IoT module by dual core application using ESP32. Measure Control 52(7–8):970–984
3. Allafi I, Iqbal T (2017)Design and implementation of low-cost web server using ESP32 for real time photovoltaic system monitoring. In: IEEE electrical power and energy conference (EPEC), pp 1–5
4. Bipasha Biswas S, Tariq Iqbal M (2018) Solar water pumping system control using a low cost ESP32 microcontroller. In: 2018 IEEE Canadian conference on electrical & computer engineering (CCECE), pp 1–5
5. Abdullah AH et al (2018) Development of ESP32-based Wi-Fi electronic nose system for monitoring LPG leakage at gas cylinder refurbish plant. In: 2018 international conference on computational approach in smart systems design and applications (ICASSDA), pp 1–5

6. Pravalika V, Rajendra Prasad C (2019) Internet of things based home monitoring and device control using Esp32. Int J Recent Technol Eng (IJRTE), 2277-3878, 8(1S4):58–62
7. Narendar Singh D, Ashwini G, Uma Rani K (2018) IOT based recommendation for crop growth management system. Int J Recent Technol Eng (IJRTE), 2277-3878 7(4):290–293
8. van Duijsen P, Woudstra J, van Willigenburg P (2018) Educational setup for power electronics and IoT. In: 2018 19th international conference on research and education in mechatronics (REM), pp 147–152
9. Rudrawar O, Daga S, Chadha J, Kulkarni PS (2018) Smart street lighting system with light intensity control using power electronics. In: 2018 technologies for smart-city energy security and power (ICSESP), pp 1–5
10. Ray PP (2016) A survey on IoT cloud platform. Future Comput Inform J 1(1–2):35–46
11. Agarwal P, Alam M (2019) Investigating IoT middleware platforms for smart application development, pp 1–14. arxiv:1810.12292
12. https://thingspeak.com Accessed on 1 Oct 2019
13. Benyezza H, Bouhedda M, Djellout K, Saidi A (2018) Smart irrigation system based ThingSpeak and Arduino. In: 2018 international conference on applied smart systems (ICASS. Medea, Algeria, pp 1–4
14. https://www.sunrom.com/p/step-up-dc-dc-based-on-xl6009. Accessed on 1 Oct 2019

Chapter 17
Intrinsic Time Decomposition Based Adaptive Reclosing Technique for Microgrid System

Shubham Ghore, Pinku Das, and Monalisa Biswal

Introduction

Uninterrupted power supply to microgrid is a future plan and can be possible with the wide application of battery energy storage system (BESS), adaptive control, monitoring and protection schemes so that total interruption time either due to abnormal causes or maintenance can be avoided. To maintain the stability, single-pole operation provision is provided to modern digital relays. Under single or double-phase fault, faulty phases can be operated with a preplanned manner to maintain the continuity of supply through other healthy phases. But in a microgrid, such a practice increases the burden on BESS to maintain the continuity of supply to the load connected to the affected phase after intentional isolation by breaker opening from both sides. BESS is indeed installed for both storage and emergency period, but uncalled utilization may deteriorate the operating life of BESS. It is hardly true that the breaker reclosing system takes attempt only at exact fault clearance time. The unwanted burden on BESS due to the conventional setting of a reclosing system reduces battery life.

As the conventional reclosing scheme has fixed dead times therefore even if the fault gets cleared much before the predefined second dead time, BESS has to supply the critical load just because the exact instant of fault clearance is not known. This causes unnecessary outage and wastage of energy as shown in Fig. 17.1.

S. Ghore (✉) · P. Das · M. Biswal
Department of Electrical Engineering, NIT Raipur, Great Eastern Road, Raipur 492010, Chhattisgarh, India
e-mail: shubhamghore@gmail.com

P. Das
e-mail: pinku.das776@gmail.com

M. Biswal
e-mail: monalisabiswal22@gmail.com

© The Editor(s) (if applicable) and The Author(s), under exclusive license to Springer Nature Singapore Pte Ltd. 2021
A. K. Singh and M. Tripathy (eds.), *Control Applications in Modern Power System*, Lecture Notes in Electrical Engineering 710,
https://doi.org/10.1007/978-981-15-8815-0_17

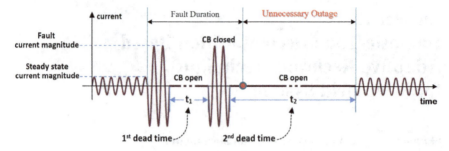

Fig. 17.1 Operating sequence of a typical autorecloser

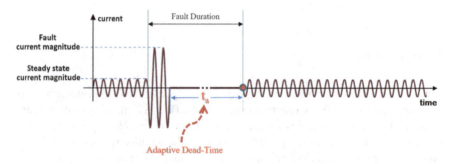

Fig. 17.2 Operating sequence of adaptive autorecloser

To overcome the above problem, the line adaptive reclosing scheme is implemented [1–7]. This can detect the exact instant of fault clearance and reclose the breaker or in other words, it modifies the fixed dead time of autorecloser to an adaptive dead time. This restores the grid supply minimizing outage time and reduces the burden on BESS. The operation of this scheme is shown in Fig. 17.2.

Microgrid System with Bess

In this section, the details of the microgrid system considered for the study are provided. The standard IEEE 13-bus system as shown in Fig. 17.3 is used for the analysis of the proposed scheme. The IEEE 13-bus system is modified by adding the Hybrid DG-BESS system. PV-BESS system is connected at bus-13 and the DFIG wind generator is connected at bus-10.

The notations and details of the studied system are as follows.

- G1, G2 are the three-phase voltage sources.
- DFIG is the Doubly Fed Induction Generator.
- BESS is the battery energy storage system consisting of battery and inverter.
- PV is the photovoltaic system consisting of photovoltaic panel and inverter.

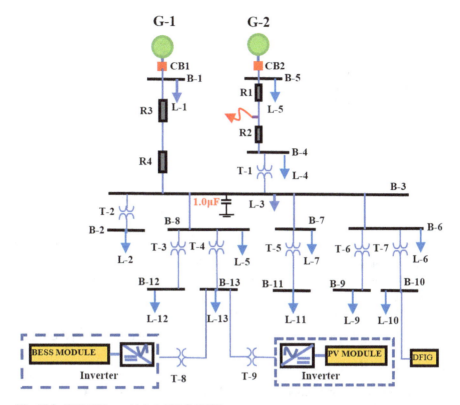

Fig. 17.3 IEEE 13 bus with hybrid DG-BESS system

- B-1, B-2, etc. are the system buses.
- T-1, T-2, etc. are the transformers.
- R-1, R-2, etc. are the relays connected at the ends of the transmission line.
- L-1, L-2, etc. are the loads connected at the bus-bars.
- μF shunt capacitor is connected at bus-3.

This test system is shown in Fig. 17.3 consists of 13 buses with hybrid DG-BESS. The distribution system is fed from a utility supply of 132 kV and the local plant distribution system operates at 11 kV. The capacitance of the overhead line and all cables are neglected. It consists of two three-phase voltage sources with rated short circuit MVA of 100 and designed for a frequency of 50 Hz.

Proposed Technique

In the proposed method, the faulty phase first is isolated from the microgrid by the single-pole operation of the corresponding circuit breaker. The fault current and voltage signal are obtained by the line side CT and PT. At the instant of fault occurrence, the faulty phase current and voltage show a transition from their steady-state values depending on the fault type and impedance. When the corresponding CB is operated, current falls to zero however a very small voltage is present in the line due to mutual coupling between the lines. At the instant fault gets cleared, the isolated phase shows a small transition in its voltage magnitude. All three cases viz., fault occurrence, CB operation, and fault clearance can be detected by using the Intrinsic Time Decomposition (ITD) algorithm.

The Intrinsic Time Decomposition (ITD) method is implemented for the adaptive reclosing technique of the microgrid. Since the ITD method is specifically designed for application of real-time signals or non-stationary signals of arbitrary origin which are obtained from complex time–frequency energy (TFE) signals system with underlying dynamics that change on the real-time scenario. ITD method doesn't presume any prior prediction on the content of the signal under analysis and response time of ITD is fast and it can be implemented in real-time or non-stationary signals [8]. Due to the underline advantage discussed above of the ITD method over other methods such as Fourier transform, Wavelet transform, Empirical mode decomposition (EMD), etc. adaptive reclosing is implemented with the help of ITD method.

ITD method decomposes a signal into two parts:

- First part consist of rotating components contains instantaneous amplitude and frequency
- Other part consists of a monotonic trend.

For understanding mathematically consider a real-time TFE signal $X(t)$ for which operator ζ is defined to extract the baseline signals from $X(t)$. Mathematically $X(t)$ can be decomposed as

$$X(t) = \zeta X(t) + (1 - \zeta)X(t) = B(t) + H(t) \tag{17.1}$$

where $B(t) = \zeta X(t) =$ baselinecomponent

$$H(t) = (1 - \zeta)X(t) = \text{rotatingcomponent}.$$

$X(t)$ is a real-time signal for $t \geq 0$ and let ξ_k for which $k \geq 1$ represents local extreme points of $X(t)$. Consider the TFE signal at ξ_k be $X(\xi_k)$ and its baseline at ξ_k be $B(\xi_k)$. Now let us assume that H(t) and B(t) is defined on the interval [0,ξ_k] and $X(t)$ is defined for the interval [0, ξ_{k+2}]. Now baseline signals operator τ is defined as a piecewise linear function on the interval [ξ_k, ξ_{k+1}] between two extrema as

$$\zeta X(t) = B(t) = B(\xi_k) + \left(\frac{B(\xi_{k+1}) - B(\xi_k)}{X(\xi_{k+1}) - X(\xi_k)}\right)(X(t) - X(\xi_k)) \tag{17.2}$$

where,

$$B(\xi_{k+1}) = \alpha\left[X(\xi_k) + \left(\frac{\xi_{k+1} - \xi_k}{\xi_{k+2} - \xi_k}\right)(X(\xi_{k+2}) - X(\xi_k))\right] + (1-\alpha)X(\xi_{k+1})$$
(17.3)

and $t \in (\xi_k, \xi_{k+1})$.

Generally, α lies in the range of (0, 1) and is kept at 0.5. Since if the magnitude of α is kept low the baseline signal component will be less and if α is kept close to 1, then it results in higher baseline amplitude. Baseline signal $B(t)$ is now constructed in such a manner that signal $X(t)$ maintain its monotonicity between its extrema. Once the baseline signal has been extracted, the remaining portion of the signal can be represented as rotating components. The rotating component can be computed as

$$H(t) = (1 - \tau) * X(t) = \beta X(t) = X(t) - B(t)$$
(17.4)

where β is a rotating operator.

After the first decomposition, a baseline component ($B^1(t)$) and a rotating component ($H^1(t)$) are generated, where $H^1(t)$ represents a high-frequency component of $X(t)$. Then the obtained baseline signal is used as a new data source and the above-mentioned decomposition process is continued for 'p' times until $B^p(t)$ turns into a monotonic function [9, 10].

Mathematically, it can be represented as,

$$X(t) = B(t) + H(t)$$
$$= \zeta X(t) + \beta X(t)$$
(17.5)

$$= \zeta[\zeta X(t) + \beta X(t)] + \beta X(t)$$
(17.6)

where $X(t) = \zeta X(t) + \beta X(t)$

$$= \zeta[(1 + \beta)X(t)] + \beta X(t)$$

$$= [\zeta(1 + \beta) + \beta] * X(t)$$

$$= [\beta(1 + \zeta) + \beta^2] * X(t)$$

$$= [\beta \sum_{k=0}^{p-1} \zeta^k + \zeta^p] * X(t)$$
(17.7)

$$= H^1(t) + H^2(t) + H^3(t) + \cdots + H^p(t) + B^p(t)$$
(17.8)

Now the original TFE signal has been decomposed as a sum of several rotating components and a monotonic component.

Simulation and Results

Various unsymmetrical fault cases are simulated and tested with the proposed method using MATLAB and PSCAD/EMTDC software. ITD algorithm successfully detected the fault occurrence, CB operation and fault clearance as shown in the simulation output. Once the exact time of fault clearance is detected, the dead time of the autorecloser can be modified to reconnect the isolated phase after synchronization.

The microgrid model is simulated for 2 s in PSCAD/EMTDC software for different fault conditions. The sequence of events performed in the microgrid model to test the proposed method are as follows:

- At $t = 1.3$, a fault is created in the line
- At $t = 1.33$, the faulty section is isolated by CB
- At $t = 1.6$, fault is cleared.

The above sequence of operation is reflected in voltage and current signals shown in Fig. 17.4. Because of single-pole switching, the other healthy phases remain live and as a result, a measurable voltage is induced in the faulty line as well. This induced voltage is negligible during the fault period. But as soon as fault gets cleared, it shows noticeable magnitude which can be measured by line side PT.

The exact time of fault clearance is detected by the ITD algorithm and the resulting baseline and rotating components are as shown in Fig. 17.5.

The proposed method is tested for LG faults with different fault resistance, fault location and fault inception time. Results for different fault scenarios are shown in Fig. 17.6, Fig. 17.7 and Fig. 17.8 respectively. Index 'ε' represents the sum of

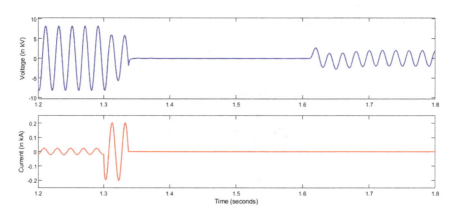

Fig. 17.4 Current and voltage signals as measured by line side CT and PT

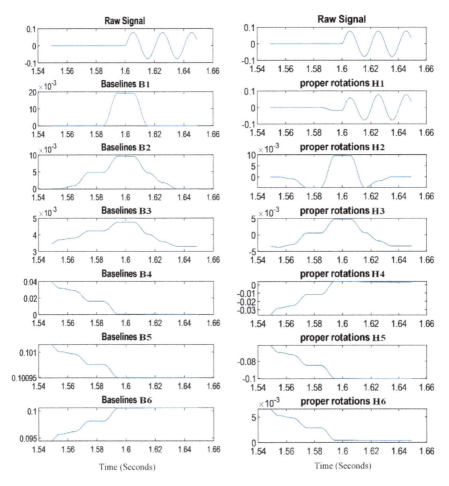

Fig. 17.5 Baseline and rotating components as obtained by the ITD algorithm at the time of fault clearance

energies of the first three baseline components of signals obtained using the ITD algorithm. Using both voltage and current data, fault occurrence and clearance can be differentiated. ε_1 and ε_2 represent the index ε during fault occurrence and clearance respectively.

Index ε_2 shows a substantial rise in magnitude as soon as the fault is cleared. The information form index ε_2 is used to modify the fixed dead time to an adaptive value.

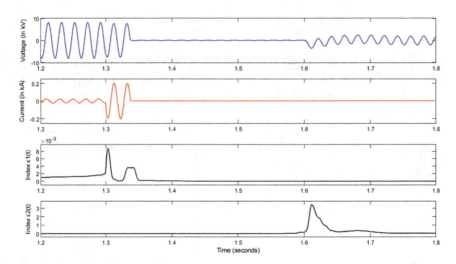

Fig. 17.6 LG fault at 5 km from bus B-5 with 2 Ω fault resistance and initiated at $t = 1.3$ s

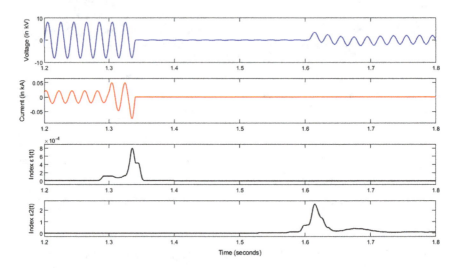

Fig. 17.7 LG fault at 15 km from bus B-5 with 100 Ω fault resistance and initiated at $t = 1.3$ s

Conclusion

The paper proposes a novel intrinsic time decomposition-based (ITD) detection logic to determine the exact fault initiation and clearance time so that an adaptive reclosing attempt can be taken by digital relay. This method is suitable for real-time implementation as gives results in less than one cycle delay. Through conventional fault detection and classification approach, the exact faulty phase is first detected and

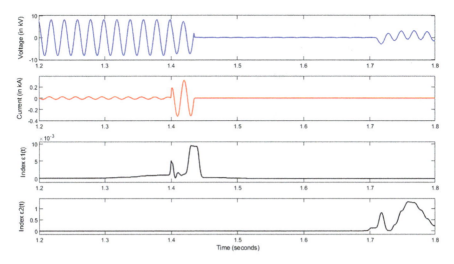

Fig. 17.8 LG fault at 10 km from bus B-5 with 10 Ω fault resistance and initiated at $t = 1.4$ s

then the proposed index is used to calculate the adaptive dead-time to reclose the breaker of only the faulty phase. Unlike conventional reclosing scheme, the proposed adaptive reclosing scheme doesn't check for the fault clearance by blindly reclosing after fixed dead times therefore it avoids the risk associated with an unsuccessful reclosing attempt. The performance of the proposed method has been evaluated using PSCAD simulations. The method delivers satisfactory results in the presence of capacitor switching, nonlinear load switching, with the presence of high resistance and high impedance fault. From the obtained simulation results and analysis, it can be concluded that the proposed method is suitable for implementation in the real system. The use of the proposed method reduces the outage time and associated energy losses, thereby making the grid more resilient to faults.

References:

1. Seo H (2017) New adaptive reclosing technique using second-order difference of THD in distribution system with BESS used as uninterruptible power supply. Int J Electr Power Energy Syst 90:315–322
2. Seo H (2017) New adaptive reclosing technique in unbalanced distribution system. Energies 10(7):1004
3. Seo H (2017) New configuration and novel reclosing procedure of distribution system for utilization of BESS as UPS in smart grid. Sustainability 9(4):507
4. Ahn S, Kim C, Aggarwal R, Johns A (2001) An alternative approach to adaptive single pole auto-reclosing in high voltage transmission systems based on variable dead time control. IEEE Trans Power Delivery 16:676–686
5. Vogelsang J, Romeis C, Jaeger J (2016) Real-time adaption of dead time for single-phase autoreclosing. IEEE Trans Power Delivery 31:1882–1890

6. Zhalefar F, Dadash Zadeh M, Sidhu T (2017) A high-speed adaptive single-phase reclosing technique based on local voltage phasors. IEEE Trans Power Delivery 32:1203–1211
7. Park J, Seo H, Kim C, Rhee S (2016) Development of adaptive reclosing scheme using wavelet transform of neutral line current in distribution system. Electric Power Comp Syst 44:426–433
8. An X, Jiang D, Chen J, Liu C (2011) Application of the intrinsic time-scale decomposition method to fault diagnosis of wind turbine bearing. J Vib Control 18:240–245
9. Martis R, Acharya U, Tan J, Petznick A, Tong L, Chua C, Ng E (2013) Application of intrinsic time-scale decomposition (ITD) to EEG signals for automated seizure prediction. Int J Neural Syst 23:1350023
10. Hu A, Yan X, Xiang L (2015) A new wind turbine fault diagnosis method based on ensemble intrinsic time-scale decomposition and WPT-fractal dimension. Renew Energy 83:767–778

Chapter 18
Design of Energy Management System for Hybrid Power Sources

Akanksha Sharma, Geeta Kumari, H. P. Singh, R. K. Viral, S. K. Sinha, and Naqui Anwer

Introduction

In this day and age, the expanding requirement for energy and the variables, for example, expanding energy costs, constrained stores, and ecological contamination, drives the sustainable power source to be the most appealing energy source [1]. The sustainable power source assets are energy and a defining moment throughout the entire existence of the power in India. India is a vast nation depending upon the power for its day to day uses. 95% of the individuals are honoured with the force provided by both centre and state governments. The administration has arranged system intends to create sun, wind, and hydro power plants to produce capacity to decrease the power cut [2]. Energy Consumption is expanding quickly and because of the depletion of the ordinary wellsprings of energy and their significant expenses, the Renewable

---

A. Sharma (✉) · G. Kumari · H. P. Singh · R. K. Viral · S. K. Sinha
Department of Electrical & Electronics Engineering, Amity School of Engineering and Technology, Amity University Uttar Pradesh, Noida, India
e-mail: akanksha.sharma2610@gmail.com

G. Kumari
e-mail: 10geetagupta@gmail.com

H. P. Singh
e-mail: hpsingh2@amity.edu

R. K. Viral
e-mail: rkviral@amity.edu

S. K. Sinha
e-mail: sksinha6@amity.edu

N. Anwer
Teri School of Advanced Studies, New Delhi, India
e-mail: naquianwer@gmail.com

Energy sources (RESs) are increasing a lot of intrigues. These RESs, similar to small hydro, wind, sun, and biomass, are an easy and more reliable option for the regular energy assets as the network power isn't available to certain remote territories. India is a developing nation where a huge piece of the populace lives in remote and detached zones. Thus, in such regions, the jolt utilizing the close by accessible RESs would prompt their general advancement. These sources incorporate sun based, wind, biomass, miniaturized scale hydro and so on. Utilization of these assets has ecological advantages as well as financial advantages like-Energy security, giving openings for work and giving an economical and clean condition. With high monetary just as mechanical development, the vitality request in India is additionally developing at a quicker rate [3]. Microgrid can reduce both cost and revenue for the customers. Microgrids are a developing section of the vitality business, speaking to a change in outlook from remote station power plants towards increasingly limited, dispersed age—particularly in urban areas, networks and grounds. The ability to disengage from the bigger framework makes microgrid strong, and the capacity to direct adaptable, equal tasks licenses conveyance of administrations that make the lattice increasingly serious [4]. Islanding is the condition where a distributed generator (DG) keeps on fuelling an area despite the fact that electrical network power is not, at this point present. Islanding can be hazardous to utility labourers, who may not understand that a circuit is as yet controlled, and it might forestall programmed re-association of gadgets. Moreover, without exacting frequency control, the harmony among burden and age in the islanded circuit will be abused, prompting irregular frequencies and voltages. Consequently, appropriated generators must identify islanding and quickly detach from the circuit; this is referred to as hostile to islanding. By "islanding" from the network in crises, a Microgrid can both keep serving its included burden when the framework is down and serve its encompassing network by giving a stage to help basic administrations from facilitating people on call and legislative capacities to offering key types of assistance and crisis cover [5].

Proposed Models

One of the most noteworthy wellsprings of maintainable power sources like solar, wind as well as small hydro. On account of the arbitrary idea of both the wind and sun arranged energy, they are not continually available and depend upon step by step or standard climatic conditions. There are numerous crossover frameworks; however, the most widely recognized framework is the sunlight and wind. The proposed model comprises of solar, wind and hydro integrated model which are improved the power quality and voltage regulation at different loads.

Solar System

The sun-powered energy is one of the most known wellsprings of sustainable power source. Before, the cost was an impediment to the utilization of the sun-based energy in light of the fact that the cells were costly, yet now the cost of the cells is moderate and can be bought no problem at all. The energy originated from the sun along with the energy which is originated from the framework can be very well utilized in terms of the time which is spent on the lightning and the heating and also during the creation of hydrogen for the creation of electrical force [6]. The most vital issue which is going towards the development method of vitality driven from the sun is the weather conditions specially the winter season when lack of sunlight is there. The issue is dealt with by utilizing Maximum Power Point Tracking (MPPT) structure. The standard furthest reaches of MPPT are to get the most preposterous force point described as the best extent of vitality accessible from solar module in various environment conditions. The silliest force is changed by modifying the sun-controlled radiation, the barometrical temperature along with sun fuelled temperature of the cell [7]. A model of a sun set up cell has been sorted out as for MATLAB/Simulink which is formed, comprising of the breeze and hydro model both to plot a hybrid framework, in order to achieve the best energy. The model is accustomed to changing the radiation at various day time events for considering the impact on cell radiation and studying distinction in power by 1000 W/m^2 radiation (Fig. 18.1).

The I-V characteristics of diode for a single module are

$$I_{\text{diode}} = I_{\text{saturation}} [exp\left(\frac{V_{\text{diode}}}{V_{\text{Thermal}}}\right) - 1] \quad (18.1)$$

$$V_{\text{Thermal}} = \frac{K T_{\text{cellTemp.}}}{q} * nI * N_{\text{cellnumber}} \quad (18.2)$$

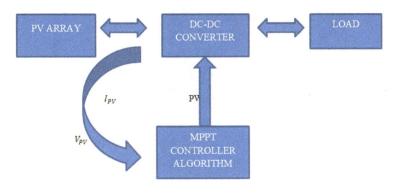

Fig. 18.1 The proposed solar energy connection [8]

where,

I_{diode}	Current of diode in Amp.
V_{diode}	Voltage of diode in Volt.
$I_{saturation}$	Saturation current of diode in Amp.
nI	ideal factor of diode, it is close to 1.0
K	Boltzmann constant = 1.3806e−23 J K^{-1}.
Q	electron charge = 1.6022e−19 C.
T	Temperature of cell (K).
N_{cell}	cells number connected in series in a module.

Wind Energy System

The mathematical displaying of wind vitality change framework incorporates wind turbine elements and generator demonstrating. An examination of the accessible writing on the framework execution appraisals for wind energy frameworks has demonstrated that very small work has been accounted for in this particular field. The analysts, for the most part, investigated the zones like provincial wind energy evaluation, wind speed dissemination capacities, financial parts of wind vitality and local wind energy strategies [9].

The basic condition of wind turbine is given by

$$P_{wind} = \frac{1}{2}C_{powercoeff.}(\lambda, \beta)\rho A V^3 \qquad (18.3)$$

where

ρ	air density in kg/m^3
$C_{powercoeff}$	power coefficient.
A	area of the rotor blades in m^3
V	average speed of the wind in m/s.
λ	Ratio of the tip speed.

The theoretical greatest estimation of coefficient of power $C_{powercoeff.}$ is 0.593. The wind turbine Tip Speed Ratio (TSR) is characterized as:

$$\lambda = \frac{R\omega}{V} \qquad (18.4)$$

where

R Turbine radius in m.
ω angular speed of wind in rad/s.
V average speed of the wind in m/s.

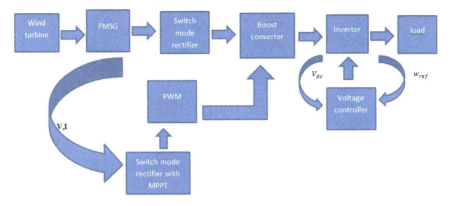

Fig. 18.2 Block diagram of wind turbine [11]

The energy generated by wind can be obtained by [10]

$$Q_w = \text{Power} \times (\text{Time})(\text{kWh}) \tag{18.5}$$

where

Q_w energy generated by wind.
P power of wind (Fig. 18.2).

Small Hydro Energy System

The target of the assessment is absolutely surveying the present hydro power status improvement in the country and making conditions of headway. The probable and the introduced limit, imaginative status, strategies and administrative help in the growth of hydro power and the strategy for the incitement of Small Hydro Power (SHP) is been researched comprehensively. The most primary stage for the showing up of hydro is calculation of steam rate. The stream rate can be resolved if the drainage area of the conduit is known in development to the precipitation data (month to month, step by step, and hourly). Drainage areas are the areas which storm the water stream into the conduit. Different parameters are having various types of effects on the SHP age and to examine these effects, an assessment has been performed [12].

The basic equation for the hydro system mechanical power is:

$$P = \eta \rho g Q H \tag{18.6}$$

Fig. 18.3 Small hydro power plant block diagram [13]

where

- P mechanical power at the turbine shaft.
- η Turbine hydraulic efficiency.
- ρ density of water.
- g acceleration due to gravity.
- Q volume flow rate passing through the turbine.
- H head of water (Fig. 18.3).

Integrated Models

A model of sun arranged imperativeness is organized on MATLAB/Simulink with breeze and hydro models both to shape hybrid structure, encouraging the three units getting the most ludicrous vitality. There are tremendous issues standing up to the blend of power age process on account of the possibility of the sun arranged wind and hydro (Fig. 18.4).

Results and Discussion

Solar Energy Model

To avow the relationship between the irradiance and the yield voltage got from the close by close planetary framework so as to empower the Simulink to show with the right affiliation, a few readings have been taken at various loads.

Without Load

Without load, the maximum power is 10,000 W and then after it is constant. Figure 18.5 displays a graph between power and time.

18 Design of Energy Management System for Hybrid Power Sources 203

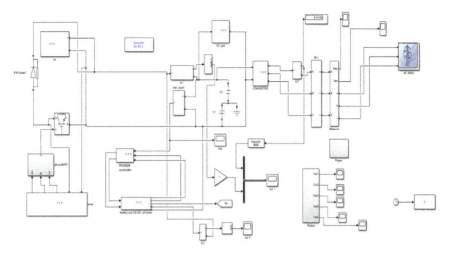

Fig. 18.4 Proposed model of integrated system

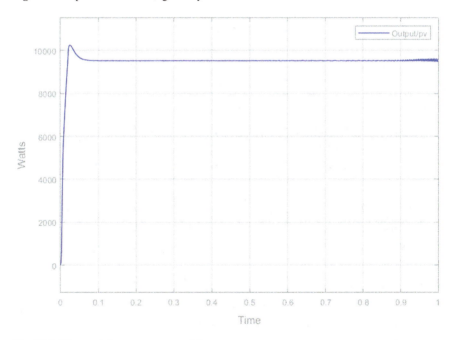

Fig. 18.5 The graph between power and time

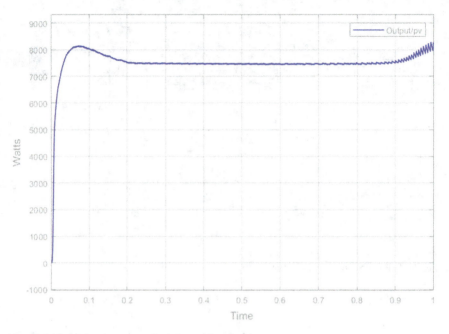

Fig. 18.6 Graph between power and time with *R* load

With R Load

When connect the R load, take the value of V_{rms} 220 V and QL = 0 var, QC = 0 var at nominal frequency 60 Hz. The output of power 8000 W then after down and then constant at 0.9 s then it again rise. Figure 18.6 displays a graph between power and time with *R* load.

With RLC Load

In RLC load graph, take V_{rms} 220 V and QL = 100 var, QC = 100 var at nominal frequency 60 Hz. The power output is 8000 W. Figure 18.7 shows the graph between power and time with RLC load.

With Dynamic Load

With dynamic load, nominal frequency 60 Hz. The power output above to 10,000 W and then constant over time. Figure 18.8 shows the graph between power and time with dynamic load.

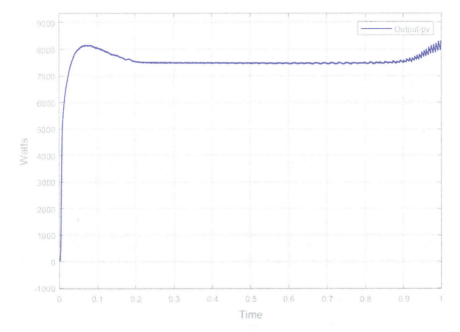

Fig. 18.7 The graph between power and time with RLC load

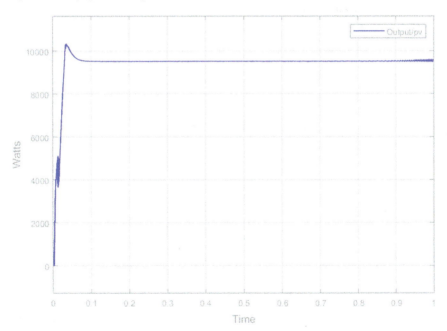

Fig. 18.8 The graph between power and time with dynamic load

Wind Energy Model

Without Load

In wind energy graph without load, the power at initial stage is 0 at 0.5 s then after it will rises upto 16,000 W at $t = 1$ s. Figure 18.9 shows the variation of output without load.

With R Load

With R load take V_{rms}=220 V and QL = 0 var, QC = 0 var. The power output 0 watts throughout at $t = 0.5$ s. then it rises few values at $t = 0.9$ s. then power 16,000 W at $t = 1$ s. Figure 18.10 shows the variation of output with R load.

With RLC Load

With RLC load take V_{rms}=220 V and QL = 100 var, QC = 100 var and maximum power 16,000 W at $t = 1$ s. Figure 18.11 shows the variation of output with RLC load.

With Dynamic Load

With dynamic load, nominal frequency 60 Hz. The power output 15,000 W at $t = 1$ s. There are some fluctuations at $t = 0.6$–0.9 s. Figure 18.12 shows the output variation of dynamic load with wind speed.

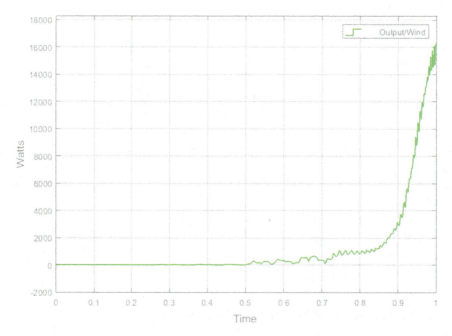

Fig. 18.9 The variation of output without load

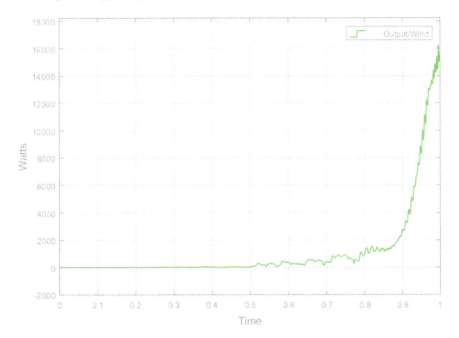

Fig. 18.10 The variation of output with R load

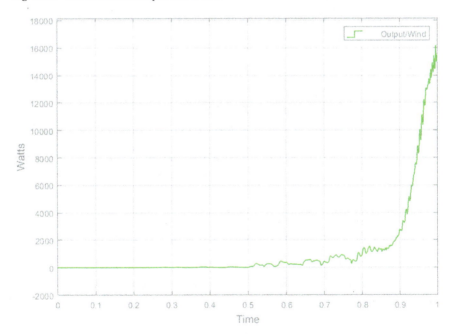

Fig. 18.11 The output variation with RLC load

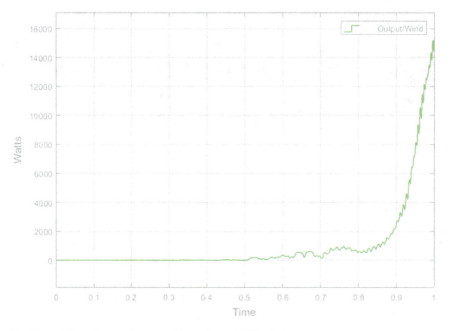

Fig. 18.12 The variation of output of dynamic load with wind speed

Hydro Energy Model

Without Load

In hydro power the condition of power have some fluctuation at $t = 0–1$ s and the maximum power is 1200 W. Figure 18.13 shows the output variation without load.

With R Load

In R load take $V_{rms} = 220$ V and QL $= 0$ var, QC $= 0$ var and the maximum power is 1200 W. Figure 18.14 shows the variation of output with R load.

With RLC Load

With RLC load take $V_{rms} = 220$ V and QL $= 100$ var, QC $= 100$ var and the maximum power is 1200 W. Figure 18.15 displays the output variation with RLC load.

With Dynamic Load

With dynamic load, the values of three phases are 0.0007477, 36.17 and 14.15. The maximum power is 1200 W. Figure 18.16 displays the variation of output with dynamic load.

18 Design of Energy Management System for Hybrid Power Sources 209

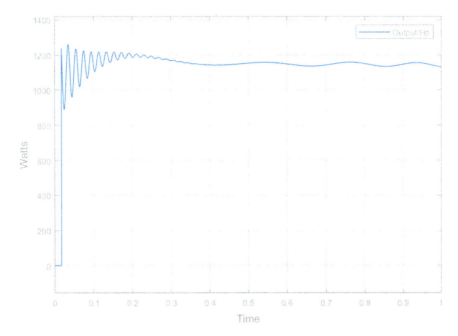

Fig. 18.13 The output variation without load

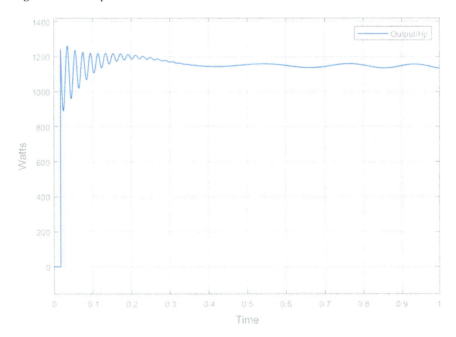

Fig. 18.14 The variation of output with R load

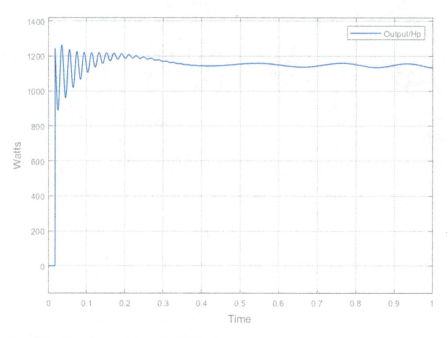

Fig. 18.15 The output variation with RLC load

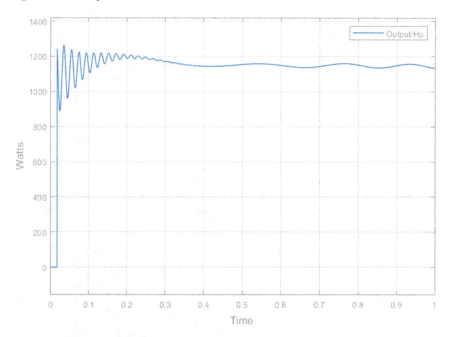

Fig. 18.16 The variation of output with dynamic load

18 Design of Energy Management System for Hybrid Power Sources

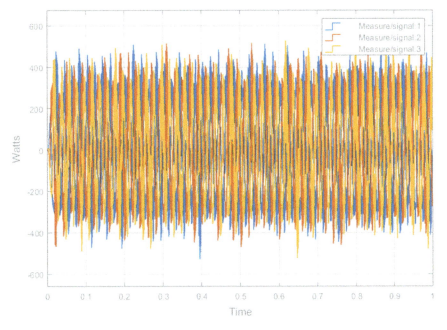

Fig. 18.17 The variation of three phases without load

Voltage Outputs

Without Load

The values of three phase's power after the converter are:

Phase	Values
A	33.13
B	−280.3
C	46

There is some variation of power at $t = 0$–1 s. The maximum power is 280.3 W with negative phase (Fig. 18.17).

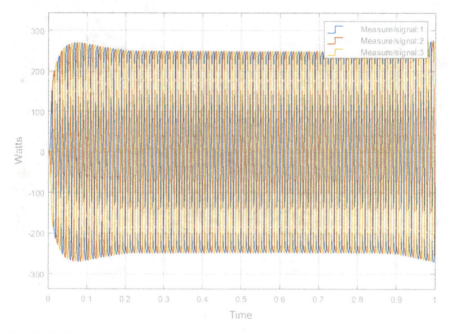

Fig. 18.18 The variation of three phases with RLC load

With RLC Load

The values of three phase's power after the converter are:

Phase	Values
A	−81.61
B	−254.8
C	273.1

To get maximum power output is 273.1 W and other phase give negative power (Fig. 18.18).

Current Outputs

Without Load

The values of three phase's power after the converter are:

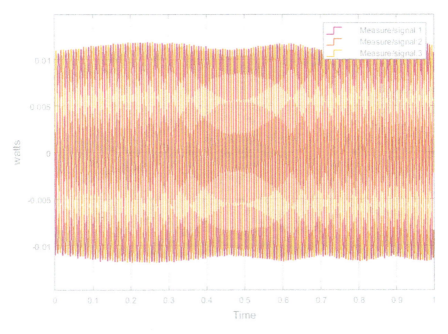

Fig. 18.19 The variation of three phases output without load

Phase	Values
A	0.01128
B	−0.006773
C	−0.00471

There is some variation of power at $t = 0\text{--}1$ s. The power is very small without load (Fig. 18.19).

With RLC Load

The values of three phase's power after the converter are:

Phase	values
A	−15.97
B	−52.73
C	57.34

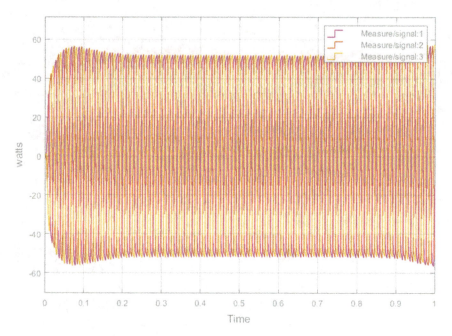

Fig. 18.20 The variation of three phases output with RLC load

There is variation of power throughout the time. With RLC load power get sinusoidal. The maximum power is 57.73 watts with negative phase (Fig. 18.20).

Conclusions

In this exploration, ideal vitality the executive's framework is planned and created for a framework associated conveyed vitality framework. The reason for the proposed calculation is to plan the age so as to limit the age cost, outflow, and to improve the usage productivity of capacity frameworks. The results can be summarised with different cases of load

- Without load maximum voltage is 46.0 V and current is min. value
- With R load max. Voltage is 272.2 V and current is 56.23 A
- With RLC load max. Voltage is 273.1 V and current is 57.34 A
- With dynamic load max. Voltage is 392.9 V with negative phase and current is min.
- The rating of battery is 200 V, 6.5 A-h and it is made up of Ni-Metal Hydride which are rechargeable
- PI controller input is V_{dc} and SOC (state of charge) and output are switch s_1 and s_2

- In this paper, take different types of loads like R, RLC and dynamic and get power, with the help of these loads the voltage regulation improved.

References

1. Ozdal MO, Hakki AI A new energy management technique for PV/wind/grid renewable energy system. Int J Photoenergy 19. https://doi.org/10.1155/2015/356930
2. Srinivasa Gopinath GS, Meher MVK (2018) Electricity a basic need for human beings. AIP Conf Proc 040024. https://doi.org/10.1063/1.5047989
3. Akanksha S, Singh HP, Sinha SK, Naqui A, Viral RK (2019) Renewable energy powered electrification in Uttar Pradesh state. In: 3rd International conference on recent developments in control, automation & power engineering (RDCAPE)
4. Microgrid, Webpage: https://en.wikipedia.org/wiki/Microgrid
5. Islanding, Webpage: https://en.wikipedia.org/wiki/Islanding
6. Likkasit C, Maroufmashat A, Ali E, Hong-ming Ku, Fowler M (2018) Solar-aided hydrogen production methods for integration of renewable energies into oil and gas industries. Energy Convers Manag 168:395–406
7. Talaat M, Farahat MA, Elkholy MH (2019) Renewable power integration: experimental and simulation study to investigate the ability of integrating wave, solar and wind energies. Energy 170:668–682
8. The configuration of the proposed solar energy connection, Webpage: https://www.researchgate.net/figure/MPPT-controller-in-a-PV-system_fig17_235761440
9. Fathima Hina A, Prabaharan N, Palanisamy K, Akhtar K, Saad M, Jackson JJ (2018) Hybrid-renewable energy systems in microgrids integration, developments and control. Woodhead Publishing. Joseph P. Hayton
10. Binayak B, Raj PS Kyung-Tae L, Sung-Hoon A (2014) Mathematical modeling of hybrid renewable energy system: a review on small hydro-solar-wind power generation. Int J Precis Eng Manuf Green Technol 1(2):157–173
11. Block diagram of PMSG based variable speed wind turbine, Webpage: https://www.researchgate.net/figure/Block-diagram-of-PMSG-based-variable-speed-wind-turbine_fig1_261048553
12. Mukesh MK, Nilay K, Bani AA (2015) Small hydro power in India: current status and future perspectives. Renew Sustain Energy Rev 51:101–115
13. Small hydro power plant block diagram, Webpage: https://www.mathwork0s.com/help/slcontrol/gs/bode-response-of-simulink-model.html

Chapter 19
Control and Coordination Issues in Community Microgrid: A Review

Seema Magadum, N. V. Archana, and Santoshkumar Hampannavar

Introduction

Today's Aggregate Technical and Commercial (AT&C) losses and scarcity of power in India, raises the question against the Indian power sector. The challenges of meeting load demand, fast depletion of conventional sources, development in industrial sectors, crawling development in renewable sectors are the heavy loop holes in the dream of economic and power independency of the country. These problems have to be addressed with advanced techniques and strategies for the development of the country.

The environmental concern has led the world to work towards the renewable sources like solar, hydro and wind power. The Distributed Generation (DG) systems with these heterogeneous sources have the power to mitigate the problem of green house effect along with power deficiency issues. The remote generation and transmission losses can be reduced by generating, transmitting and distributing the energy locally. The encouragement is needed to utilize the renewable sources, DG systems for local generation and consumption.

Microgrid is the small area of the grid, where consumption, generation, DG systems and renewable sources are maintained locally, so that it forms a healthy connection with the utility grid. The main interest of collaboration of DC &AC sources lies in the fact that there is increased usage of DC loads and sources in much dominated AC systems. This hybrid DC/AC grid, with the help of converters creates an efficient and reliable interconnecting system with main grid. The microgrid can

---

S. Magadum (✉) · S. Hampannavar
REVA University, Bangalore, India
e-mail: seemamagadum@reva.edu.in

N. V. Archana
NIEIT, Mysuru, India

be collaborated with main grid for power exchange in either direction. The utility grid can import power from microgrid with excess power and even microgrid will be able operate in isolated mode by retaining its individual properties.

In regard to achieve more reliability multi microgrid or community microgrid is gaining much importance. In community microgrid system, group of microgrids are interconnected with interlinking converters. Each microgrid in community gird can operate either in isolated or grid connected mode. The interconnection between the neighboring microgrids may provide the channel of power exchange within the community grid during emergency conditions. Community microgrid may have both ac and dc microgrids. This cluster of microgrids improves performance of individual microgrid systems in the economic and reliability prospective [1].

The gap between the smart grid and microgrid can be filled by the concept of community microgrid. As smart grid encourages the decentralized operation, smart usage of electricity usage, reliable grid maintenance adding the strength to the conventional system, where the microgrid system strengthens the system by local generation, transmission and distribution system. The coordination among the adjacent microgrids can be obtained with the help of numerous control and communication techniques. These features make the system to be capable of adopting many smart grid features [2]. The implementation of community microgrid requires suitable techniques to control and coordinate the power exchange in different mode of operation.

Issues in Multi Microgrid Operation

Due to intermittent nature of DERs and technical issues in the interconnection with the utility grid may face different challenges:

Power Sharing

Microgrids should respond to load variations and maintain the frequency/voltage at the point of common coupling. The system aims to have seamless exchange of power between the utility and multi microgrid.

Seamless Transfer of Mode

The system should be capable of exchanging power between microgrids and utility grid. Islanding operation should not affect the remaining part of the grid. Once fault clears the system should reconnect in short time lapse without disturbing the system stability [3, 4]. The many existing control techniques fails to achieve seamless voltage and power sharing between the different operating modes of the system due to poor disturbance rejection [5]. However the change of operation mode may results in the failure of the controller and also induces the large transients [6] and that will affect the power quality.

Islanding

In case of intentional islanding part of grid is isolated from the remaining part of the grid. Islanding operation is mostly not preferred in multi microgrid system to avoid the load shedding. The noncritical loads may shed to balance the generation and loads in an independent entity.

Power Quality

The voltage and frequency stabilization and reduction in distortion of injecting current is the major challenge due to nonlinear loads and instable operation of Distributed Generation (DG) [7] existed in the system.

Self Healing Technique

Self healing nature of network enables the system to respond, detect the fault and analyze the conditions and reset the system with less time lapse and human interference. The only affected part of system will get isolate to avoid the spreading of fault for large area and to avoid power interruption. The intelligent agents are adopted in the system to make the network as self healing system in different operating conditions [8]. The maintenance and reliability of the system can be enhanced with the help of these intelligent agents in real time operations [9].

Literature Review on Control Strategies

The large number of DG units causes several operational and technical issues, such as system stability, power quality, network voltage and low inertia. To overcome the mentioned issues, many advanced control techniques are adopted in microgrid system so that it can be capable to operate either in utility connected or isolated mode in concern with the stability and reliability of the network.

The community microgrid which includes clusters of microgrid should be capable of monitoring frequency, power and voltage in different mode of operation. System should be capable of providing reference voltage and frequency for islanded operation in order to address the need of critical loads. As smart grid feature, the microgrid should adjust to abnormal situations without affecting the stability. Microgrid should operate in grid following mode or grid forming mode. The grid following mode can adopts Current Source Inverters (CSI) while in later condition Voltage source Inverters (VSI). Different controlling algorithms are adopted on the operating modes.

The microgrid operation faces challenges such as frequency, voltage and power flow management, load sharing and coordination issues. Hierarchical control techniques are introduced in microgrid to address the above issues [1, 6, 10, 11]. The system includes three levels as voltage control, primary and secondary or/and tertiary control.

The voltage control addresses the power quality issues and provides the reference grid voltage. Control loop may consists the combination of voltage and current control or inner-loop voltage control [12]. To regulate current through the inductor and voltage across capacitor, the linear/nonlinear control loops, feedback, feed-forward and virtual impedance can be adopted in the control loop. The primary control in microgrids aims to enhance power sharing, frequency and voltage control, stability and system performance in the presence of both nonlinear and linear loads. The control techniques in primary loop may include communication based or without communication techniques. The communication less control techniques are more reliable and improves the system performance as it utilizes the local measurements [1]. The secondary and tertiary control measures the frequency and voltage at the PCC. It regulates the power sharing, frequency, voltage and reactive power. The efficient synchronization of microgrid to utility grid can be achieved with the help of these control techniques.

Inner-Loop Control Techniques

Research is progressing in the direction to address various aspects in the microgrid system mainly frequency and voltage maintenance, power management issues in different mode of operation, smooth mode switching and self healing etc. Numerous techniques are proposed by researchers to resolve these issues as listed in (Table 19.1).

Proportional Integral (PI) controller is effectively adopted for reactive power and real power control. It provides the active and reactive current references to the controller [13, 14]. The performance of PI controller can be improved by utilizing cross coupling term and feed-forward voltage. It results in zero steady state error in dq frame of reference; however controller fails to retain stability in distorted electrical conditions and may result in weak dynamics performance of system. Use of multiple PI controllers in unbalanced or nonlinear load condition to avoid harmonics can overcome by the adoption of proportional resonant (PR) controller. The controller can be designed both in abc frame and $\alpha\beta$ frame. PR controller tuned at fundamental frequency regulates both negative and positive sequence components and provides the zero steady state error. The controller at harmonic frequency may able to mitigate voltage harmonics. It also maintains the voltage regulation in addition to the harmonic reduction [15]. PR controller faces the problem of accurate tuning and it is sensitive to the frequency variations. Linear Quadratic Regulator (LQR) controller improves the dynamic response of the system [16, 17]. LQR is independent of load characteristics and network parameters [9]. The controller is simple to implement and effective in both transient and steady state conditions of the microgrid incorporating multiple DERs [18] but tracking accuracy during load variation of the controller is poor. Linear Quadratic Integrator (LQI) controller improves the dynamic response and reduces the steady state voltage error. In the disturbed condition, integral term of the controller reduces the error, which provides an accurate voltage tracking with zero steady state error. The controller faces the issue to track voltage in the normal operation. Deadbeat Controller (DB) controls the current of an inverter. It improves the dynamic performance with current control technique and reduces the harmonic components [19, 20]. This controller is sensitive to network parameters. Model Predictive controller manages the disturbances and nonlinearities of the system by predicting the controlled variables by using its present states [21]. It also minimizes current tracking error. This controller is sensitive to the parameter variation [22].

Hysteresis controller is simple, easy to implement and have good dynamic response. The controller has fast response and produces the control signal if error between measured and reference signal exceeds the limits and produces the switching signals for an inverter [23]. Controller should be designed to reduce THD so that output current contains fewer ripples. In addition to that the design filter adds to the difficulty, owing to the output randomness. The practical difficulty arises to maintain switching frequency for inverters as depends on load variation. H-infinity (H-∞) controller is robust and easy to implement. It has robust performance under unbalanced load condition. It offers less THD and less tracking error [24]. An advantage of H-∞ controller is it reduces the effect of disturbance on the output. The controller is mathematically difficult to understand and have slow dynamics. Repetitive Controller RC consists of integral controller, resonant controller and proportional control in parallel combination. The controller adopts the internal model principle to reduce the error in a dynamic system [25, 26]. In the system with nonlinear load, the low pass filter can be accommodated with controller to reduce the harmonics in the output voltage. Neural Network (NN) consists of 3 parallel interconnected layers such as input, hidden and output layers to process the task. NN is a systematic technique transfer the biological nerve system with considerable time delay. These algorithms are self learning and feasible to provides robust performance for different operating conditions and grid disturbances and validated in [27]. Fuzzy controller deals with linguistic values, which ranges 1 for completely true and 0 for completely false. The fuzzy controller is adopted to enhance the tracking performance and to reduce the overshoot as achieved in [28]. Sliding Mode Control (SMC) facilitates the strong control action during the disturbances [29]. It provides the robust performance for different range of operating points. The controller has advantages such as good disturbance rejection, insensitive to parameter variations and easy implementation.

Table 19.1 Merit and demerit of inner-loop control techniques

Control techniques	Merits	Demerits
Proportional integral	Easy to implement Results in Zero steady state error in dq frame	Poor performance under disturbed conditions Leads to Steady state error in an unbalanced system
Proportional resonant	Robust controller, manages to get Zero steady state error, and implementation is easy	Affects by frequency variations, High THD Requirement of accurate tuning
Linear quadratic regulator	Dynamic response is fast, easy to implement, reliable tracking performance	Voltage tracking error in disturbance, results in voltage tracking errors in normal condition
Linear quadratic integrator	Dynamic response is fast, easy to implement, reliable tracking performance	Results in voltage tracking errors in normal condition, difficult to design the model
Deadbeat controller	Good harmonic control, results in better transient response	Additional filter circuit, depends on network parameters
Model predictive controller	Accurate current control with low THD, suitable for nonlinear systems	Additional filter circuit, depends on network parameters, complex in design
Hysteresis controller	Easy to implement, fast transient response, inherent current control	Suitable for low power levels, harmonics problems
H-infinity(H-∞) controller	Low THD, suitable for linear and nonlinear loads, negligible tracking error	Slow dynamics, mathematical complexity
Repetitive controller	Zero steady state error, robust performance	Slow dynamics, mathematical complexity
Neural network	Good performance	Slow dynamic response
Fuzzy controller	Independent of parameter variation, suitable for nonlinear loads	Slow method
Sliding mode control (SMC)	Robust performance, low THD	Difficult to design. Chattering phenomenon

Control Methods for Power Sharing

Research on Primary loop/secondary loop deals with power management for microgrid system. The system should be capable of meeting required load demand either in grid connected or islanded mode. Smart paradigm concentrates to meet the critical load in disturbed condition without losing quality and stability of the system. Approach towards addressing these critical factors of microgrid operation called for suitable techniques either on the base of communication system or communication less system. The communication based control techniques are used to provide accurate reference points for power sharing and stability control. As it depends on the communication system, techniques are not reliable and flexible. Many techniques are in practice such as, average load sharing, centralized control, distributed control, consensus-based droop, peak-value-based current sharing, master–slave, circular chain and angle droop control.

The master–slave control incorporates the coordination techniques between the converters [30]. In [31] the master converter acts as a Voltage Source Inverter (VSI) which produces the controlled voltage and Current Source Inverter acts as slave inverters by receiving the command signal from the master inverter. The inverters are continuously updated with current reference values as a weighted average current in the average load sharing control [32, 33]. A centralized control manages the equal current for all DG units [34]. In peak-value based current sharing control, peak current of VSI is used to set as current reference for converters. This method is effective for smooth mode transfer and proper power sharing [35, 36]. The converters are assumed to be operated in chain links, to provide the current reference for the inverters from preceding inverter in a circular chain control technique [37].The distributed control can be designed in between the high and low bandwidth local controllers, to improve the reliability by reducing the communication lines [38]. Angle droop control adopts the communication line to determine the phase angle reference for controlling the active power [10, 39]. Efficient load sharing among the inverters can be achieved by using the modified voltage angle control loop [40]. Voltage angle loop along with integral control can achieve a better steady state performance. The issues of reactive power sharing and dependency on parameter variation can be reduced by adopting the consensus-based droop control with sparse communication network under line impedances variations [41].

The high bandwidth communication system are not suitable in the case of the system having large number of DG units in remote places, it will increase the infrastructure cost. Popularly droop based techniques are used in microgrid operation to eliminate the limitations of communication based techniques. Droop techniques are more reliable and flexible but there are some limitation such as dependency on network impedance, frequency and voltage variations and effect of nonlinear loads.

Power/Frequency (P/f) Droop Control technique can be utilized in an islanded microgrid operation and it suits for high voltage transmission lines [42–44]. It is effective to obtain plug and play features. The P/f droop control is to be adopted based on the system characteristics, as network with large synchronous machine will have inertia, but converter based microgrid system lack the inertia. The conventional droop function can be modified to achieve better performance such as (Table 19.2):

Active and reactive power sharing accuracy is monitored from central controller to synchronize the system, which may results in stability issues.

Power/voltage (P/V) droop control focuses on dispatchable DG units and can be results in good response compare to P/f droop control in low-voltage distribution network as observed in [14, 45]. The technique addresses the issue of reactive power sharing and more suitable for low-voltage system. The P/V droop technique can be adopted in the microgrid with large number of DG systems where inductance is

Table 19.2 Modifications in conventional droop function

Conventional function	$\omega_i = \omega^* - Kf(P_i - P_i^*)$ $V_i = V^* - Kv(Q_i - Q_i^*)$	ω-angular frequency P_i-measured active power P_i^*-reference active power Kf-frequency droop coefficient Kv-voltage droop gain Q_i-measured reactive power Q_i^*-reference reactive power
To improve dynamics and reduce transients	$\omega_i = \omega^* - Kf * P_i - \widehat{Kfd}\dfrac{dP_i}{dt}$ $V_i = V^* - Kv * Q_i - \widehat{Kvd}\dfrac{dQ_i}{dt}$	$\widehat{Kfd}$ and $\widehat{Kvd}$—adaptive transient droop gains
To avoid coupling issues	$\omega_i = \omega^* - Kf(P_i - Q_i)$ $V_i = V^* - Kv(P_i + Q_i)$	Kf-frequency droop coefficient Kv-voltage droop gain
To reduce power sharing error	$\omega_i = \omega^* - Kf * P_i - Kv * Q_i$ $V_i = V^* - Kv * Q_i + \dfrac{K_c}{S}(P_i - P_{\text{avg}})$	K_c-integral term P_{avg}-average steady state active power

much low compared to resistance. The conventional droop function relates the active power with voltage and reactive power with frequency as shown in Eqs. (19.1) and (19.2).

$$\omega_i = \omega^* + Kf(Q_i - Q_i^*) \qquad (1)$$

$$V_i = V^* - Kv(P_i - P_i^*) \qquad (2)$$

The system dynamics can be improved by adopting derivative term as follows:

$$\omega_i = \omega^* + Kf * Q_i + K_{q,d}\dfrac{dQ_i}{dt} \qquad (3)$$

$$V_i = V^* - Kv * P_i - K_{p,d}\dfrac{dP_i}{dt} \qquad (4)$$

The technique proves better control over reactive power sharing but may results in the reliability issues during fault conditions.

Voltage Based Droop (VBD) control is applied to low-voltage islanded microgrids with majority of renewable energy sources [46]. The technique results in the seamless transition between the islanded and grid mode of operation and reduces the switching transients as achieved in [47]. The control techniques need not to be modified for different operating modes. The VBD divides P/V droop techniques as Pdc/Vg with constant power band and V_g/V_{dc} droop control. The Vg/Vdc control can be expressed as:

$$V_g^* = V_g^0 + Kv(V_{dc} - V_{dc}^0) \tag{5}$$

where V_{dc}^0 and V_g^0 are reference DC ink voltage and terminal voltage. The method is flexible to adjust power during voltage deviation and need to be verified for its stability margin.

The V/I droop method can be adopted for the improvement of power sharing and system performance with direct and quadrature axis voltages. The load sharing capability of converter can be improved by adopting modified droop control method. The controller uses minimum output current tracking control in low-voltage Microgrid systems [48]. Jin et al. [49] monitors power distribution by sending a command signal with the help of demand droop control and also a normalization technique in an autonomous hybrid microgrid by merging the information on both sides of the interfacing converter. Augustine et al. [50] presents low-voltage dc microgrid with the Droop Index (DI) method for dc–dc converter operation. The proposed technique can regulate the load current and also reduce the circulating current by calculating the virtual resistance Rdroop based on the converters voltage. Many of these techniques compromise in the aspects of power quality due to integration aspects (Table 19.3).

The important aspects of practical implementation of microgrid are its operating capabilities under both utility connected mode and in isolated mode. Because of nonlinear nature of load, inverter characteristics and also because of highly inductive property of the system, the many discussed techniques fail to achieve a seamless transfer between clusters of grids.

Table 19.3 Merit and demerit of power sharing

Control techniques	Merits	Demerits
P/V droop	Preferred for low line, Easy to implement, no requirement of communication system	Slow dynamic response, sensitive to network parameters, Poor active power control
P/f droop	Easy to implement Suitable for high and medium voltage transmission line, no requirement of communication system	Slow dynamic response, sensitive to network parameters, Poor reactive power control
Voltage based droop	Suitable for the system with renewable source, suitable for large resistive network, good power balance method	Difficult to implement, voltage variation under load variation
V/I droop	Fast dynamic response Good power sharing, suitable for system with DG units	Voltage variation under load variation, oscillation problem for low droop coefficients

Conclusion

Numerous control technologies are adopted in many research works for the coordinated operation of microgrid with main grid to achieve basic aspects of grid interconnection issues. The inner-loop control techniques are concentrated to regulate the frequency and voltage, whereas power sharing techniques are adopted for coordinated operation. The paper highlights the merits and demerits of most of microgrid techniques. Any one of the technique will not provide the complete solution to all microgrid concerned issues. Combination of inner-loop control and power sharing techniques can provide the better solution for major concerned issues of microgrid operation. As most of work discussed is applied to microgrid/hybrid microgrid, the control techniques may need up gradation to achieve the coordination aspects of the multi microgrid with the utility grid.

References

1. Che YL, Shahidehpour M, Alabdulwahab A, Al-Turki Y (2015) Hierarchical coordination of a community microgrid with AC and DC microgrids. IEEE Trans Smart Grid 3042–3051
2. Ng EJ, Ramadan A (2010) El-Shatshat.: Multi-microgrid control systems (MMCS). IEEE, pp 1–5
3. Rocabert J, Azevedo G, Candela I, Teoderescu R, Rodriguez P, Etxebarria-Otadui I (2010) Microgrid connection management based on an intelligent connection agent. IEEE 3028–3033
4. Heredero-Peris D, Chill´on-Ant´on C, Pag‘es-Gim´enez M, Gross G, Montesinos-Miracle D (2013) Implementation of grid-connected to/from off-grid transference for micro-grid inverters. IEEE 840–845
5. Mohamed YA-RI, Radwan, AA (2011) Hierarchical control system for robust microgrid operation and seamless mode transfer in active distribution systems. IEEE Trans Smart Grid 352–362
6. Huang P-H, Xiao W, El Moursi MS (2013) A practical load sharing control strategy for DC microgrids and DC supplied houses. IEEE 7124–7128
7. Hu J, Zhu J, Platt G (2011) Smart grid—the next generation electricity grid with power flow optimization and high power quality. IEEE 1–6 (2011)
8. Pashajavid E, Shahnia F, Ghosh A (2015) A decentralized strategy to remedy the power deficiency in remote area microgrids. IEEE, 2015
9. Shahnia F, Chandrasena RPS, Rajakaruna S, Ghosh A (2014) Primary control level of parallel distributed energy resources converters in system of multiple interconnected autonomous microgrids within self-healing networks. IET Gener Transm Distrib 203–222
10. Majumder R, Chaudhuri B, Ghosh A, Majumder R, Ledwich G, Zare F (2010) Improvement of stability and load sharing in an autonomous microgrid using supplimentary droop control loop. IEEE Trans Power Syst 796–808
11. Liu X, Wang P, Loh PC (2011) A hybrid AC/DC microgrid and its coordination control. IEEE Trans Smart Grid 278–286
12. Li Y, Vilathgamuwa DM, Loh PC (2004) Design, analysis, and real-time testing of a controller for multibus microgrid system. IEEE Trans Power Electron 1195–1203
13. Guerrero JM, Vásquez JC, Teodorescu R (2009) Hierarchical control of droop–controlled DC and AC microgrids—a general approach towards standardization. IEEE 4305–4310
14. Majumder R (2014) A hybrid microgrid with DC connection at back to back converters. IEEE Trans Smart Grid 251–259

15. Nian H, Zeng R (2011) Improved control strategy for stand-alone distributed generation system under unbalanced and non-linear loads. IET Renew Power Gener 323–331
16. Mahmud MA, Hossain MJ, Pota HR, Roy NK (2014) Nonlinear distributed controller design for maintaining power balance in islanded microgrids. IEEE
17. Raju P, Jain T (2014) Centralized supplementary controller to stabilize an Islanded AC microgrid. IEEE
18. Rana MdM, Li L (2015) Controlling the distributed energy resources using smart grid communications. IEEE 490–495
19. Xueguang Z, Wenjie Z, Jiaming C, Dianguo X (2014) Deadbeat control strategy of circulating currents in parallel connection system of three-phase PWM converter. IEEE Trans Energy Convers 406–417
20. Kim J, Hong J, Kim H (2016) Improved direct deadbeat voltage control with an actively damped inductor-capacitor plant model in an Islanded AC microgrid. Energies
21. Hu J, Zhu J (2015) Model predictive control of grid-connected inverters for PV systems with flexible power regulation and switching frequency reduction. IEEE Trans Ind 587–594
22. Cortés P, Kazmierkowski MP, Kennel RM, Quevedo DE, Rodríguez J (2008) Predictive control in power electronics and drives. IEEE Trans Ind Electron 4312–4324
23. Rahim NA, Selvaraj J, Krismadinata (2007) Implementation of hysteresis current control for single-phase connected inverter. IEEE
24. Hornik T, Zhong QC (2013) Parallel PI voltage H-∞ current controller for the neutral point of a three-phase inverter. IEEE Trans Ind Electron 1335–1343
25. Hara S, Yamamoto Y, Omata T, Nakano M (1988) Repetitive control system: a new type servo system for periodic exogenous signals. IEEE Trans Autom Control 659–668
26. Jin W, Li Y, Sun G, Bu L (2017) H-infinity repetitive control based on active damping with reduced computation delay for LCL-type grid-connected inverters. Energies
27. Hatti M, Tioursi M (2009) Dynamic neural network controller model of PEM fuel cell system. Int J Hydrogen Energy 5015–5021
28. Hasanien HM, Matar M (2015) A fuzzy logic controller for autonomous operation of a voltage source converter-based distributed generation system. IEEE Trans Smart Grid 158–165
29. Chen Z, Luo A, Wang H, Chen Y, Li M, Huang Y (2015) Adaptive sliding-mode voltage control for inverter operating in islanded mode in microgrid. Int J Electr Power Energy Syst 133–143
30. Chen JF, Chu C (1995) Combination voltage controlled an current controlled PWM inverter for UPS parallel operation. IEEE Trans Power Electron 547–558
31. Pei Y, Jiang G, Yang X, Wang Z (2004) Auto-master-slave control technique of parallel inverters in distributed AC power systems and UPS. IEEE
32. Roslan AM, Ahmed KH, Finney SJ, Williams BW (2011) Improved instantaneous average current-sharing control scheme for parallel-connected inverter considering line impedance impact in microgrid networks. IEEE Trans Power Electron 702–716
33. Sun X, Lee YS, Xu D (2003) Modeling, analysis, and implementation of parallel multi-inverter systems with instantaneous average-current-sharing scheme. IEEE Trans Power Electron 844–856
34. Siri K, Lee C, Wu TF (1992) Current distribution control for parallel connected converters. IEEE Trans Aerosp Electron Syst 829–840
35. Chen CL, Wang Y, Lai JS, Lee YS, Martin D (2010) Design of parallel inverters for smooth mode transfer microgrid applications. IEEE Trans Power Electron 6–15
36. Chen Q, Ju P, Shi K, Tang Y, Shao Z, Yang W (2010) Parameter estimation and comparison of the load models with considering distribution network directly or indirectly. IEEE
37. Wu TF, Chen YK, Huang YH (2000) 3C strategy for inverters in parallel operation achieving an equal current distribution. IEEE Trans Ind Electron 273–281
38. Prodanovic M, Green TC (2006) High-quality power generation distributed control of a power park microgrid. IEEE Trans Ind Electron 1471–1482
39. Pota HR, Hossain MJ, Mahmud M, Gadh R (2014) Control for microgrids with inverter connected renewable energy resources. IEEE (2014)

40. Majumder R, Ghosh A, Ledwich G, Zare F (2008) Control of parallel converters for load sharing with seamless transfer between grid connected and Islanded modes. IEEE (2008)
41. Lu LY, Chu CC (2015) Consensus-based droop control for multiple DICs in isolated microgrids. IEEE Trans Power Syst 2243–2256
42. Barklund E, Pogaku N, Prodanovic M, Hernandez-Aramburo C, Green TC (2008) Energy management in autonomous microgrid using stability-constrained droop control of inverters. IEEE Trans Power Electron 2346–2352
43. Chandorkar MC, Divan DM, Adapa R (1993) Control of parallel connected inverters in standalone AC supply systems. IEEE Trans Ind Appl 136–143
44. Marwali MN, Jung JW, Keyhani A (2004) Control of distributed generation systems—Part II: Load sharing control. IEEE Trans Power Electron 1551–1561
45. Au-Yeung J, Vanalme GM, Myrzik JM, Karaliolios P, Bongaerts M, Bozelie J, Kling WL (2009) Development of a voltage and frequency control strategy for an autonomous LV network with distributed generators. IEEE
46. Vandoorn TL, Meersman B, De Kooning JDM, Vandevelde L (2013) Transition from Islanded to grid-connected mode of microgrids with voltage-based droop control. IEEE Trans Power Syst 2545–2553
47. Pashajavid E, Shahnia F, Ghosh A (2015) A decentralized strategy to remedy the power deficiency in remote area microgrids. IEEE
48. Lee C-T, Chuang C-C, Chu C-C, Cheng P-T (2009) Control strategies for distributed energy resources interface converters in the low voltage microgrid. IEEE 2022–2029
49. Jin C, Loh PC, Wang P, Mi Y, Blaabjerg F (2010) Autonomous operation of hybrid AC-DC microgrids. IEEE ICSET (2010)
50. Augustine S, Mishra MK, Lakshminarasamma N (2015) Adaptive droop control strategy for load sharing and circulating current minimization in low-voltage standalone DC microgrid. IEEE Trans Sustain Energy 132–141
51. Zou H, Mao S, Wang Y, Zhang F, Chen X, Cheng L (2019) A survey of energy management in interconnected multi-microgrids. IEEE Access 72158–72169
52. Saleh M, Esa Y, Hariri ME, Mohamed A (2019) Impact of information and communication technology limitations on microgrid operation. Energies 1–24
53. Zhou J, Zhang J, Cai X, Shi G, Wang J, Zang J (2019) Design and analysis of flexible multi-microgrid interconnection scheme for mitigating power fluctuation and optimizing storage capacity. Energies 1–21
54. Lee W-P, Choi J-Y, Won D-J (2017) Coordination strategy for optimal scheduling of multiple microgrids based on hierarchical system. Energies 1–18

Chapter 20
Optimal Generation Sizing for Jharkhand Remote Rural Area by Employing Integrated Renewable Energy Models Opting Energy Management

Nishant Kumar and Kumari Namrata

Introduction

In developing countries such as India, growths depend on the energy they consume or generate that affects the economy of these countries. Major cities and Industries are fed with the most of electricity generated resulting shortage of energy needed by rest of the small, scattered villages far away from cities, located in dense forest and hilly regions. Energy shortage and environment protection concerns are increasing in this era, both clean energy and new fuel technologies are being actively researched and pursued. Collectively the renewable energy, (micro-hydropower, biomass and solar energy) is converted into electrical energy, which can be directly delivered to the public power grid or used for isolated loads. G. K Singh reviewed the Photovoltaic Energy Generation [1]. R. K. Akikur, R. Saidur, H. W. Ping, K. R. Ullah studied stand-alone system and hybrid systems of solar with off grid to electrify remote villages [2]. It is essentially worthy that economic benefits and speedup utilization of distributed energy resources necessitate the optimal models development to meet the demand through these renewable energy resources [3]. Due to unpredictability of the availability of renewable resources, research preferred the integrated renewable energy Model (IREM) in comparison of hybrid systems or stand-alone to increase reliability of those system [4]. Further, IRE model are reliable, non-polluting, and more effective in reduction in the cost of operation and maintenance [5]. Due to

---

N. Kumar (✉)
Chaibasa Engineering College, Kelende, Jharkhand, India
e-mail: krnishant125@gmail.com

K. Namrata
National Institute of Technology, Jamshedpur, Jamshedpur, Jharkhand, India
e-mail: namarata.ee@nitjsr.ac.in

© The Editor(s) (if applicable) and The Author(s), under exclusive license to Springer Nature Singapore Pte Ltd. 2021
A. K. Singh and M. Tripathy (eds.), *Control Applications in Modern Power System*, Lecture Notes in Electrical Engineering 710,
https://doi.org/10.1007/978-981-15-8815-0_20

involvement of each component's mathematical models, the modeling of an IRE model or system claims to be a complex process. S. Rajanna, and R. P. Saini, has modeled such IRE system to supply power and verified the optimal sizing using PSO [6].

Appropriate and effective load demand management is needed for efficient IREM modeling. Energy Management (EM), a method traditionally used which offers feasible solution to improve efficiency of IRE Model by altering, adjusting operating time and appliance quantity, applying Load shifting Technique during peak periods to off peak periods. However, this method provides daily load curve which is uniform.

Study Area Minutiae

Jharkhand with 24 districts covering an area of 79,714 km^2, belongs to the northern India. Chatra district in Jharkhand has 12 blocks (Chatra, Kunda, Hunterganj, Pratappur, Lawalong, Gidhor, Pathalgada, Simaria, Tandwa, Itkhori, Kanhanchatti and Mayurhand) out of which Hunterganj block has been considered for study purpose. Hunterganj is well nourished by geographical assets consists of sufficient water bodies being the part of Upper and lower Hazaribagh Plateau and northern scarp. Hunterganj stands on the banks of Lilajan River, with 60% of area covered with forest and rain fed agriculture. Economically, the entire study area relies on agriculture, forest, animal husbandry and marginal workers. Agriculture residue and forest foliage are the resources available in considerable amount for cooking and feedstock for the cattle.

Hunterganj block located at 24.4341° N 84.8149° E [7] has the highest percentage of un-electrified villages as compared to rest of the blocks in Jharkhand. There are 270 villages with residing population of 187590, in which 250 villages are inhabited and electrified and 20 villages remains uninhabited and un-electrified further segregating the focused area depending upon locality and resources a gram panchayat LENJWA was considered with 5 villages 3 electrified and 2 un-electrified. Both areas are at the different part of the study area, consisting total household of 480 with a population of 2908. The details of the area are highlighted in Table 1.

Load Minutiae

The analysis of the total load demand, done considering the population size and the countable households of the entire study area. The different sectors considered for determining the load demand are domestic load, community load, commercial load and small industry loads, which are listed in details in Table 2. The total load demand estimation of each village of that area was made on the basis of minimum desired hourly load profile. The total annual load demand has been estimation based

Table 1 Glance of case study area

Parameters	Area
Name of Gram Panchayat	LENJWA
Populations (Nos.)	2908
Electrified villages (Nos.)	3
Un-electrified villages (Nos.)	2
Household (Nos.)	480
Households population (Nos.)	960
Latitude and longitude of study area	24.4619° N, 84.7159° E

Table 2 Minutiae of load in area

Appliances	Power Rating (Watts)	Cost ($)	Switching Points	Number in Used	Energy utilization (KWh)
LED lighting (HH)	10	5.50	2/HH	960	9.60
Ceiling Fan (HH)	50	27	2/HH	960	48.00
LED TV (HH)	60	115	1/HH	480	28.80
Refrigerator (HH)	42	430	1/HH	480	20.16
LED lighting (SC)	10	5.50	10 per village	30	0.30
Ceiling Fan (SC)	50	27	12 per village	36	1.80
LED lighting (H)	10	5.50	6	6	0.06
Ceiling Fan (H)	50	27	6	6	0.30
Refrigerator (H)	42	430	1	1	0.042
LED lighting (CH)	10	5.50	1 per village	3	0.30
Ceiling Fan (CH)	50	27	1 per village	3	0.15
Street Light (Vill)	20	11	1 per 20 HH	24	0.48
Water Pump Motor (Vill)	3675	176.92	1 per village	3	11.025
Flour Mills Motor (Vill)	3750	253.846	1 per 3 village	1	3.75
Grain thrasher Motor (Vill)	5000	315.38	1 per 3 village	1	5.00
Saw Mills Motor (Vill)	5000	315.38	1 per 3 village	1	5.00

on collected data from survey. MCMR equipment is considered for this survey the detail profile (rating and cost) are listed in the Table 2. The estimated total energy requirement of the focused area is 287669.64 kWh/yr.

Resource Assessment Minutiae

Solar Energy

The solar radiation data was taken from the Homer software and verified data obtained from NASA Surface meteorology and Solar Energy: RETScreen Data, by providing longitude and latitude of the study area. Annual energy from solar available for the area is 1944 KWh/m^2/year. The solar output power from the PV panel (P_{PV}) can be calculated from following Eq. (20.1) [8];

$$P_{PV} = S_R * \eta_{PV} * A_{PV} \tag{20.1}$$

where S_R is solar radiance, η_{PV} is efficiency of panel, and A_{PV} is Area of a panel.

Micro-hydro

Low volumetric flow rate with high head height or low head with high volumetric flow rate both conditions are suitable for micro-hydro generate to operate. Based on the analysis done in the study area, rough idea of availability of water streams types and water runoff rate has been derived to estimate the micro-hydropower potential. For predicting direct runoff volume the Soil Conservation Service Curve Number (SCS-CN) approach is used for a given rainfall data. The topo-sheet (maps) or contour maps provides the require data like areas, the watershed length and elevation difference of the water bodies for estimation of discharge rate. The monthly rainfall data is taken from the HOMER Software supported by NASA Surface meteorology and Solar Energy: RETScreen Data by providing longitude and latitude of the study area. Actual water runoff (Q_d) has been calculated using following Eq. (20.2) [8]:

$$Q_d = \frac{(I - 0.2S)^2}{(I + 0.8S)} \text{ when } I > 0.2S \tag{20.2}$$

$$S = \frac{25400}{CN} - 254 \tag{20.3}$$

where I in (mm) monthly recorded rainfall; S in (mm) max potential of retention in the watershed (mm); CN is curve number of runoff for hydrological sail (CN =

40–58; CN = 50–55 for forest which are dense and agriculture land, respectively) [10]. The following Eq. (20.5) is used to calculate the discharge rate;

$$T_P = 0.6 T_C + \sqrt{T_C} \qquad (20.4)$$

$$Q(t) = \frac{0.0208 * A * Q_d}{T_P} \qquad (20.5)$$

where; $Q(t)$ is discharge rate (m³/s); T_P is peak runoff time (hr); T_C is concentration time (hr); H_{net} is watershed head (m); A is watershed Area (hectare). Adapting methodology mentioned above, the discharge rate of the focused area estimated as 367 l/s (50% dependability). Equation (20.6) represents the output power of micro-hydro (P_{MH}) plant;

$$P_{MH} = 9.81 * Q(t) * h_{net} * \eta_{MH} * \rho_w \qquad (20.6)$$

where h_{net} is net height, ρ_w is density of water, and η_{MH} is efficiency of plant.

Biogas

Biogas is also produced from the manure of the cattle. The biogas potentials are totally dependent on the population of cows, goats, buffalos, ox, and sheep. The dung produce from the each animal was assumed as 10 kg/day dung by cow/horse/ox, 15 kg/day dung by buffaloes and 1 kg/day dung by goat/sheep. Based on the collected data the biogas potential of the area is 217 m³/day. Considering only 80% of the available waste is used to produce electricity only. The period of operation of the Biogas generator is operated 10 h per day and scheduled based on the peak hours, where demand is greatest. The output of biogas power (P_{BG}) can be calculated from following Eq. (20.7);

$$P_{BG} = \frac{\text{Gas yeild}(m^3/\text{day}) * CV_{BG} * \eta_{BG}}{860 * (\text{operating hours/day})} \qquad (20.7)$$

CV_{BG} in (kcal/kg) is calorific value, and η_{BG} is efficiency.

Biomass

Biomass often refers to the residue obtained from the agricultural and forest. The biomass potential is estimated considering the efficiency of collection as 75% and collected forest waste average as 165 kg/ha/yr. It is proposed and recommended to use a biomass gasifier-based power generation system to convert 80% of the total biomass potential into electrical energy. Based on the collected data the biomass potential of the area is 147 ton/yr. The hourly output from the biomass generator (P_{BM}) can be calculated from following Eq. (20.8);

$$P_{BM} = \frac{\text{Fuel availbility(ton/yr)} * CV_{BM} * \eta_{BM} * 1000}{365 * 860 * (\text{operating hours/day})} \quad (20.8)$$

CV_{BM} (kcal/kg) is calorific value, and η_{BM} is efficiency. The biomass generators are operated 9 h/day and scheduled based on the peak hours, where demand is greatest.

Battery Storage Backup

The IRE system is backed with the battery storage and bi-directional converter set. The battery storage system is always tagged with Discharging Limit (DL) before optimal sizing.

$$S\_C_{min} = DL * B\_C \quad (20.9)$$

The discharging and charging of the battery bank (i.e., State of charge SC) is examined through given Eqs. (20.10, 20.11):

$$SC_{t+1} = SC_{t+1} * (1-\epsilon) + (P_{DC}(t) + \eta_b * P_{AC} - P_{DM}(t)) * \eta_{BC} \quad (20.10)$$

$$SC_{t+1} = SC_{t+1} * (1-\epsilon) - \frac{(P_{DC}(t) + \eta_b * P_{AC} - P_{DM}(t))}{\eta_{BC}} \quad (20.11)$$

where S_C_{Min} is minimum storage capacity and B_C is initial battery capacity of storage. The constraint considered while designing the battery system while charging and discharging are in following Eq. (20.12). Energy storage should be in the limits under stated; i.e., it should not exceed maximum limit and should not less than the minimum limit.

$$SC_{min}(t) \leq SC(t) \leq SC_{max}(t) \quad (20.12)$$

Model and Objective Function Formation

An integrated renewable energy model shown Fig. 1 has been proposed to generate electricity, which can be used to supply the study area described in section 'Load Minutiae.' This model is been optimized providing constraints to minimize the

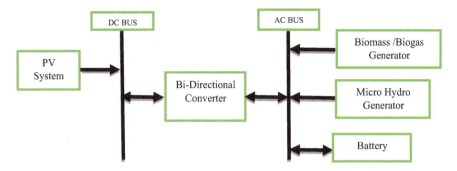

Fig. 1 IRE model

total overall generation cost using Particle swarm optimization and Jaya algorithm supported by technical parameter as Energy Index = 1 given below Eq. (20.13):

$$\text{Min TONC}_G = C\_C_{NV} + O\_M_{NV} + F\_C_{NV} + R\_C_{NV} + A\_C_{NV} \quad (20.13)$$

where C_C_{NV} is capital cost net value, O_M_{NV} is operation and maintenance net value, F_C_{NV} is fuel cost net value, R_C_{NV} replacement cost net value, and A_C_{NV} is appliance cost net value.

This study considered Technical Parameter and the Cost parameters to analysis and development of the IRE System. The above mentioned Cost parameters are calculated based on the no of equipment used and the net value of it. As the project have a life span of project life is 20 years, battery and converter has life span of 5 and 10 years, respectively; therefore, repairing or replacement cost depends on the given Eqs. (20.14, 20.15):

$$Z = \frac{N}{L} - 1 \quad (20.14)$$

$$C\_E(\text{Rs/kwh}) = \frac{\text{TONC}_G}{\sum_{t=1}^{8760} E_G(t)} \quad (20.15)$$

$EG(t)$ is the total energy required annually (kwh) generated from the renewable generators in IRE Model; TONC_G is total overall net cost of generation; L is life span of component (Batt/Converter); N is life span of project; Z is replacement frequency.

Energy Management

Energy management concept provides a numerous option and essential strategies to utilize load demand efficiently considering IRE System. As numerous investment option are highlighted in few previous studies like low, medium high ratings and

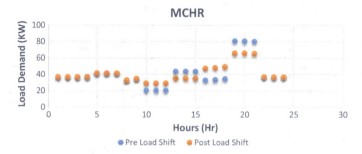

Fig. 2 Load curve pre- and post-load shifting

investment, so the best scenario considered for present study is Medium Cost high Rating (MCHR). Out of different Energy management concept Peak Load Shifting [11, 12] offers a best need solution, as the loads considerable can be segmented into two Shift-able Load (Water Pump, Saw Mills, Flour Mills,) and Non Shift-able Loads (Lights, Fan, Refrigerator, Street Lights, TV, Radio). The strategy adapted in present study Load Shifting at Peak time, i.e., Shift-able loads operating time are shifted from the peak curve to fill valley in the load curve, equations are illustrated as follows (Eqs. 20.16, 20.17):

$$\text{DM}(t)_{\text{WLS}} = L_{\text{NS}} + \sum_{I=1}^{24} L(t)_S * T_{\text{ONT}} \qquad (20.16)$$

$$\text{DM}(t)_{\text{LS}} = L_{\text{NS}} + \sum_{I=1}^{24} L(t)_S * T_{\text{OFFT}} \qquad (20.17)$$

where; $\text{DM}(t)_{\text{WLS}}$ and $\text{DM}(t)_{\text{LS}}$ is load demand pre-load shifting and post-load shifting, respectively, L_{NS} is Load non shift-able, L_S is Load shit-able, T_{ONT} are the operating time of shift-able loads at peak time. The below graph in Fig. 20.2 reflects the peak shifting concept used considering the biogas and Biomass operating time as it provides the fixed operating time as per load curve of the study Area.

Solution Adaptive Technique

Adaptive Technique 1: Particle Swarm Optimization (PSO)

Particle Swarm optimization is based on natural phenomena as swarm intelligence. A population based optimization a very powerful searching technique. It is most favorable techniques due to its implementation is easy. Steps follows as it assign random values to the size of the population and each particle searches the best solution in the assigned areas updating the velocity of best solution among the size. If the Objective function is to minimize whose Y_k^i be solution or position of the solution of kth term of ith iteration, then next position would be updated by Eq. (20.18) [13]:

$$Y_{k+1}^i = Y_k^i + v_{k+1}^i \qquad (20.18)$$

For updating the Velocity of each particle follows the stated Eq. (20.19):

$$v_{k+1}^i = w_k v_{k+1}^i + c_1 r_1 \left(P_k^i - Y_k^i\right) + c_2 r_2 \left(P_k^g - Y_k^i\right) \qquad (20.19)$$

where v is velocity, P_k^g best solution c_1 and c_2 (1.25, 1.25) cognitive and scaling value, r_1 and r_2 values randomly obtained between (0–1) and weightage of velocity ($w = 0.8$).

Adaptive Technique 2: JAYA Algorithm

Jaya algorithm is uprising as handy algorithm can be opted for optimization for solving constrained optimization objective functions. This algorithm provides the desired solution or best solution avoiding worst solution. Let $g(x)$ be the objective function. Be 'n' as number of design variables; 'm' as population size. $g(x)_{best}$ and $g(x)_{worst}$ be the best candidate and worst among the population size, respectively [14]. If $Y_{j,k,i}$ is the solution of the *j*th variable for the *k*th particle, in *i*th iteration, so to modify the solution below Eq. (20.18) is followed [15]:

$$Y'_{j,k,i} = Y_{j,k,i} + r1_{j,i}\left(Y_{j,\text{best},i} - \left|Y_{j,k,i}\right|\right) - r2_{j,i}\left(Y_{j,\text{worst},i} - \left|Y_{j,k,i}\right|\right) \qquad (20.20)$$

Results and Discussion

Optimization of the study area is analyzed considering the MCHR scenario using PSO and JAYA algorithm, which is portrayed into the Table 3. The economic and technical parameter based results observed under post- and pre-peak load shifting. The result obtained from PSO and JAYA algorithm clear states that it's more suitable to use these searching algorithm for the formulated objective functions. The result from both are almost same and very minute difference which can be neglected are detailed into Table 4. Amount of $ 9860 and $ 11076 are savings in terms of TONC and 0.0017 ($/KWh) and 0.0015 ($/KWh) savings in Cost of Energy. Technical and Economic observations are similar for both strategy used but Jaya algorithm converges early than the PSO algorithm.

Table 3 Optimal Sized IRE Model Minutiae for MCHR

Parameters	PSO		Jaya	
	Pre-load Shifting	Post-load Shifting	Pre-load Shifting	Post-load Shifting
Solar (kW)	46	35	43	32
Micro-hydro (kW)	14	14	14	14
Biogas (kW)	26	26	26	26
Biomass (kW)	23	23	23	23
Battery (Nos.)	99	82	120	95
TONC ($)	103190	93330	105370	94294
CE ($/kWh)	0.0223	0.0206	0.0225	0.0210

Table 4 Comparison of economic and technical result on MCHR observing both strategy

Parameters	PSO			Jaya		
	Pre-load shifting	Post-load shifting	Savings	Pre-load shifting	Post-load shifting	Savings
Solar (kW)	46	35	11	43	32	9
Micro-hydro (kW)	14	14	–	14	14	–
Biogas (kW)	26	26	–	26	26	–
Biomass (kW)	23	23	–	23	23	–
Battery (Nos.)	99	82	17	120	95	15
TONC ($)	103190	93330	9860	105370	94294	11076
CE ($/kWh)	0.0223	0.0206	0.0017	0.0225	0.0210	0.0015

Conclusions

The motive of the this study was to model an IREM system supported with Battery bank, whose sizing can be done optimally by minimizing the TONC and CE which engage the distributed renewable sources such as Biomass, Biogas, Solar (PV), and micro-hydro present in the focused area. So, the IREM with battery bank can supply the LENJWA panchayat villages which are un-electrified. The comparison of two searching algorithm is enhanced and an optimal solution to electrify the rural similar to study area is proposed where the loads are considered based on the MCHR appliances.

References

1. Singh GK (2013) Solar power generation by PV (photovoltaic) technology: a review. Energy pp 1–13. 27 Feb 2013
2. Tien W, Kuo K-C (2010) An analysis of power generation from municipal solid waste (MSW) incineration plants in Taiwan. Energy 35:4824e30
3. Raj Kumar DKK (2012) Optimal planning of distributed generation systems in distribution system: a review. Renew Sustain Energy Rev 16:5146–5165. 9 May 2012
4. Upadhyay S, Sharma MP (2014) A review on configurations, control and sizing methodologies of hybrid energy systems. Renew Sustain Energy Rev 38:47–63. 20 Mar 2014
5. Cano A, Francisco J, Higinio S, Fernandez M (2014) Optimal sizing of standalone hybrid systems based on PV/WT/FC by using several methodologies. J Energy Inst 87(1):330–340
6. Rajanna S, Saini RP (2016) Modeling of integrated renewable energy system for electrification of a remote area in India. Renew Energy 90:175–187
7. Jharkhand Census (2015) 2011 Census C.D. Block Wise Primary Census Abstract Data (PCA). Jharkhand—District-wise CD Blocks. Registrar General and Census Commissioner, India. Retrieved 20 Nov 2015
8. Protogeropoulos C, Brinkworth BJ, Marshall RH (1997) Sizing and techno-economical optimization for hybrid solar photovoltaic/wind power system with battery storage. Int J Energy Res 21:465–479
9. Michael AM, Ojha TP (2013) Principles of agriculture engineering. Jain Brothers
10. Kusakana K (2016) Optimal scheduling for distributed hybrid system with pumped hydro storage. Energy Convers Manag 111:253–260
11. Shina A, De M (2016) Load shifting for reduction of peak generation capacity requirements in smart grid. In: 1st IEEE international conference on power electronics, intelligent control and energy system (ICPEICES-2016)
12. Uuemaa P, Kilter J, Valti J, Drovtar I, Rosin A, Puusepp A (2013) Cost-effective optimization of load shifting in the industry by using intermediate storages. 4th IEEE PES Innovative Smart Grid Technologies Europe (ISGT Europe), October 6–9, Copenhagen
13. Hakimi SM, Moghaddas-Tafreshi SM (2009) Optimal sizing of a stand-alone hybrid power system via particle swarm optimization for Kahnouj area in south-east of Iran. Renew Energy 34:1855–1862
14. Venkata Rao R, Waghmare GG (2016) A new optimization algorithm for solving complex constrained design optimization problems. Eng Optim
15. Venkata Rao R (2016) Jaya: a simple and new optimization for solving constrained and unconstrained optimization problems. Int J Indus Eng Comput 7:19–34

Chapter 21
IoT-integrated Smart Grid Using PLC and NodeMCU

Kumari Namrata, Abhishek Dayal, Dhanesh Tolia, Kalaga Arun, and Ayush Ranjan

Introduction

Energy transmission is an indispensable part of the energy sector and its production has no value until the energy reaches the destination for the final consumer. The humongous amount of energy produced in the power station is to be transferred over many kilometers to reach load centers to serve the needs of customers using transmission lines and towers. Although our country has sufficient energy production capacity still a huge portion of its population has a narrow supply of electricity majorly due to lack of proper transmission infrastructure.

In a bulk electric system such as the national grid, it is necessary to control the voltage because there is an inverse relationship between voltage and current. As electric usage rises, such as on a hot day when everybody's air conditioner is running, the current draw on the system rises. This current rise causes the voltage to

---

K. Namrata · A. Dayal · A. Ranjan
Department of Electrical Engineering, National Institute of Technology Jamshedpur, Jamshedpur, Jharkhand, India
e-mail: namrata.ee@nitjsr.ac.in

A. Dayal
e-mail: adayal487@gmail.com

A. Ranjan
e-mail: ayushranjan9911@gmail.com

D. Tolia (✉) · K. Arun
Department of Electronics and Communication Engineering, National Institute of Technology Jamshedpur, Jamshedpur, Jharkhand, India
e-mail: dstolia07@gmail.com

K. Arun
e-mail: kakarun1908@gmail.com

© The Editor(s) (if applicable) and The Author(s), under exclusive license to Springer Nature Singapore Pte Ltd. 2021
A. K. Singh and M. Tripathy (eds.), *Control Applications in Modern Power System*, Lecture Notes in Electrical Engineering 710,
https://doi.org/10.1007/978-981-15-8815-0_21

drop. The drop-in voltage, if severe enough, can cause problems and even damage to electrical equipment that requires a steady voltage source. Usually, at peak time, when the temperature soars the highest, the industries consume more energy, thereby drawing more load current and hence the voltage drops. This voltage drops occurring causes the drop also at the lower distributaries resulting in low fluctuating voltages at the residential areas or even a blackout in case of the peak voltage. Because of the problems, this can cause, voltage regulation is used on the national grid system. This is accomplished through the use of standalone voltage regulators or load tap changers that are incorporated into a power transformer. These devices will raise or lower the voltage each time it changes by 5/8 of a per cent to maintain a steady voltage as current draw rises and falls.

Existing Systems

Smart Electricity Grids are new to the twenty-first century. But ironically, there is already a lot of research done on them. Let me mention a few of these:

In 2017, delivery system load management was proposed by organizing several bands of combined air-conditioners. This helped deal with the complex load by synchronizing compound groups of Virtual Power Storage Space Scheme (VPSSS) [1]. The same year, a Bluetooth Low Energy-based Home Energy Management system was proposed which was maintained by Artificial Neural Network (ANN). This smart grid comprises an electrical network superimposed by a communication system [2]. Another system proposed was an Intelligent Residential Energy Management System (IREMS). It merged demand-side management (DSM) with a special algorithm based on mixed-integer linear programming whose main purpose is the reduction in energy expenditure for residential consumers [3]. The year 2017 also witnessed a Building Management System (BMS) whose function was to monitor and coordinate the self-moving and electrical apparatus of a building using the consolidation of IPv4 and IPv6, Power over Ethernet (PoE) and Internet of Things (IoT) [4].

With 2018, the efficient use of Vehicle-to-Grid (V2G) was made feasible for Plug-in Electric Vehicles (PEVs) with two energy management strategies all the way taking care of energy imbalance in associated microgrids [5]. But in 2016, coalition-based game theory was used in Cyber-Physical Systems (CPSs) to develop an energy-efficient smart grid [6]. Another system proposed dealt with Coordinated Multi-Point (CoMP) transmission driven by active energy management for smart grid power [7]. The year 2016 also saw a system developed keeping the smart microgrid ecosystem in mind, called the Intelligent Dynamic Energy Management System (I-DEMS). The proposed system assembles the total of the severe load exact demands and also dispatches power to convenient loads which in turn helps microgrids be consistent, self-sustainable and ecologically responsive [8].

The year 2017 saw multiple customers, compound micro grid based Home Energy Management System with storage (HoMeS) was introduced [9]. But in 2015, a

Unified Energy Management System (UEMS) representation based on Distributed Location Marginal Pricing (DLMP) was proposed [10]. Come 2017 and another system proposed aimed to decrease overall production expense all the while keeping constraints related to the thermal and stimulating system in check and conclude the excellent dispatching of low voltage microgrids [11].

With 2014, a Smart Home Energy Management System (SHEMS) structural design was proposed which could monitor energy production and consumption simultaneously [12]. Another system proposed has a Home Plug answerable to a Power Line Communication (PLC) to supervise a photovoltaic (PV) unit [13]. The same year, an excellent Maximum Power Point Tracking (MPPT) based system was presented for the creation of photovoltaic (PV) power pump constant. This causes a rise in the PV operation and transfers the impressive PV energy produced into storage space during the beat break period [14].

The onslaught of 2013 saw a Smart Home Energy Management System (SHEMS). Its functionalities included mechanism learning algorithm, sense and announcement technology [15]. Another system proposed devised a Markov decision process (MDP) out of the preparation difficulty of many energy systems and gave two estimated resolution methods [16]. The same year, a Home Energy Management System (HEMS) with good air conditioning and heating prediction technique was introduced which took note of distinctiveness of thermal appliances and consumer handiness in a smart home situation [17].

Proposed System

By considering all the problems, we came up with the idea of establishing an Internet of Things (IoT) network among all the heavy- load appliances and the Regional-level power grid. PLCs would regulate the behavior of these appliances as per the directions are given by the grid. Thus, enabling us to distribute power efficiently to everyone without causing a complete power cut in any area.

Block Diagram

Our area of work would include three divisions: (i) Intermediate Substation, (ii) Distributing Substation (33KV/440 V), (iii) Home.

Since the Intermediate Substation would know when the load on the electricity grid is exceeding the limit, hence, we would put NodeMCU (Master) here. All the appliances in the Sub-Station would have NodeMCU (Mid-Level). Master and Mid-Level NodeMCU would communicate using MQTT (Message Queue Telemetry Transport) protocol where the master would publish signals under a unique event name and all the Mid-Level devices would subscribe to it. This is also shown in Fig. 21.1.

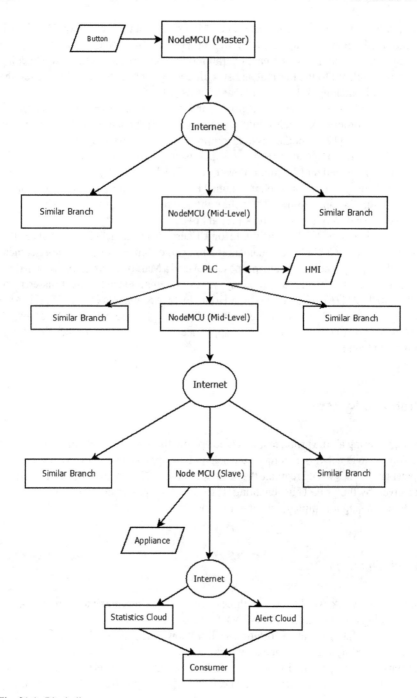

Fig. 21.1 Block diagram

Hence, when Master would declare excessive load acting on the grid, this data would be published to the cloud as an event. All Mid-Level devices subscribed to this event would receive the warning and would start regulating the activities of the PLC under them.

At Distributing Substation level, the Node MCU (Mid-Level) will send different feeds to Arduino-interfaced Relay module which further transmit the feed to the PLC input port.

We will receive a 3.3 V DC output from the output port of PLC which will act input to a second node MCU (Mid-Level). This one will again broadcast signals under a unique event name which will be received by NodeMCU (slaves) at the Home level.

At Home level, NodeMCU (slaves) will regulate the activities of appliances under them. Before regulation, an alert would be sent to the consumer. These NodeMCU (slaves) will also collect power consumption data of their appliances which can be made available to the consumer on-demand.

Hardware Implementation

We sectionalize the construction into three parts: (A) Intermediate Substation, (B) Distributing Substation (33KV/440 V), (C) Feeders.

Intermediate Substation

In the Intermediate Substation control room, the in-charge will be monitoring the load condition across the grid. On observing high load, he would press the button- 1 which in turn makes the NodeMCU (Master) publish an alert on the cloud platform. Do note, the NodeMCU (Master) needs to be connected to the Internet. When the load on the grid normalizes, the in-charge will press button-2 which in turn makes the NodeMCU (Master) publish a relax command on the cloud platform.

At Distributing Substation (33 kV/440 V)

The NodeMCU (Mid-Level) will receive a particular load level from the Intermediate Substation. One of the five digital pins will receive data from the cloud such that each pin will represent a different mode of working of Substation.

We assign five corresponding pins or input levels in PLC. These pins in PLC get activated by signals transmitted from NodeMCU at sub-station level boosted by an Arduino–Relay interface from 3.3 to 24 V.

Initially, we divide our locality into 4 divisions, each of which is controlled by a NodeMCU, so there will be four NodeMCU initially each of which controls five

Table 21.1 Load levels with their associated pins and voltage levels

Load level	Pins assigned on node MCU	Associated voltage (V)
Very high	D0	0.000
High	D1	0.825
Medium	D2	1.650
Low	D5	2.475
Very low	D6	3.300

Fig. 21.2 Final Model of Distributing Substation with our Analytics Web site on Laptop

devices. We divide the voltage into 5 levels varying from 0 to 3.3 V. Refer to Table 21.1 for details regarding the pin configuration, load level and associated voltage.

Output ports are taken from PLC for transferring data from substation level to equipment via [NodeMCU (Mid-Level) to cloud to NodeMCU (Slave)] interface. So, every NodeMCU will be connected to output ports of PLC as:

- NodeMCU1: Y001-Y005
- NodeMCU2: Y006-Y010
- NodeMCU3: Y011-Y015
- NodeMCU4: Y016-Y020.

Note: We have also added an HMI with PLC here, to ease troubleshooting at the Sub-station level (see Fig. 21.2).

Feeders

The NodeMCU (slaves) on receiving their load level would start manipulating the activities of the device under them (see Fig. 21.3). Before a complete power cut-off

Fig. 21.3 Final Model of Feeders and their respective output devices whose operations we manipulate

or any other regulation, an alert email will be sent to the consumer by the NodeMCU (slaves). They will also be collecting power consumption details of every high-load device and uploading them to the cloud platform. Here, informative graphs would be plotted based on the collected data which will be made available to the user on demand.

It is important to note that our system would only work for "Smart Devices" i.e., devices which have a microcontroller already embedded in them and also have the capability to connect to the Internet.

Results and Discussions

The following Table 21.2 shows the observations and results for the proposed system when different commands are given from the Intermediate Substation.

The following Table 21.3 shows how a decrease in voltage affects the temperature of an Air Conditioner.

Table 21.2 Active pins and voltage being given out by different NodeMCUs

NodeMCU/voltage given when different pins are active	D0	D1 (shifts values vertically every minute)	D2 (shifts values vertically every minute)	D5 (shifts values vertically every minute)	D6
NodeMCU 1 (V)	0	0.825	1.65	2.475	3.3
NodeMCU 2 (V)	0	0.825	1.65	3.300	3.3
NodeMCU 3 (V)	0	0.825	3.30	3.300	3.3
NodeMCU 4 (V)	0	3.300	3.30	3.300	3.3

Table 21.3 The voltage being given out by NodeMCU and temperature

NodeMCU output voltage to air conditioner (V)	The temperature of the air conditioner
0.000	Device off
0.825	Fixed at 30 °C
1.650	Between 30 and 27 °C
2.475	Between 30 and 24 °C
3.300	User defined

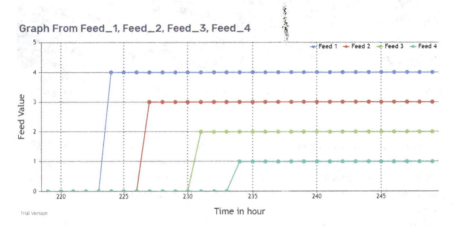

Fig. 21.4 Graph depicting various feed values being given by different feeders

We have also developed a Web site through which an operator can monitor the workings of distributing substation remotely. The graphs below in Figs. 21.4, 21.5 and 21.6 displays various feed levels of different feeders, change in cost per unit electricity with time and change in feed level with change in total load on the grid, respectively.

Conclusion

In this work, a smart grid by creating an IoT network of every heavy-load appliance has been implemented so that they can communicate with the intermediate substation directly. As the load on the grid changes, variation in the level of operation of every high-load device is witnessed. The status of this system can be seen either on the HMI situated at the distributing substation or on the dedicated Web site. By employing this work, big power corporations can save lots of electricity and reduce power cuts, thus keeping their profit and customers happy. This idea, if implemented on a large scale, can dynamically transform the power production and distribution industry, not only in India but also abroad.

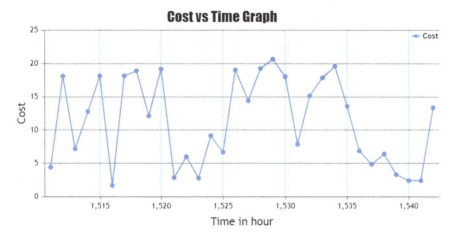

Fig. 21.5 Graph depicting varying cost value with respect to time. The variability will depend on total load faced by the grid

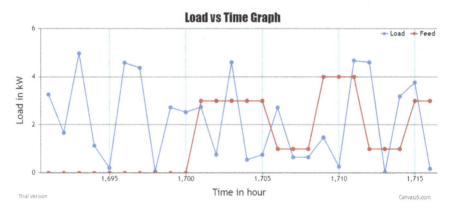

Fig. 21.6 Graph depicting the change in Feed value with respect to the change in Load faced by the grid

References

1. Meng K, Dong ZY, Xu Z, Hill DJ, Zheng Y (2017) Coordinated dispatch of virtual energy storage systems in smart dstribution networks for loading management. IEEE Trans Syst Man Cybern Syst 49(4):776–786
2. Collotta M, Pau G (2017) An innovative approach for forcasting of energy requirements to improve a smart home management system based on BLE. IEEE Trans Green Commun Netw 1(1):112–120
3. Arun SL, Selvan MP (2018) Intelligent residential energy management system for dynamic demand response in smart buildings. IEEE Syst J 12(2):1329–1340
4. Minoli D, Sohraby K, Occhioggrosso B (2017) IoT considerations, requirements, and architectures for smart buildings—energy optimization and next-generation building management systems. IEEE Internet Things J 4(1):269–283

5. Kumar Nunna HSVS, Battula S, Doolla S (2018) Energy management in smart distribution systems with vehicle-to-grid integrated microgrids. IEEE Trans Smart Grid 9(2):4004–4016
6. Kumar N, Zeadally S, Misra SC (2016) Mobile cloud networking for efficient energy management in smart grid cyber-physical systems. IEEE Wirel Commun 23(5):100–108
7. Wang X, Zhang Y, Chen T, Giannakis GB (2016) Dynamic energy management for smart-grid-powered coordinated multipoint systems. IEEE J Sel Areas Commun 34(5):1348–1359
8. Venayagamoorthy GK, Sharma RK, Gautam PK, Ahmadi A (2016) Dynamic energy management system for a smart microgrid. IEEE Trans Neural Netw Learn Syst 27(8):1643–1656
9. Mondal A, Misra S, Obaidat MS (2017) Distributed home energy management system with storage in smart grid using game theory. IEEE Syst J 11(3):1857–1866
10. Wang K, Ouyang Z, Krishnan R, Lei S, Lei H (2015) A game theory-based energy management system using price elasticity for smart grids. IEEE Trans Industr Inf 11(6):1607–1616
11. Bracco S, Brignone M, Delfino F, Procopio R (2017) An energy management system for the savona campus smart polygeneration microgrid. IEEE Syst J 11(3):1799–1809
12. Han J, Choi CS, Park WK, Lee I, Kim S-H (2014a) Smart home energy management system including renewable energy based on ZigBee and PLC. IEEE Trans Consum Electron 60(2):198–202
13. Han J, Choi CS, Park WK, Lee I, Kim S-H (2014b) PLC-based photovoltaic system management for smart home energy management system. IEEE Trans Consum Electron 60(2):184–189
14. Hsieh HI, Tsai CY, Hsieh GC (2014) Photovoltaic burp charge system on energy-saving configuration by smart charge management. IEEE Trans Power Electron 29(4):1777–1790
15. Hu Q, Li F (2013) Hardware design of smart home energy management system with dynamic price response. IEEE Trans Smart Grid 4(4):1878–1887
16. Xu Z, Jia QS, Guan X, Shen J (2013) Smart management of multiple energy systems in automotive painting shop. IEEE Trans Autom Sci Eng 10(3):603–614
17. Jo HC, Kim S, Joo SK (2013) Smart heating and air conditioning scheduling method incorporating customer convenience for home energy management system. IEEE Trans Consum Electron 59(2):316–322

Chapter 22
Fuzzy Model for Efficiency Estimation of Solar PV Based Hydrogen Generation Electrolyser

Sandhya Prajapati and Eugene Fernandez

Introduction

Electrolysis has become an important method for extracting hydrogen from water. This proven technology is based on the production of oxygen and hydrogen with the direct current flow through the water for splitting it. The electrolysis can be driven by renewable energy sources like solar and wind, both have now become the attractive technology to extract a large amount of hydrogen production without causing any emission unlike other resources based on fossil fuel or nuclear [1–11].

The hydrogen thus obtained using electrolysis purest form of hydrogen, after drying out the produced hydrogen the impurities have been removed in the gaseous form. The most suitable form of hydrogen for low-temperature fuel cell is produced from electrolysis, such purity levels offer a great advantage against production by other processes that may require fossil fuels. The processing cost ranges between US$ 3.5–16/kg of hydrogen, depending on the electrolysis system size [11].

The efficiency of hydrogen production lies in the low range of 2–6% with renewable energy sources. This necessitates a large space area of the PV array installation. The optimal load matching of the voltage and solar power is necessary for the higher efficiencies of up to 12%. Figure 22.1 shows the basic layout of a typical solar PV electrolysis scheme [12]. The coupling strategies may be direct, with a dc-dc converter or a dc-ac-dc converter.

S. Prajapati (✉) · E. Fernandez
Department of Electrical Engineering, Indian Institute of Technology Roorkee, Roorkee 247667, Uttarakhand, India
e-mail: sandhyaprajapati92@gmail.com

© The Editor(s) (if applicable) and The Author(s), under exclusive license to Springer Nature Singapore Pte Ltd. 2021
A. K. Singh and M. Tripathy (eds.), *Control Applications in Modern Power System*, Lecture Notes in Electrical Engineering 710, https://doi.org/10.1007/978-981-15-8815-0_22

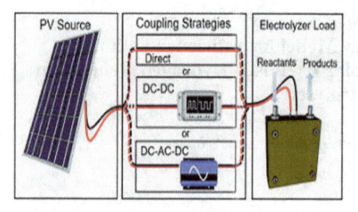

Fig. 22.1 Basic layout of a typical solar PV electrolysis scheme [12]

The paper develops a fuzzy set-based model for simulating the efficiency of a solar PV based water electrolysis unit which employs direct coupling of the PV array and electrolyser. The model is developed from experimental data provided in [13].

Solar Hydrogen System

In a direct connected solar PV-electrolyser, there are only two essential parts:

(a) The solar PV module array assembly.
(b) The water electrolyser.

Each component is discussed below:

Solar PV Module Array Assembly

The solar PV array can be made up of a series–parallel combination of smaller modules to give the required terminal voltage and current needed by the electrolyser unit. The solar PV characteristics represented by V–I curve are nonlinear in nature and hence the controlling equations are given as:

$$V = V_{th} \times \log\left(\frac{I_{ph} - I}{I_s} + 1\right) - R_s I$$

$$I_{ph} = I_s \left(\exp\left(\frac{V_{oc}}{V_{th}}\right) - 1\right) \tag{22.1}$$

The voltage and saturation current of the PV module is given as the coordination of an optimum powerpoint (I_{op}, V_{op}):

$$V_{th} = \left(\frac{V_{op} + R_s I_{op} - V_{oc}}{\ln\left(1 - \frac{I_{op}}{I_{sc}}\right)} \right)$$

$$I_s = \frac{I_{sc}}{\exp\left(\frac{V_{oc}}{V_{th}}\right) - \exp\left(\frac{R_s I_{sc}}{V_{th}}\right)} \tag{22.2}$$

where

R_s is the series resistance of the module (Ω), I_{ph} and V_{th} = are the photocurrent (A) and thermal voltage (V), saturation current is I_s in (A), V_{op} and I_{op} are the optimum voltage and current of the module, V_{oc} and I_{sc} are the open-circuit voltage and short circuit current of the module in volt and amp, respectively.

Water Electrolyser

This is a mono cell consisting of two electrodes—cathode and anode immersed in the electrolyte. It is powered by the solar PV array. The electrolyte consists of an aqueous alkaline solution (KOH solution) with a concentration of 27%. The equation represents the hydrogen production from the water with electrolytic:

$$H_2O(l) + \text{Electric Energy} \rightarrow H_2(g) + O_2(g) \tag{22.3}$$

The anode and cathode reactions are as follows:

Anode: $2OH(aq) \rightarrow \frac{1}{2}O_2(g) + H_2O(l) + 2e$ $\quad E_{\text{rev\_an.25°}} = 1.299\ V$
Cathod: $2H_2O(l) + 2e^- \rightarrow H_2(g) + 2OH^-(aq)$ $\quad E_{\text{rev\_ca.25°}} = 0\ V$ $\quad$ (22.4)

The electrolyser model is represented with a temperature-dependent voltage source with nonlinear resistor in series that depends upon current and temperature. The current injected in electrolyser is proportional to hydrogen production and is recorded for specific times during the day. The hydrogen production is proportional to the *adaptive efficiency* for the solar PV-electrolyser system and is given by Eq. (22.5). Here V_{dc} and I_{dc} are the direct-coupled voltage and current, respectively.

$$\eta_{ae} = \left(\frac{V_{dc} \times I_{dc}}{V_{op} \times I_{op}} \right) \qquad (22.5)$$

where:
V_{dc} = directly coupling voltage, V
I_{dc} = directly coupling current, A.

The Fuzzy Model for the System

The fuzzy logic is based on the decision making depending upon the non-numeric information. The capability of fuzzy logic is widespread for various purposes like interpretation, data utilisation, representation, recognition and manipulation. The fuzzy logic utilises the non-numeric information for rule and fact for fuzzy set formation instead of the numeric values used in mathematics. To assign the numeric values the fuzzification is used, which assigns the degree of membership for the respective fuzzy set in the range of [0 1] [14]. A fuzzy set model was developed in MATLAB® Fuzzy toolbox environment for the study of the efficiency of the solar PV-electrolyser. The fuzzy model uses two parameters—the radiation and temperature as input parameters and the electrolyser efficiency as the output parameter.

The membership functions use the asymmetrical triangular membership and these are shown in Fig. 22.2a–c: The temperature range is split into different range categories T_l (temp—low), T_m (temp—medium) and T_h(temp—high) Likewise, the radiation levels ranges are R_l (radiation—low), R_{lm}(radiation—low medium), R_m (radiation—medium), R_h (radiation—high) The output efficiency is categorised into five categories P_l (efficiency—low), P_{ll} (efficiency—low little), P_{ml} (efficiency—medium little), P_m(efficiency—medium) and P_h (efficiency—high) (Fig. 22.3).

The SOM (small of maxima) defuzzification method has been used.

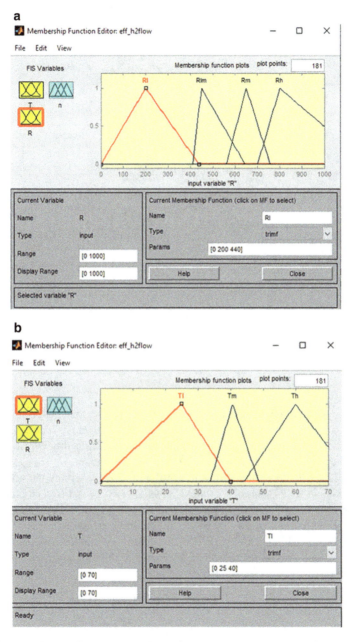

Fig. 22.2 a Membership functions for *Solar Radiation* (input). b Membership functions *Temperature* (input). c Membership functions *Efficiency* (output)

Fig. 22.2 (continued)

Results

Forty-eight observational sets of readings were taken for the model development from the experimental results of [13]. Figure 22.4 shows the results obtained. It may be seen from the results that a very good fit has been obtained between the actual data and the simulated data points using the developed model.

Figure 22.5a–d shows the application of the model to examine the effect of the magnitude of the solar radiation on the system efficiency. Four different temperatures have been tested 10 °C, 20 °C (low temperatures), 40 °C (medium temperature) and 60 °C (high temperature). The variations inefficiency are shown. It is observed at up to radiation levels of 400–450 W/m^2, the efficiency remains virtually constant. However, beyond this range, there is an increase in the efficiency for higher solar radiation levels.

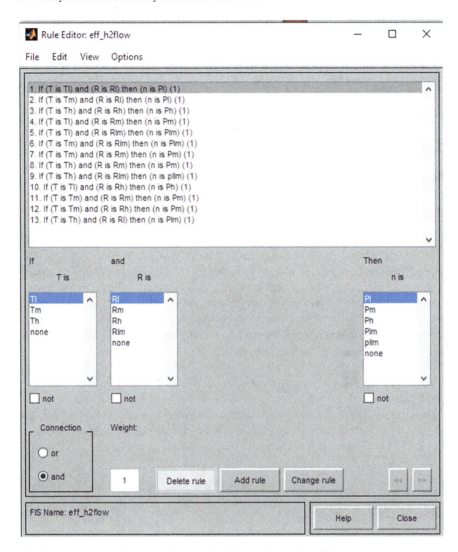

Fig. 22.3 gives a snap shot of the developed *If-Then* rules by the toolbox

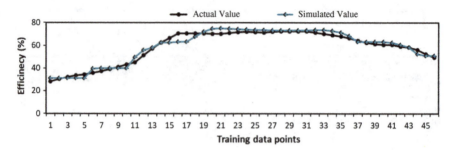

Fig. 22.4 Results of the simulation using the model

Conclusion

The paper presents a fuzzy model developed for predicting the efficiency of a solar PV-hydrogen electrolyser. Two input variables are used for a system of given specifications, namely, solar radiation and temperature. The output is the system efficiency in terms of the power drawn by the electrolyser and the solar PV input. The scheme uses a solar PV hydrogen production with direct coupling and is useful as a simple scheme in remote or rural areas where technical complexity in hardware is not desired. The results of the simulations based on the model designed from an experimental database shows good accuracy for use in further simulation. As an example, the simulated results for efficiency at different solar radiations and temperatures are shown graphically.

As an extension of this study, the model can be used to predict the hydrogen flow rate and the effects of changes in alkaline (KOH) concentration levels on efficiency. These will be considered for further work using the developed model.

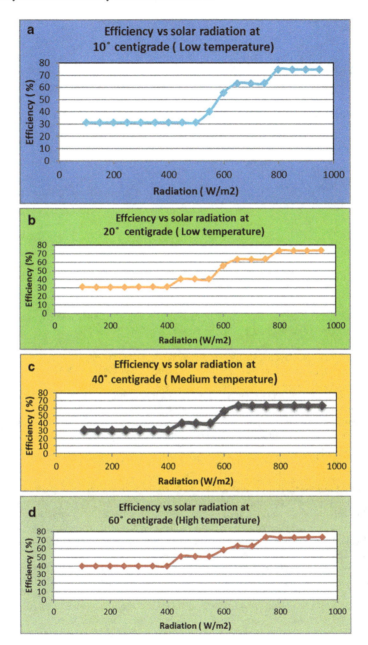

Fig. 22.5 a Efficiency versus solar radiation at 10 °C. b Efficiency versus solar radiation at 20 °C. c Efficiency versus solar radiation at 40 °C. d Efficiency versus solar radiation at 60 °C

Appendix

Specifications of the system used.

Parameters of module	Symbol	Value
Area of each module	A_{mod}	1.003×0.462 m^2
Peak power	P_{max}	55 W
Peak power voltage	V_{max}	17.5 V
Peak power current	I_{max}	3.14 A
Short circuit current	I_{sc}	3.5 A
Open-circuit voltage	V_{oc}	22.2 V

References

1. Balat M (2008) Potential importance of hydrogen as a future solution to environmental and transportation problems. Int J Hydrog Energy 33:4013–4029
2. Plass HJ, Barbir F, Miller HP, Veziroglu TN (1990) Economics of hydrogen as a fuel for surface transportation. Int J Hydrog Energy 15:663–668
3. Granovskii M, Dincer I, Rosen MA (2006) Life cycle assessment of hydrogen fuel cell and gasoline vehicles. Int J Hydrog Energy 31:337–352
4. Granovskii M, Dincer I, Rosen MA (2007) "Exergetic life cycle assessment of hydrogen production from renewables. J Power Sources 167:461–471
5. Elgowainy A, Gaines L, Wang M (2009) Fuel-cycle analysis of early market applications of fuel cells: Forklift propulsion systems and distribute power generation. Int J Hydrog Energy 34:3357–3570
6. Martin KB, Grasman SE (2009) An assessment of wind-hydrogen systems for light duty vehicles. Int J Hydrog Energy 34:6581–6588
7. Leaver JD, Gillingham KT, Leaver LHT (2009) Assessment of primary impacts of a hydrogen economy in New Zealand using UniSyD. Int J Hydrog Energy 34:2855–2865
8. Hajimiragha A, Fowler MW, Canizares CA (2009) Hydrogen economy transition in Ontario-Canada, considering the electricity grid constraints. Int J Hydr Energy 34:5275–5293
9. Stolzenburg K, Tsatsami V, Grubel H (2009) Lessons learned from infrastructure operation in the CUTE project. Int J Hydrog Energy 34:7114–7124
10. Shayegan S, Pearso PJG, Hart D (2009) Hydrogen for buses in London: a scenario analysis of changes over time in refueling infrastructure costs. Int J Hydrog Energy 34:8415–8427
11. Ursu A, Gandıa LM, Sanchis P (2012) Hydrogen production from water electrolysis: current status and future trends. In: Proceedings of the IEEE, vol 100, no. 2, February 2012, pp 410–426
12. Sriramagiri GM, Luc W, Jiao F, Ayers K, Dobsona KD, Hegedus SS (2019) Computation and assessment of solar electrolyzer field performance: comparing coupling strategies. Sustain Energy Fuels 3:422–430
13. Djafour A, Matoug M, Bouras H, Bouchekima B, Aida MS, Azoui B (2011) Photovoltaic-assisted alkaline water electrolysis: basic Principles. Int J Hydrogen Energy 36:4117–4124
14. https:// en.wikipedia.org/wiki

Chapter 23
Temperature-Dependent Economical and Technical Aspect of Solar Photo Voltaic Power Plant

Subhash Chandra

Nomenclature

A	Area of land
INOCT	0.8' kW/m^2
G	Radiation 'W/m^2'
T_{NOCT}	47 °C
$T_{a,NOCT}$	20 °C
α	Voltage temperature coefficient
c-Si	Crystalline silicon
m-Silicon	Multicrystalline silicon
a-Si	Amorphous silicon
η_i	Monthly average electrical efficiency
STC	Standard test conditions
T_m Module	Temperature °C
T_a Ambient	Temperature °C
V_{oc}	Open circuit voltage of module 'Volt'
V_m	Maximum power point voltage 'Volt'
β	Power temperature coefficient
η_0	Standard efficiency
η_{el}	Temperature-dependent electrical efficiency

S. Chandra (✉)
Electrical Engineering Department, GLA University, Mathura 281406, India
e-mail: subhash.chandra@gla.ac.in

© The Editor(s) (if applicable) and The Author(s), under exclusive license to Springer Nature Singapore Pte Ltd. 2021
A. K. Singh and M. Tripathy (eds.), *Control Applications in Modern Power System*, Lecture Notes in Electrical Engineering 710,
https://doi.org/10.1007/978-981-15-8815-0_23

Introduction

India is a country where the government has taken good initiatives to increase renewable energy. They are focusing especially on solar-based electricity, i.e., out of 175 GW by 2022, 100 GW will be produced by solar-based power plants. They may be rooftop, building integrated, ground-mounted or solar thermal based. As per available data in February 2020, India has 34 GW installed capacity w.r.t. to the target of 100 GW. India is a country with a very high population density therefore land scarcity is always a problem in the present scenario. To generate power from solar PhotoVoltaic (PV), land is a very essential prime requirement [1, 2]. In the era of progress in solar cell material technology, solar cells and solar PV modules are manufactured from various types of technologies like c-Si, mc-si, amorphous si, thin-film, and Heterojunction technology [3]. Although c-si, mc-si are the two most mature technology and used widely, however, they suffer from the problem of power loss in high-temperature zones. India is a country with several kinds of temperature zones hence to install the PV based projects with conventional module technology leads to economical and technical barriers [4]. It seems less significant to compare the power drop of PV modules of small ratings but since nowadays multiples modules are connected to achieve a large amount of power so in the strings meaningful voltage drop is observed. This voltage drop makes the system less stable and more sensitive [5, 6]. Researchers are continuously searching the option for more efficient and less temperature-sensitive materials for making solar cells. In this sequence various types of materials based like amorphous si, thin-film exist but suffer from low-efficiency problems in contrast to the better performance at high temperatures. The hetrojunction (HIT) technology-based solar cells are also in the developing phase and yet to be commercialized. The rejection or selection of technology should not be a blind procedure, it should be judged on the available natural parameters at the local site [7–10].

To compare the various available technology PV modules, keeping the technical and economical aspects in focus, an analysis is done for a 50 kWp power plant shown in Fig. 23.1. For which required land area and open-circuit voltage drop of a string is estimated in local climate conditions.

State of Art

Electrical efficiency of solar cells is different than the standard efficiency which is indicated by the manufacturer. It is a function of temperature and related by

$$\eta_{el} = \eta_o[1 - \beta(T_m - T_a)] \qquad (23.1)$$

The module temperature is a crucial parameter and estimated by various approaches. The most widely used approach is

Fig. 23.1 String of 50 KW system situated at university roof

$$T_\mathrm{m} = T_\mathrm{a} + \left[\frac{T_\mathrm{NOCT} - T_\mathrm{a,NOCT}}{0.8}\right] \cdot G \qquad (23.2)$$

Efficiency and power temperature coefficient of various materials based on solar cells is shown in Fig. 23.2.

The power temperature coefficient which is the product of voltage and current temperature coefficient is mainly affected by a drop in voltage rather than rise in current. This voltage drop is material and temperature-dependent and its variation produces the instability in DC link voltage and leads to variation in maximum power point voltage. Therefore it is most important to choose the appropriate module technology in order to minimize the oscillation of maximum power point voltage between these two values. Unfortunately, c-Si and mci-Si are widely used due to their high efficiency while thin-film technology is a better solution for this kind of problem because of its low voltage drop at high temperatures. Another side thin-film technology needs a larger area to produce the same amount of power in the same conditions. Thus a technology known as HIT having efficiency near to c-Si and much less voltage

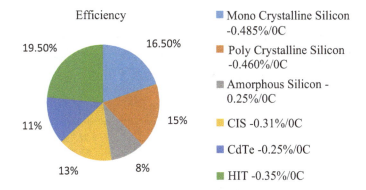

Fig. 23.2 Power temperature coefficient and efficiency and of various materials

Fig. 23.3 Pretit GL 100 data logger

temperature coefficient is promoted for the manufacturing of Solar PV modules. To compare the effect of module temperature on various module technologies in context to voltage and efficiency drop, Pretit data logger of GL 100 series is used for recording the module temperature which is used in 50 kWp system. Measuring conditions are set according to the module like to record the data, sampling is set. Monthly average values of the module temperature are used for the analysis purpose. This instrument is shown in Fig. 23.3. With the help of collected data, monthly electrical efficiencies are calculated for various types of technologies by using Eq. 23.1.

Then to find the required area for 50 kWp system in Mathura, annual average values of electrical efficiency is calculated by

$$\eta_{\text{elavg}} = \sum_{i=1}^{12} \frac{\eta_i}{12} \tag{23.3}$$

Considering 6 h solar radiation daily, the required land area for each PV technology is

$$A = x \times \frac{kW_{\text{stc}}}{G \times \eta_{\text{elavg}}} \tag{23.4}$$

where x ranges from 1.5 to 2 depend upon several factors.

The impact of temperature on open circuit voltage is observed taking 25 °C as a reference.

$$V_{\text{oc max}} \text{ (w.r.t. minimum ambient Temperature)} = V_{\text{oc}} + \Delta V_{\text{oc1}}$$
$$V_{\text{oc min}} \text{ (w.r.t. Maximum module Temperature)} = V_{\text{oc}} - \Delta V_{\text{oc2}}$$

where $\Delta V_{\text{oc1}} = \alpha \cdot (25 - T_{a\,\text{min}})$ and $\Delta V_{\text{oc2}} = \alpha \cdot (T_{m\,\text{max}} - 25)$

Thus V_{oc} will oscillate between two boundary values, i.e., $V_{\text{oc max}}$ and $V_{\text{oc min}}$ with variation in temperature, i.e., $V_{\text{oc min}} < V_m < V_{\text{oc}} < V_{\text{oc max}}$.

% Voltage instability with respect to HIT $= \dfrac{[\Delta V_{\text{ocCry}} - \Delta V_{\text{oc HIT}}] \times 100}{\Delta V_{\text{oc HIT}}}$ (23.5)

Results and Discussions

Monthly average values of global radiation G, ambient temperature T_a, and wind speed, V_s are plotted on both axes. Minimum average 300 W/m^2 radiation and Maximum average 800 W/m^2 radiation and temperature 14–32 °C is in the month of December and May, respectively. Here it is important to note that radiation drops suddenly in the month of July and August due to cloudy conditions while temperature does not. The module temperature ranges from 30 to 60 °C as shown in Fig. 23.4.

Here it important to observe the variation in monthly average electrical efficiency of various PV module technology. The maximum drop in the efficiency is seen in the MAY due to the highest ambient temperature of the local site. In context to this, it is important to see that c-Si has max drop in efficiency while amorphous c-Si gives the least variation. Although electrical efficiency is radiation dependent also so average radiation values are used to calculate monthly average electrical efficiency for each month. Annual variation for c-Si, m-Si, A-Si, thin-film, and HIT technology is 14.78, 13.2, 7.57, 10.41, and 18.03%, respectively as shown in Fig. 23.5. Key points to be noted here are with reference to HIT, c-Si 18%, m-Si 25%, a-Si 58%, and thin film 42.3% is less efficient for the considered case.

The well-proven adverse effect in efficiency and area required for PV installation is shown in Fig. 23.6. The monthly average area required maximum in the month of JANUARY and minimum in the month of MAY for each technology. Although in these technologies, the minimum area is required if PV module are used based on HIT technology provided that other parameters remain the same. The required area for c-Si, m-Si, A-Si, thin-film, and HIT technology is 828, 906, 1621, 1179, and 680 m^2, respectively. Required land area is a crucial parameter for a country

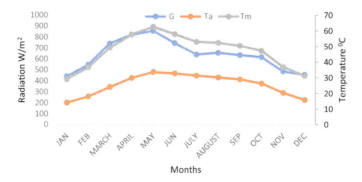

Fig. 23.4 Annual variation in radiation, ambient and module temperature

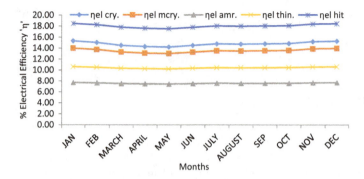

Fig. 23.5 Annual variation in electrical efficiency of various types of module technology

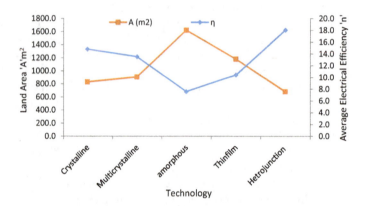

Fig. 23.6 Required land area w.r.t. electrical efficiency for various types of module technology

like India where population density is too high; however, rooftops of government buildings, canals, and non-agricultural land are suitable for the projects. Somewhere it provides passive cooling for buildings hence reduces the electricity consumption indirectly.

The crucial impact of temperature variation does not only affect the efficiency but significantly contributes to voltage drop also as shown in Fig. 23.7. This voltage variation becomes a very effective and key role player in larger rating systems. The technical parameters estimated by neglecting this effect may lead to the failure of appliances connected with the system. Moreover the maximum power point voltage will oscillate more if the temperature effect is not taken care of properly. With respect to V_{oc} at 25 °C (which is indicated by the manufacturer), maximum difference between V_{oc} max and V_{oc} min is observed in c-Si technology therefore it provides much instability in system voltage while the least difference is obtained for thin-film so it seems better suitable. But another side thin-film technology needs a larger area. With respect to HIT, c-Si 38%, and m-Si 31% provides more instability in open-circuit voltage while, a-Si 28% and thin film 28% provide less voltage instability for the

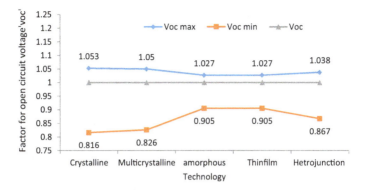

Fig. 23.7 Boundary values of V_{oc} for maximum module and minimum ambient temperature

considered case. Hence a PV module based on HIT gives the better trade-off between area and voltage instability, are a good option.

Conclusion and Future Scope

Above-detailed analysis carried for a 50 kWp system, which is installed on the roof and m-Si technology-based PV modules is utilized for the purpose. In the local climate conditions, the behavior of available techniques is studied and analyzed which concludes that large adverse effects between area and voltage instability exist for c-Si and a-Si techniques. The study suggests that for high-temperature zones c-Si, m-Si are less suitable due to their high power temperature coefficient. Another side choosing low-temperature coefficient material based a-Si and thin film are also not preferable due to their less efficiency. Hence a PV module based on HIT gives better trade-off between area and voltage instability, are a good option.

In continuation of the above work, PV modules of different techniques can be placed in outdoor conditions and their module temperature can be noted. By this process, one can collect the data of temperature from different modules that can further be used for analysis. This will conclude the practice area required for power plant and voltage drop of the string.

References

1. http://mnre.gov.in/file-manager/grid-solar/100000MW-Grid-Connected-Solar-Power-Projects-by-2021-22.pdf
2. Romero-Fiances I et al (2019) Analysis of the performance of various pv module technologies in Peru. Energies 12(1):186

3. Chandra S et al (2020) Material and temperature dependent performance parameters of solar PV system in local climate conditions. Mater Today Proc
4. De Prada-Gil M et al (2016) Technical and economic comparison of various electrical collection grid configurations for large photovoltaic power plants. IET Renew Power Gener 11(3):226–236
5. Edalati S, Ameri M, Iranmanesh M (2015) Comparative performance investigation of mono-and poly-crystalline silicon photovoltaic modules for use in grid-connected photovoltaic systems in dry climates. Appl Energy 160:255–265
6. Chandra S, Agrawal S, Chauhan DS (2018) Effect of ambient temperature and wind speed on performance ratio of polycrystalline solar photovoltaic module: an experimental analysis. Int Energy J 18(2)
7. Ye J Y et al (2014) Effect of solar spectrum on the performance of various thin-film PV module technologies in tropical Singapore. IEEE J Photovolt 4(5):1268–1274
8. Chandra S, Agrawal S, Chauhan DS (2018) Soft computing based approach to evaluate the performance of solar PV module considering wind effect in laboratory condition. Energy Reports 4:252–259
9. Abdallah A et al (2016) Performance of silicon heterojunction photovoltaic modules in Qatar climatic conditions. Renew Energy 97:860–865
10. Chandra S et al (2017) Experimental investigation of optimum wind speed for material dependent temperature loss compensation in PV modules. In: 2017 3rd international conference on condition assessment techniques in electrical systems (CATCON). IEEE

Chapter 24
An Efficient Optimization Approach for Coordination of Network Reconfiguration and PV Generation on Performance Improvement of Distribution System

Sachin Sharma, Khaleequr Rehman Niazi, Kusum Verma, and Tanuj Rawat

Introduction

The growing needs of electricity demand and significant depletion of fossil fuel, there has been a keen interest in renewable energy generation over the previous decade. Therefore, the accommodation of distributed generation (DG) into the distribution system is very popular among the power researchers. The DG's placement in optimal positions and with optimal sizes will offer numerous benefits to the power system, such as line load reduction, active power loss and, mitigation of reactive power requirements and enhancement of the voltage profile. Several researchers have suggested different optimization techniques to solve the above problems such as artificial intelligence, and hybrid intelligent techniques [1]. In the past literature, the main objective fulfilment is the minimization of network loss [2]. Moreover, the many regional power sectors are suffering from the high power delivery loses and that objective is somewhat directly associated with the economy. For example, in the Indian power sector, the major challenge is of high power loss that is approximately 23% in 2018. Consequently, India's finance commission estimated that power purchase costs increased from about ₹680 billion to around ₹1160 billion, which affects all stakeholders of power sectors [3]. Therefore, DGs accommodation problem is developed by using analytical [4], meta-heuristic [5–9], and heuristic methods [10].

In addition to the beneficial effects of incorporating optimal DGs into a power system, network reconfiguration can also be seen as another solution to reducing network losses in the distribution system. The system reconfiguration can be described as a technique that manages the open/close positions of tie switches and

---

S. Sharma (✉) · K. R. Niazi · K. Verma · T. Rawat
Malaviya National Institute of Technology, Jaipur, India
e-mail: sachineesharma@gmail.com

sectionalizes switches to determine the best optimal configuration of the system. The optimal configuration is subject to the considered objective functions and constraints. However, distribution networks are typically mesh-structured but run in a radial configuration to efficiently reduce the network faults. Therefore, the objective of network reconfiguring is to discover an optimal configuration while maintaining the radial nature of the distribution network that optimizes certain objective functions. The finding of a large number of switching combinations while maintaining the radiality of distribution networks makes the reconfiguration problem a non-differentiable and highly complex optimization problem. In the recent past literature, numerous meta-heuristic optimization techniques were developed to address the issue of network reconfiguration. In 1975, Merlin and Back carried out the first analysis of network reconfiguration for the minimization of active power loss [10]. Over the previous years, numerous studies have been devoted to resolving the network reconfiguration problem by adding certain modifications in meta-heuristic techniques and to investigate new optimization techniques for optimizing various objectives such as system stability indices, voltage profile, etc. The literature on network reconfiguration shows several optimization techniques to solve the various objective problem such as GA [11], PSO [12], ACO [13], tabu search algorithm [14]. The reconfiguration of the distribution network and the optimum integration of the DGs are generally studied separately. However, the combination of these two problems together adds benefits to the whole distribution network. In Ref. [13], authors address the problem of minimizing the network loss, improving the balance of the feeder loading balance, and enhancing the system's voltage profile using ant colony optimization (ACO). In [15], the authors solved the same problem with the bang-big crunch optimization algorithm.

In this paper, dynamic network reconfiguration and optimal accommodation of multiple PVs are investigated by new physical phenomenon based optimization techniques. The water evaporation optimization (WEO) algorithm is developed to solve the objective of minimize the network losses of complete one day.

Problem Formulation

The optimum arrangement of reconfiguration of complex network and multiple PVs for the distribution system is designed to reduce energy losses. In this paper, firstly, the size and location of multiple PVs are optimally determined and then with that optimal size and location, the optimal configuration of the network are determined to reduce the energy loss and minimize the node voltage deviations. Therefore, the network loss minimization of complete one day is mathematically expressed as:

$$\min(F) = \sum_{T_m=1}^{24} P_L^{T_m} \tag{1}$$

where, $P_L^{T_m}$ symbolize the active power loss of the network for a time T_m period. The network loss is calculating with the help of Eq. (2) and is taken from [7].

$$P_L^{T_m} = \sum_{i=1}^{N_a} \sum_{j=1}^{N_a} \alpha_{ij}^{T_m} \left(P_i^{T_m} P_j^{T_m} + Q_i^{T_m} Q_j^{T_m}\right) + \beta_{ij}^{T_m} \left(Q_i^{T_m} P_j^{T_m} - P_i^{T_m} Q_j^{T_m}\right) \forall T_m \quad (2)$$

$$\alpha_{ij}^{T_m} = r_{ij} \cos\left(\delta_i^{T_m} - \delta_j^{T_m}\right) / V_i^{T_m} V_j^{T_m} \quad (3)$$

$$\beta_{ij}^{T_m} = r_{ij} \sin(\delta_i^{T_m} - \delta_j^{T_m}) / V_i^{T_m} V_j^{T_m} \quad (4)$$

where, $V_i^{T_m}, V_j^{T_m}, P_i^{T_m}, P_j^{T_m}, Q_i^{T_m}, Q_j^{T_m}, r_{ij}$ and $\delta_i^{T_m}$ represents the voltage of respective nodes, network real and reactive power, the branch resistance and angle of ith and jth node for the T_m time period respectively.

Constraints:

$$I_{f\ell,ij}^{T_m} \le I_{a,ij}^{mx} \quad \forall T_m, i, j \quad (5)$$

$$P_i^{T_m} = V_i^{T_m} \sum_{j=1}^{N_a} V_j^{T_m} Y_{ij} \cos\left(\theta_{ij} + \delta_j^{T_m} - \delta_i^{T_m}\right) \quad \forall T_m, i \quad (6)$$

$$Q_i^{T_m} = -V_i^{T_m} \sum_{j=1}^{N_a} V_j^{T_m} Y_{ij} \sin\left(\theta_{ij} + \delta_j^{T_m} - \delta_i^{T_m}\right) \forall T_m, i \quad (7)$$

$$V^{mn} < V < V^{mx} \quad (8)$$

where, limits of current flow through feeders are presented in Eq. (5) [7]. The system power balance is presented by Eqs. (6) and (7). Equation (8) represents the maximum and minimum limit of node voltage.

Here, $I_{f\ell,ij}^{T_m}, I_{a,ij}^{mx}$ are the flow of real current through branch and extreme limit of current through the line between nodes ith and jth.

Network Reconfiguration

In this paper, to establish a feasible individual with the modification of the infeasible individuals, the theory of graph-based rule has been used and is taken from [16]. Therefore, the terminology used in the rule-based radiality check constraints of network reconfiguration is expressed as follows:

Loop vector (LV) is the combination of switches in a distribution network graph that constitute a closed path.

Common branch vectors (CBV) are the combination of switches common to any of the LV of a distribution network graph.

Prohibited group vector (PGV) is the combination of branch vectors from which, if one switch is opened, then internal DNG nodes are insulated.

Therefore, to check the radial constraints following rules are used and in ref [16], these rules are explained in more detail.

Rule 1: Every candidate switch will have to correspond to their respective LV.

Rule 2: Choose only a single switch from one CBV.

Rule 3: All the elements of the CBV of the PGV cannot be used to form an entity at the same time.

To retain diversity at each point of the evolution process, the individuals are created under the guidance of these rules by the random selection of n numbers of individual switches to confirm the radial configuration. Rule 1 and 2 prevent the islanding of outer and inner nodes of the distribution network graph. Rule 3 is used to protect the islanding of the principal interiors node of distribution network graphs. Therefore, these rules ensure that individuals follow the feasible radial topologies. Table 1 shows all the vectors based on the above discussion.

Table 1 Information on CBV, PGV, and LV	Common branch vectors (C)	Prohibited group vector (P)	Loop vectors (L)
	C12: (33)	P5: (C14, C15, C45)	L1: (2, 3, 4, 5, 6, 7, 33, 20, 19, 18)
	C23: (9, 10, 11)	P7: (C12, C15, C25)	L2: (8, 9, 10, 11, 35, 21, 33)
	C14: (3, 4, 5)	P8: (C23, C25, C35)	L3: (9, 10, 11, 12, 13, 14, 34)
	C35: (34)	P57: (C12, C14, C25, C45)	L4: (22, 23, 24, 37, 28, 27, 26, 25, 5, 4, 3)
	C25: (8)	P78: (C12, C15 C23, C35)	L5: (25, 26, 27, 28, 29, 30, 31, 32, 36, 17, 16, 15, 34, 8, 7, 6)
	C15: (6,7)	P578: (C12, C14, C23, C35, C45)	–
	C45: (25, 26, 27, 28)	–	–

Solution Methodology

The main aim of the paper is to investigate the new physical phenomenon based WEO algorithm to determine the dynamic optimal configuration of the network with multiple PVs accommodation problem. The WEO initiates the evaporation of a small number of water molecules applied to the solid surface of varying wettability that can be analyzed through simulation of molecular dynamics [17]. The molecules of water and solid surfaces with variable wettability are allowed to treat as algorithm individuals and search space of the problem respectively. As per the molecule dynamics, the diminishing surface wettability shifts the movement of water from the monolayer to a phase of sessile droplets. This behavior is consistent with how the actual model varies with others as the algorithm progresses. This reducing wettability will reflect on the objective minimization of the problem. The rate of evaporation of the water molecules is considered with the most effective measures to update the individuals whose pattern of modification is in strong arrangement with the global and local searching ability of optimization algorithm. Therefore, these molecular properties help the WEO to achieve significantly better-searching behavior and a simple structure of the algorithm. The WEO has used two evaporation phases such as monolayer and droplet to search the global and local solutions for the optimization problem and both the phases are expressed as follows:

For the monolayer evaporation phase the existing fitness is scaled to the period of -3.5 to -0.5 in each iteration and reflects the equivalent substrate energy vector as follows:

$$Z^i_{\text{sub,iter}} = \frac{(Z_{\max} - Z_{\min})(\text{fitn}^i_{\text{iter}} - \min(\text{fitn}))}{(\max(\text{fitn}) - \min(\text{fitn}))} + Z_{\min} \qquad (9)$$

$$\text{MOE}^{i,j}_{p,\text{iter}} = \begin{cases} 1 & \text{if rand}^{i,j} < \exp(Z^i_{\text{sub,iter}}) \\ 0 & \text{if rand}^{i,j} \geq \exp(Z^i_{\text{sub,iter}}) \end{cases} \qquad (10)$$

where, Eqs. (9) and (10) shows the energy vector of the substrate $Z^i_{\text{sub,iter}}$ and probability matrix of the monolayer evaporation phase (*MOE*).

For the droplet evaporation phase the existing fitness is scaled to the periods of $-50°$ to $-20°$ in each iteration and reflects the equivalent contact angle vector as follows:

$$\sigma^i_{\text{iter}} = \frac{(\sigma_{\max} - \sigma_{\min})(\text{fitn}^i_{\text{iter}} - \min(\text{fitn}))}{\max(\text{fitn}) - \min(\text{fitn})} + \sigma_{\min} \qquad (11)$$

$$\text{DOE}^{i,j}_{p,\text{iter}} = \begin{cases} 1 & \text{if rand}^{i,j} < J(\sigma^i_{\text{iter}}) \\ 0 & \text{if rand}^{i,j} \geq J(\sigma^i_{\text{iter}}) \end{cases} \qquad (12)$$

$$J(\sigma_{\text{iter}}^i) = \frac{1}{2.6}\left(\frac{2}{3} + \frac{\cos^3(\sigma_{\text{iter}}^i)}{3} - \cos(\sigma_{\text{iter}}^i)\right)^{-\frac{2}{3}} * (1 - \cos(\sigma_{\text{iter}}^i)) \quad (13)$$

where, Eqs. (11) and (12) shows the vector of the contact angle (σ_{iter}) and probability matrix of droplet evaporation phase (DOE), respectively.

For updating the water molecules:

$$G = \text{rand.}\left(\text{WAM}_{\text{iter}}^T[\text{permut1}(i)(j)] - \text{WAM}_{\text{iter}}^T[\text{permut2}(i)(j)]\right) \quad (14)$$

$$\text{WAM}_{\text{iter}+1} = \text{WAM}_{\text{iter}} + G \times \begin{cases} \text{MOE}_{\text{p,iter}} & \text{iter} \leq \text{miter}/2 \\ \text{DOE}_{\text{p,iter}} & \text{iter} > \text{miter}/2 \end{cases} \quad (15)$$

where, $\text{WAM}_{\text{iter}+1}^T$, $\text{WAM}_{\text{iter}}^T$ are the next and current set of water molecules. permut is the permutation function used in Eq. (14).

Following steps are performed to initializing the water molecules for determining the network loss of the considered problem:

I. Read the distribution system data and run the newton-rephson load flow analysis on the radial configuration of the base case for calculating the network loss.
II. Load the optimal size of PVs at the optimal location that is determined in first step.
III. Then in the second step re-configure the distribution network to absorb that PVs with an optimal radial configuration.
IV. Therefore, to determine the radial configuration, choose any single basic loop, and replace the opened switch on either side of any adjacent switch.
V. Verify the radial topology using LV, PGV, and CBV if the radial constraints are broken then choose the next sequence of switches.
VI. If the combination of switches does not violate the radial constraints then determine the network loss using newton-rephson load flow analysis.
VII. Repeat the above steps until the minimum network loss is achieved.

After initializing the water molecules, the monolayer and droplet evaporation phases are applied and then updating the water molecules as per the Eqs. (14) and (15). But, after updating the water molecules again check the radial constraints using the rule-based method.

Result Discussion

In order to validate the problem of network reconfiguration and optimal accommodation of multiple PVs, it is executed on the IEEE radial 33-bus distribution system. The radial distribution system IEEE 33-bus data is drawn from [18]. The per-unit limits of node voltage are considered to be 0.90–1. The following cases are framed and analyzed to show the feasibility of the developed methodology:

Case 1: Feasibility check of the WEO algorithm.

Case 2: Without any modification in the standard 33-bus distribution network.

Case 3: Multiple PVs modules are incorporated into the distribution network.

Case 4: Dynamic network reconfiguration in the presence of multiple PVs are investigated.

For the performance analysis of the WEO algorithm, it is developed for the optimal accommodation of multiple DGs problem. As the nature of all the framed cases is the same that is mixed-integer non-linear non-convex. Therefore, WEO is compared with the available literature on DGs accommodation problem and is shown in Table 2.

Table 2 shows that the searching capability of WEO is significantly better than the available meta-heuristic technique for case 1. Moreover, the case 2 is used for the reference purpose. In this case, no modification is done in the standard 33-bus distribution network. The network loss of the complete one day for case 2 is 1.3566 MW and is shown in Table 3. In case 3, three PVs are optimally integrated into the standard distribution network for complete one day. The network loss of case 3 is 1.0976 MW that is significantly reduced than case 2. However, in case 4, dynamic network reconfiguration is done with the effect of multiple PVs. In this case, the network loss is further reduced to 1.0820 MWh. This case 4 is best among all the

Table 2 Comparison of WEO with the available literature

Literature	Optimal nodes (sizes in MW)	Network loss (MW)
TLBO [19]	10 (0.8246), 24 (1.0311), 31 (0.8862)	0.0755
PSO [20]	8 (1.1543), 13 (0.9708), 32 (0.9438)	0.1066
GA [20]	11 (1.3328), 29 (0.8014), 30 (0.9654)	0.1082
IA [21]	6 (0.900), 12 (0.900), 31 (0.720)	0.0811
ELF [21]	13 (0.900), 24 (0.900), 30 (0.900)	0.0743
WEO	14 (0.832), 24 (0.790), 30 (0.985)	0.0716

Table 3 Impact of all the farmed cases on network losses

Case	Network Losses (MWh)
Case 2	1.3566
Case 3 (PV)	1.0976
Case 4 (PV + NR)	1.0820

Table 4 Optimal configuration of switches for case 4 (dynamic reconfiguration)

Hours	Optimal selection of Switches
1	19, 21, 12, 23, 30
2	18, 21, 12, 23, 17
3	18, 21, 12, 23, 17
4	20, 35, 12, 22, 17
5	20, 35, 12, 22, 17
6	20, 35, 12, 22, 17
7	20, 35, 12, 22, 17
8	18, 21, 12, 23, 17
9	2, 35, 12, 37, 16
10	20, 35, 14, 23, 30
11	2, 21, 12, 37, 30
12	2, 21, 12, 37, 30
13	2, 21, 12, 37, 30
14	19, 21, 12, 23, 30
15	20, 21, 12, 22, 17
16	20, 21, 12, 22, 17
17	18, 21, 13, 37, 16
18	18, 35, 14, 37, 15
19	20, 21, 14, 22, 36
20	18, 21, 14, 37, 15
21	20, 35, 13, 23, 31
22	19, 35, 14, 22, 30
23	20, 35, 12, 23, 29
24	20, 21, 14, 23, 31

considered cases and network loss is reduced to 20.24% from case 2 respectively. The hourly optimal configuration of the distribution network for case 4 is shown in Table 4.

Conclusion

This paper aims to improve the network losses for complete one day of distribution network by optimal accommodation of multiple PVs and dynamic network reconfiguration. To show the importance of network reconfiguration and PV accommodation four multiple cases are framed and analyzed. The hourly reconfiguration of switches for the distribution system is predicted in the presence of varying multiple PVs and load profile. Case 4 is the combination of PVs and network reconfiguration and it achieves the best result among all the framed cases. The results are found to be significantly improved and the use of the newly developed WEO algorithm increases the capability of searching global solutions.

References

1. Singh B, Sharma J (2017) A review on distributed generation planning. Renew Sustain Energy Rev 76:529–544
2. Sharma S, Niazi KR, Verma K, Rawat T (2019) Impact of multiple battery energy storage system strategies on energy loss of active distribution network. Int J Renew Energy Res (IJRER) 9(4):1705–1711
3. Sharma S, Niazi KR, Verma K, Meena NK (2019) Multiple DRPs to maximise the techno-economic benefits of the distribution network. J Eng 18:5240–5244
4. Naik SNG, Khatod DK (2014) Analytical approach for optimal siting and sizing of distributed generation in radial distribution networks. IET Gener Transm Distrib 9(3):209–220
5. Hemmati R, Hooshmand RA (2015) Distribution network expansion planning and DG placement in the presence of uncertainties. Int J Electr Power Energy Syst 73:665–673
6. Sharma S, Niazi KR, Verma K (2019) Bi-level optimization framework for impact analysis of DR on optimal accommodation of PV and BESS in distribution system. Int Trans Electr Energy Syst 29(9):e12062
7. Wang L, Singh C (2008) Reliability-constrained optimum placement of reclosers and distributed generators in distribution networks using an ant colony system algorithm. IEEE Trans Syst Man Cybern Part C (Appl Rev) 38(6):757–764
8. Prakash DB, Lakshminarayana C (2016) Multiple DG placements in distribution system for power loss reduction using PSO Algorithm. Procedia Technol 25:785–792
9. Merlin A (1975) Search for a minimal-loss operating spanning tree configuration for an urban power distribution system. In: Proceedings of 5th PSCC, vol 1
10. Segura S, Romero R (2010) Efficient heuristic algorithm used for optimal capacitor placement in distribution systems. Int J Electr Power Energy Syst 32(1):71–78
11. Mendoza J, Lopez R, Morales D (2006) Minimal loss reconfiguration using genetic algorithms with restricted population and addressed operators: real application. IEEE Trans Power Syst 21(2):948–954
12. Dahalan WM, Mokhlis H (2012) Network reconfiguration for loss reduction with distributed generations using PSO. In: 2012 IEEE international conference on power and energy (PECon), pp 823–828. IEEE
13. Falaghi H, Haghifam MR (2008) Ant colony optimization-based method for placement of sectionalizing switches in distribution networks using a fuzzy multi-objective approach. IEEE Trans Power Deliv 24(1):268–276
14. Ramirez-Rosado IJ, Domínguez-Navarro JA (2006) New multiobjective tabu search algorithm for fuzzy optimal planning of power distribution systems. IEEE Trans Power Syst 21(1):224–233
15. Sedighizadeh M, Esmaili M (2014) Application of the hybrid Big Bang-Big Crunch algorithm to optimal reconfiguration and distributed generation power allocation in distribution systems. Energy 76:920–930
16. Gupta N, Swarnkar A, Niazi KR (2010) Multi-objective reconfiguration of distribution systems using adaptive genetic algorithm in fuzzy framework. IET Gener Transm Distrib 4(12):1288–1298
17. Kaveh A, Bakhshpoori T (2016) Water evaporation optimization: a novel physically inspired optimization algorithm. Comput Struct 167:69–85
18. Baran ME, Wu FF (1989) Network reconfiguration in distribution systems for loss reduction and load balancing. IEEE Power Eng Rev 9(4):101–102

19. Sultana S, Roy PK (2014) Multi-objective quasi-oppositional teaching learning based optimization for optimal location of distributed generator in radial distribution systems. Int J Electr Power Energy Syst 63:534–545
20. Capitanescu F, Ochoa LF, Margossian H (2014) Assessing the potential of network reconfiguration to improve distributed generation hosting capacity in active distribution systems. IEEE Trans Power Syst 30(1):346–356
21. Hung DQ, Mithulananthan N (2011) Multiple distributed generator placement in primary distribution networks for loss reduction. IEEE Trans Industr Electron 60(4):1700–1708

Chapter 25
Enhancement of Hybrid PV-Wind System by Ingenious Neural Network Technique Indeed Noble DVR System

Roopal Pancholi and Sunita Chahar

Introduction

The need for clean and environment-friendly energy makes the system's approach toward the prolific generation of power with pecuniary techniques employed within it. As renewable energies make the desired possibility of power generation as per the demand of the consumers effectively. The (Photovoltaic) PV system technology has covered the vast area of the world as for fulfilling the demand of energy productively and skillfully. Since the efficacious response of the PV system depends on its maximum power tracking capability. And that follows with the technique implemented to it. The comparison was made between various ingenious techniques applied to MPPT of the PV system that was discussed by the authors of [1, 2]. The authors of [3] proposed diverse 19 techniques for maximizing the output of MPPT of a PV system. Every technique comes up with its own pros and cons and that is vastly described by the authors. The PV system makes the generation of power accessible and for this reason, many researchers focused on enhancing the output power of MPPT, with this point of view the author of [4] has realized the Neural Network Technique in the MPPT has got 97% of higher adroit response for the system. For achieving the agile and legitimate response for the system the implementation of Neural Network in a PV system is applied.

Further, in this context, we also have an ultimate RBFN and ENN neural network techniques which have proved its performance as the authors of [5] presented in their configuration. In addition to it, the NN technique facilitates the system's proficiency

---

R. Pancholi · S. Chahar (✉)
Department of Electrical Engineering, Rajasthan Technical University, Kota, India
e-mail: schahar@rtu.ac.in

R. Pancholi
e-mail: roopalpancholi96@gmail.com

© The Editor(s) (if applicable) and The Author(s), under exclusive license to Springer Nature Singapore Pte Ltd. 2021
A. K. Singh and M. Tripathy (eds.), *Control Applications in Modern Power System*, Lecture Notes in Electrical Engineering 710, https://doi.org/10.1007/978-981-15-8815-0_25

to a new level of extent as by detecting the global maximum power point of a PV system [6]. With the accession of technology and making a way for developing nation to developed nation the implicating of renewable energy will make it more dynamic effortlessly. And with this advent speculation, many researchers provides a potent solution for it, as the PV system was installed on the ship and the predictive control approach was made for tracking maximum power for the system as explained by the authors in [7], of their literature. The advantages of the controller are not only limited with the tracking of maximum power in normal condition but also in partial shading conditions occurs with the climatic changes for PV system [8]. The predictive controller approach is extended toward the prediction of the error signal and improving the efficiency of the system as authors of [9] presented. The comprehensive performance analysis of controllers specifies that the performance of the controller is also based on converters applications [10]. In [11] authors described the configuration based on PMSG Wind Turbine. As the system can overcome the distinctive mechanical and economical parameters for gaining the overall efficiency for the system.

The PMSG based Wind Turbine system offers conducive capabilities for fault ride-through in the system and with grid performance which makes the ability of system more decisive and vigorous [12]. By controlling the voltage and pitch angle of wind energy conversion system with PMSG based system can contribute to the increase in proficiency of the proposed network [13]. In the [14] authors suggested the two diverse approaches which are based on analytical and simulation work for obtaining the gain of pitch angle by PI controller in the system and for maintaining the efficacy of the network. We can also make a combination of MPPT with the pitch angle control in WECS as this is now being considered an optimal solution for fulfilling of indispensable requirements of grid-connected with it. And the controlling is made possible by Neural Network technique then it will be a milestone in controlling parameters efficaciously [15]. A high-performance of training of neural network is also found applications with the RBFN technique that can be implemented for controlling purpose of pitch angle and with this for the errorless system the fuzzy inference system gives effectiveness to the system response [16]. The DVR system provides a decisive strategy for controlling voltage and maintaining it with the reference voltage with regulating the harmonics of the system profitably [17].

DVR is an obligatory part of the power electronics circuitry and is upcoming with a new era of mitigation of all errors, due to several disturbances occurs in the network. It can be implemented on both low and medium voltage levels and with this, a convenient means of protection of high power sensitive loads from perturbations are provided [18]. The installation of DVR is made possible between the supply terminals and precarious loads which supply active power to the system during voltage sag and the sag compensation techniques were elaborated by the authors [19]. With the installation of a system in another grid network leads to controlling of parameters that will give a result-oriented performance for the conducive system. As same installing of DVR needs to be controlled or regulated in a systematic manner which is possible with the fuzzy logic control and also with ANFIS controller [20]. The perspicacious DVR can be easily applied to PV-Wind Hybrid system and also plays a vital role in improving the power quality and low voltage ride-through capability of the network efficiently [21].

The applications of Neural Networks has spread its aura widely in renewable energy for enhancement in the output of the system conducive [22]. For making the system more stable and extracting the productive performance of the system, it has become important to include the PV-Wind Hybrid system. And for making an intelligent Hybrid power system the NN method is a premium technique for escalating the response and vitality of the system [23]. Also with this the fulfillment of basic requirements for extraction of maximum output power with maintaining the prudent behavior the NN technique various methods were adopted in the literature [24].

PV-Wind Hybrid System Proposed Techniques

PV System Methodology

The (Photovoltaic) PV system consists of a double diode structure based solar cell. The merits of the proposed cell include diffusion and recombination characteristics of charge carriers with the recombination of space charge zone of the structure. Further, the proposed structure is connected with distinct series and parallel resistances. The PV system has got an indispensable world's recognition in terms of generating the desired amount of power to every type of consumer. The configuration has a potent array type structure as SunPower SPR-X20-445-COM with this 106 parallel strings and 15 series connected modules per string in the respective PV array.

As Fig. 25.1, expresses the double diode equivalent structure of solar cell and the equation which can fully satisfy the proposed configuration with its characteristics is as follows [25]:

$$I = I_{ph} - I_{sa1}\left(e^{q\frac{V+Ir_{se}}{\eta kT}} - 1\right) - I_{sa2}\left(e^{q\frac{V+Ir_{se}}{\eta' kT}} 1\right) - \frac{V + Ir_{se}}{r_p} \quad (25.1)$$

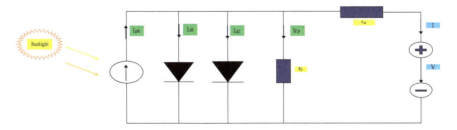

Fig. 25.1 Solar PV two-diode equivalent circuit

where, I is the output of the proposed PV module, q is electric charge (1.6×10^{-9} °C, V is the terminal voltage and k is the Boltzmann constant (1.38×10^{-23} J/K),T is the cell temperature (K) the η is an ideality factor.

Further, taking the assumption that all the cells are commensurate and likewise in operating conditions, similarly all the voltages will be multiplied by the number of series-connected cells, Ns and all the currents of the PV cell will be multiplied by, N_p, V_t is the thermal voltage of the module and I_{ph} is the charge carrier generation of PV cell in the following equations it is well stated:

$$I_{ph,field} = N_p \cdot I_{ph,cell} \tag{25.2}$$

$$V_{t,field} = N_s \cdot V_{t,cell} \tag{25.3}$$

$$r_{p,field} = \frac{N_s}{N_p} \cdot r_{p,cell} \tag{25.4}$$

Figure 25.2 expresses the characteristics, I–V and P–V of the proposed system in which the suitable array type is recommended for the lossless system. In Table 25.1, the description of the parameters with their specified values is mentioned.

With this the Boost converter, a step-up converter is employed in the PV system. This converter gives new heights and magnification to voltage in the system for enhancing the output power for the betterment of the PV system. In the converter, various electronics components provide their profitable and valuable results for output as by filtering the output voltage of the system. Figure 25.3 is a representation

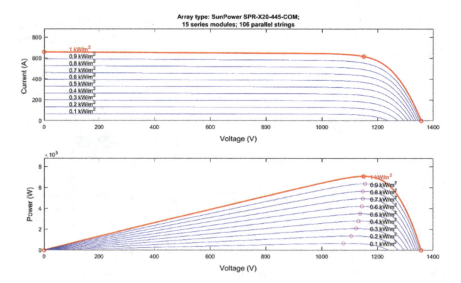

Fig. 25.2 Characteristics of the proposed PV array

Table 25.1 Parameters of PV array

Parameters	Values obtained
Maximum power	444.86 W
Open circuit voltage (V_{oc})	90.5 V
Cells per module (N_{cell})	128
Short-circuit current (I_{sc})	6.21 A
Light generated current (IL)	6.216 A

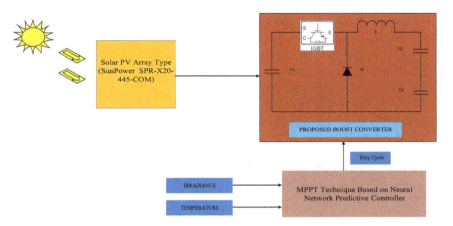

Fig. 25.3 Schematic representation of Boost Converter

of the Boost converter used for the system. Also, the Boost converter needs to convert 730 V of dc voltage to 440 Vac, 50 Hz and that is possible with the 3- level transformerless inverter. For reduction of harmonics the PV system is further connected with low-frequency transformer for stepping up the desired voltage, 440 V–30 kV with the inductive grid filter when transmitting the energy of the system to the grid-connected. With this, the PV system is connected to proposed DVR configuration and then to grid-connected within parallel to load of the system.

The Proposed Novel MPPT's Technique

The productive and profitable PV system has set a remarkable benchmark in the power generation systems. It has not only created an imperative platform for remunerative energy supply for humans but also provides sustainable development for the societies well-being. MPPT is an important part of the PV system. As the system's efficiency solely depends on it, so it has become a prime duty for choosing of the conducive neural network technique for the system's effectual performance. In the Neural Network techniques, the Model Predictive Controller is employed for tracking of the maximum output power for the system to give efficient results. The input given to the is based on irradiance of solar with the adversities comes in the temperature while tracking of power for energy generation and in return will give the pertinent duty cycle for the converter's operation.

The vigorous and quick responsive controllers of Neural Network provides various facilities to the system. The beneficial scheme of NN provides a great role in improving low voltage ride-through capability for the system. The proposed configuration offers the independency on raw material gatherings, it is self-functioning unit also with this it maintains the high efficiency of the plant [26, 27]. Offering the electricity to every type of consumer within its remunerative ways as rural and urban should be facilitated with the evergreen demand fulfilling PV generation. It has given more possibilities with the emerging of NN configuration implementing the system. In several areas, there the availability of renewable sources are present in plenty amount, for such reasons establishing of PV systems will be a great choice [28]. So, the proposed Neural Networks Model Predictive Controller offers various highlights to the system as fast processing of data, accurate prediction of network implemented, qualified and quantitative recognition of the system's performance and credibility in controlling of the plant efficiently and decisively.

The Neural Network Model Predictive Controller uses to predict the future, proposed PV plant performance and that is done in the two steps. The first step of the system is the identification step that determines the plant model of the network then the second step clarifies the controlling process for prediction of system. Figure 25.4 indicates the step 1 configuration of the Model Predictive Controller for systems maximum power tracking operation. Further, in Fig. 25.5 the controller's performance step 2 is performed by predicting the performance of the plant. In the predictive control system the performance of plant in controlling the signal and minimizing it is completely based on the predictions which are further explained in following equation over the specified horizon.

Based on performance and variables taken for controlling purpose of the plant as mentioned in Figs. 25.4 and 25.5 the optimization results follow as:

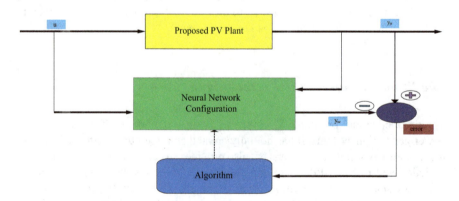

Fig. 25.4 System Identification of NN Controller

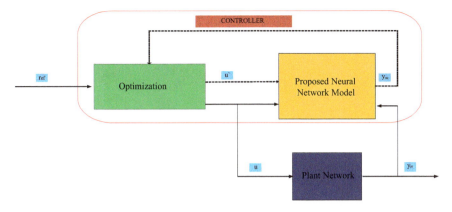

Fig. 25.5 The Controller of system used for prediction

$$J = \sum_{j=n_1}^{n_2}(r_{\text{ef}}(t+j) - y_{\text{mr}}(t+j))^2 + \rho \sum_{j=1}^{n_u}(u'(t+j-1) - u'(t+j-2))^2 \tag{25.5}$$

where, n_1, n_2, n_u are the horizons for tracking the error during system identification process. The u' signifies controlling signal, r_{ef} and y_{mr} is the response desired with that another one is the response of the proposed network model and the ρ value defines the sum of the squares has control increments on the index of performance. The block describes in Fig. 25.5 as u' that minimizes J which gives u as the input to the plant. Since, in NN the data which is needed that has to be of large amount of data for accurate outcomes for training purposes. So, we have used the data used for training the NN Predictive Controller based on irradiance and temperatures data and the obtained duty cycle for converter by MPPT is well described in Table 25.2.

Table 25.2 NN implemented training data	Irradiance (W/m^2)	Temperature (°C)	Duty cycle
	1000	45	0.9999
	900	45	0.9999
	800	40	0.9998
	700	35	0.9998
	600	40	0.9998
	500	35	0.9997
	400	40	0.9997
	300	30	0.9995
	200	20	0.9994
	100	20	0.9993

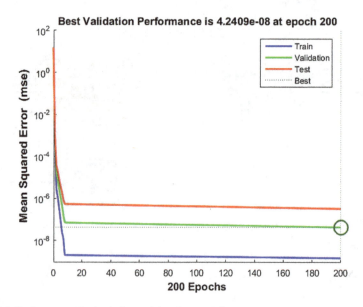

Fig. 25.6 Performance obtained after training the controller

After training of the NN Predictive Controller with required data the performance plot is obtained as shown in Fig. 25.6 with the 200 epochs.

Wind Power Plant

The PMSG based Wind Turbine system is gaining a lot of recognition around the world because of its various merits over another generator implemented in the system. Whereas the proposed configuration, wind system is coming up with many certainties and opportunities of energy generation as a clean source. The main advantage of PMSG Wind turbine that it offers a gearless system, which reduces wear and tear operations and with it, maintenance cost is also reduced. The Wind Turbine offers the maintenance at low-cost as it can be done at due timings [29]. With the elevation in technology the wind energy conversion system provides the LVRT controlling process and inculcates the system performance based on several parameters.

As for strengthening the operation and weakening the debility of system the authors of [30] proposes two techniques as active damping control and bandwidth retuning process. These measures were so taken as the Wind system is considered taken environment-friendly energy source for Modern systems. By controlling the wind turbine various parameters give benefits to the system to last up to the maximum wind speed and also reduces the mechanical loads of the turbine.

The voltage equations for PMSG based Wind Turbine System as reference rotating frame of d-q or Park's transformations, stator equations are described below [31]:

$$\frac{d}{dt}i_{ds} = \frac{1}{L_{ds}}V_{ds} - \frac{R_s}{L_{ds}}i_{ds} + \frac{L_{qs}}{L_{ds}}p\omega_r i_{qs} \qquad (25.6)$$

$$\frac{d}{dt}i_{qs} = \frac{1}{L_{qs}}V_{qs} - \frac{R_s}{L_{qs}}i_{qs} + \frac{L_{ds}}{L_{qs}}p\omega_r i_{ds-\lambda_a}\rho\frac{\omega_r}{L_{qs}} \qquad (25.7)$$

The PMSG rotor equations are as:

$$\frac{d}{dt}i_{dr} = \frac{1}{L_{dr}}V_{dr} - \frac{R_r}{L_{dr}}i_{dr} + \frac{L_{qr}}{L_{dr}}p\omega_s i_{qr} \qquad (25.8)$$

$$\frac{d}{dt}i_{qr} = \frac{1}{L_{qr}}V_{qr} - \frac{R_r}{L_{qr}}i_{qr} + \frac{L_{dr}}{L_{qr}}p\omega_s i_{dr-\lambda_a}\rho\frac{\omega_s}{L_{qr}} \qquad (25.9)$$

The Flux Linkage equations of stator are given as:

$$\Psi_{dst} = L_{dst}i_{dst} - \Psi_m \qquad (25.10)$$

$$\Psi_{qst} = L_{qst}i_{qst} \qquad (25.11)$$

The active and reactive powers of PMSG Based Wind Turbine System are given as:

$$P_s = \frac{3}{2}(V_{ds}i_{ds} + V_{qs}i_{qs}) \qquad (25.12)$$

$$Q_s = \frac{3}{2}(V_{qs}i_{ds} - V_{ds}i_{qs}) \qquad (25.13)$$

where, V, I, Ψ are denoted as voltage, current and flux of the proposed PMSG based Wind Turbine, whereas L and R denotes the respective inductance and resistance of machine. In the respective Fig. 25.7 the PMSG Turbine characteristics are shown which tells the base wind speed and graph is a plot between Turbine speed and Turbine output power. In Table 25.3, the details of Wind Turbine Plant are explained with this we have used the Wind Gust Model in plant.

Neural Network with PI Controller in WT Plant

The PMSG Based Wind Turbine system makes the system realistic and optimistic for every type of consumers. And with the controlling of the pitch angle of Wind Turbine for making the output more consistent and profitable in energy production, whether it is in terms of environment or resources. Due to pertaining such adored and enduring characteristics that have given a vast development in the field of renewable energy. As for maintaining the constant rated power in the system, the controlling

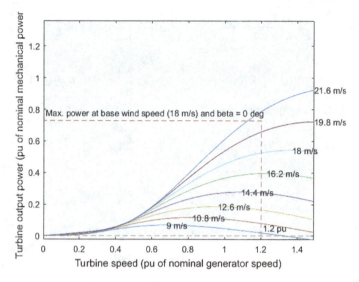

Fig. 25.7 PMSG based wind turbine characteristics

Table 25.7 Ratings of wind turbine system

Parameters	Values
Base wind speed	18 m/s
Output power	700 W
Maximum power at base wind speed	0.73
Gust start time	7 s

of pitch angle is an indispensable part in designing of the Wind turbine and that is possible with PI controller [32]. The controllability and feasibility of the system is an important issue in the case of Wind Turbine systems [33]. As the pitch angle control is possible by the PI controller and also by NN Controller for the system. The pitch angle control in PMSG based Wind Turbine System offers various advantages as it regulated the generated power up to its rated power by adjusting the angles of the blades, it gives complete control of mechanical power, also provides acceleration in the turbine speed with this it has a feature of large settling time for the system [34].

In the proposed system, PMSG Based Wind Turbine we have used NN Reference Controller with PI controller for controlling the pitch angle of the turbine. The pitch angle control is of the great need of Wind Turbine for maintaining the wind speed in a condition where it is running above the rated speed [35]. The PI Controller is coupled with Neural Network Reference Controller for obtaining the desired pitch angle control variations in the output of PMSG Based Wind Turbine system effectively. The effectual and efficacious NN Controllers efficiency is worldwide known as of obtaining the optimal power for a long time and it gives authenticate and meticulous results for the proposed network [36]. In the Model Reference Controller or NN

25 Enhancement of Hybrid PV-Wind System …

Controller, the parallel presentation of Controller is implemented in the system for productively controlling the output of the system with it of zero error approach for the system. The input to the system is presented by r(t) for the proposed network the input of the system is presented as and the output as $u_p(t)$ and $y_{po}(t)$. The proposed plant model can be expressed as follows [37]:

$$G_p(s) = \frac{y_{po}(s)}{u_p(s)} = \frac{k_p}{s^2 + a_p s + b_p} \quad (25.14)$$

where the values of

$$k_p = \frac{V_O}{L_O C_1}, \quad a_p = \frac{1}{R_1 C_1} \text{ and } b_p = \frac{1}{L_O C_1}$$

The Reference Model equation is as stated below:

$$G_m(s) = \frac{y_{m0}(s)}{r(s)} = \frac{k_m}{s^2 + a_m s + b_m} \quad (25.15)$$

Here, a_m and b_m is used in controller for tracking purpose as $y_{m0}(t)$. As shown in Fig. 25.9 the controlling purposed duty cycle expression is described as [38]:

$$u = \theta_o r + \theta_1 \frac{1}{s + \lambda} u_p + \theta_2 \frac{1}{s + \lambda} y_{po} + \theta_3 y_{po} \quad (25.16)$$

In this, $u_f = \frac{1}{s+\lambda} u$ and $y_f = \frac{1}{s+\lambda} y_{po} \cdot \frac{1}{s+\lambda}$
We can also write it as:

$$\theta_o r + \theta_1 u_f + \theta_2 y_f + \theta_3 y_{po} = \theta^T \omega \quad (25.17)$$

Since, ω is defined as $[r\, u_f\, y_f\, y_{po}]^T$, is a regression vector.

In Fig. 25.8 the structure of the proposed controller is shown, in which the desired response of the system is made near to the reference model of controller with required dynamics. The concept of the system is based on an adjustable mechanism for effective controlling of controllers parameters. And with this the key objective of system is to drive the output error to $e_1 = y_{po} - y_{mo}$. As in Fig. 25.9 the NN Reference Controller's performance represented as based on zero error concept.

The dynamic behavior of the system is described as between the pitch demand and, pitch angle. The desired pitch angle for Wind Turbine based on a PI controller can be expressed as follows:

$$\beta_{dm} = K_p e + K_i \int e\, dt \quad (25.18)$$

$$\text{Since, } e = \omega_r - \omega_m \quad (25.19)$$

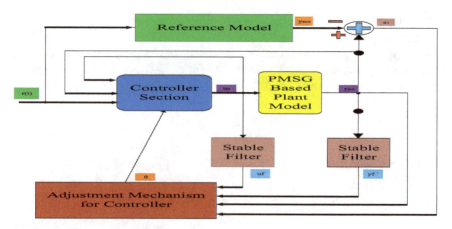

Fig. 25.8 Structure of the NN reference controller

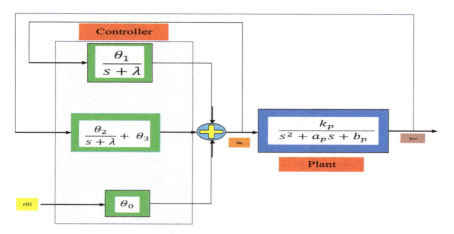

Fig. 25.9 Representation of controllers performance

Here, the output signal from PI Controller is as described in Fig. 25.10 as β_{dm}, ω_r is the reference speed of generator and ω_m is the generator speed. In Fig. 25.10 the diagrammatic presentation of the proposed scheme in the PMSG Based Wind Turbine System as of controlling the pitch angles by the Novel NN Controller and with the PI Controller for enriching the output response of the desired system. Figures 25.11 and 25.12 the proposed controllers performance is shown. The variable pitch angle values were taken to evaluate the performance of the controller as 0°, 5°, 10°, ..., 45°, based on this the controllers performance output of higher efficiency is obtained 0.9999–0.9997. In Fig. 25.11 the PI Controllers response is also mentioned as the values of $K_p = 1.86$ and $K_i = 3.118$ used for controlling the pitch angle of the wind turbine. And in Fig. 25.12 the proposed NN Based Reference Controller performance is validated after importing the necessary data in it with 300 epochs.

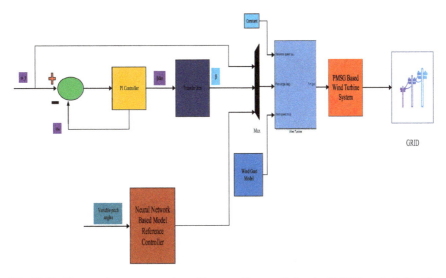

Fig. 25.10 The complete representation of the controllers applied to the PMSG based wind turbine system

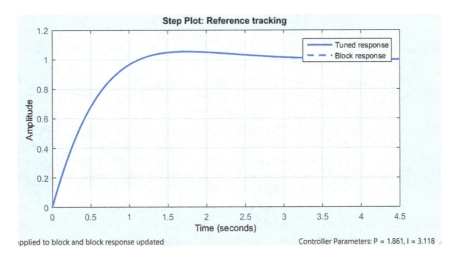

Fig. 25.11 The response plot of PI controller

DVR's Proposed Topology

The Dynamic Voltage Restorers is a type of a FACTS device which has set a revolution in the field of energy. For maintaining the consistency in the network and offering the asset in an innovative way to the system of high speed and high reliability. And making the system sustainably effective. With the concern of power quality issues

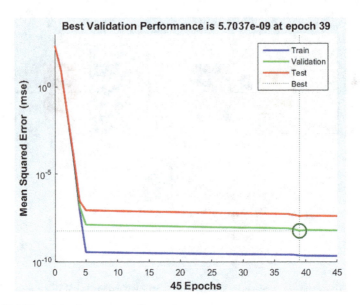

Fig. 25.12 NN model reference controller performance

taken place in the power system the DVR plays an indispensable role in improving the interruptions occurs in the system [39]. The applications of FACTS devices are widely accepted in the field of the corporate world as of planning and management and also it is linked with the people in their daily routine. FACTS technology and DVR within it provides a cost-effective solution for the problems occurred in the generation whether it is related with voltage sags and swells flickering and discrepancies caused by the harmonic distortions occurs in the system [40]. For distinctive types of uncertainties occurred in the system for a short duration of time that can be easily solved with the help of DVR imposing in the system. And with this the SRF technique gives a reliable solution with the Adaptive Neuro-Fuzzy Inference System in the network [41].

The novel controllers applied to the DVR system comes with the same issues and shortcomings of stability in wider operating regions. So, after with such experiences, the authors proposed in their literature Sugeno-type Fuzzy Controller and ANFIS generates acceptable statistics for efficient system output [42]. In the proposed network we had implemented the DVR applications in the PV-Wind Hybrid System for avoiding any transitory disruption and making the system sustainably productive. Further, the DVR is controlled by the peculiar ANFIS controller for making the performance of the DVR more result-oriented performance of the system. In Fig. 25.13 the structure of DVR is presented which consists of several beneficiary and potent components as injection transformer, energy storage device, VSC, control system and harmonic filter accompanied within it.

$$V_{dvr} = V_{load} + Z_{th} I_{load} - V_{th} \tag{25.20}$$

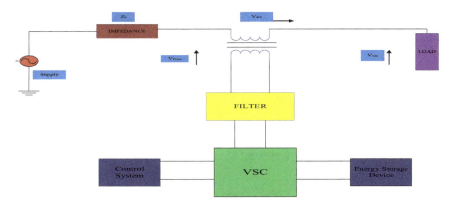

Fig. 25.13 General schematic representation of DVR

Taking load voltage, V_{load} as the reference voltage, the voltage equation can be written as follows:

$$V_{\text{dvr}} = V_{\text{load}}^{\angle 0} + z_{\text{th}}^{\angle \alpha' - \theta} - V_{\text{th}}^{\angle \delta} \tag{25.21}$$

$$\text{Since, } \theta = \tan^{-1}\left(\frac{\theta_l}{p_l}\right) \tag{25.22}$$

ANFIS Proposed Controller Technique

The Adaptive Network-Based Fuzzy Inference System connotatively equal, Adaptive Neural Fuzzy Inference System is based on a feed-forward network. It is also based on hybrid learning rule which helps it to be decisively efficient in handling various types of and also for optimizing the Fuzzy Inference System, FIS it captures the learning abilities of NN to FIS. The learning algorithm accustomed the membership functions of the Sugeno-type Fuzzy Inference System employing the input-output data as in the controllers training performance and purposes [43]. The ANFIS architecture is used to utilize the nonlinear functions of the model with on-line diagnosing the components and predicting the anarchic time series. With the diverse adaptive capabilities, the circles and squares nodes are used in the network parameters. The FIS system is based on Back Propagation NN structure and has five layers are stated below as:

$$\text{Layer 1}: O_i^1 = N_i(x_i) \tag{25.23}$$

where, $i = 1, 2, \ldots, p$.

And the node functions $N_1, N_2, \ldots N_q$ is equivalent to membership function $\mu(x)$, which is used in regular fuzzy systems.

$$\text{Layer 2}: O_i^2 = N_i(x_i) AND N_j(x_j) \tag{25.24}$$

Every node output is presented by Layer 2 and it is the product of all incoming nodes.

$$\text{Layer 3}: O_i^3 = \frac{O_i^2}{\sum_i O_i^2} \tag{25.25}$$

All nodes normalized firing strengths are compared with the addition of all firing strengths and are explained in Layer 3. In the Layer 4 Sugeno-Type FIS is presented as the proposed structure of DVR. It is representing the input and output variables of each IF-THEN rules. The consequent parameters in layer 4 are also explained.

$$\text{Layer 4}: O_i^4 = O_i^3 \sum_{j=1}^{p} P_j x_j + c_j \tag{25.26}$$

The final Layer 5 represents the defuzzification process of the system that is the sum of all inputs as:

$$\text{Layer 5}: O_i^5 = \sum_i O_i^4 \tag{25.27}$$

Layer 5 represents the systems overall equation of the output. As in the Fig. 25.14 Sugeno-type Fuzzy Inference structure is proposed in the DVR of the system with input1 as error and input 2 as d_error and output as dvr_output implemented in it. With 6 membership functions and several 15 rules were realized for making the systems result more profitable. And in Fig. 25.15 the ANFIS controller's structure is conferred that clarifies the rules presentation of Sugeno-type FIS system. It also displays that it can also remunerate source and load side complications competently.

PV-Wind Hybrid Proposed Configuration

With the concern of environmental issues toward a well sustainably developed energy system, the PV-Wind has emerged as one of the best alternatives to fulfill the crucial and imperative demand of society affordably. The combination of PV and Wind system has increased the authenticity of the network and make the system self-dependable with the natural resources efficaciously. The prime asset of renewable energies is that they can lighten up those aloof areas which are still not connected with

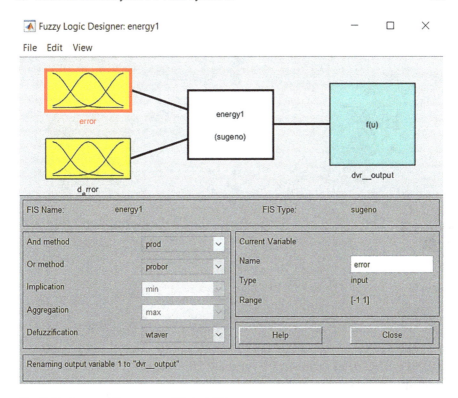

Fig. 25.14 Proposed Sugeno-type FIS for DVR

the main powers systems. And sometimes the PV-Wind system provides compatible support for both of them as due to ambivalence situation occurs in the weather [44].

A stand-alone system is described by the authors of [45], in which for improving the stability and substantiality of the system the NN technique is imposed. The quick response of the system is an obligatory requirement for any healthy network and in that condition, the Neural Network Techniques plays an imperative role in boosting the efficiency of system. The Hybrid systems should satisfy the requirement of loss of power supply probability for system and that is made possible by the decisive ANFIS technique for an adroit system [46].

In this paper, we proposed the PV-Wind hybrid system in which the MPPT of the PV system is controlled through NN based Model Predictive Controller and the PMSG Based Wind Turbine Systems pitch angle is controlled by virtuous NN based Model Reference Controller with PI controller for making a profitable system for different types of consumers. With this, the DVR implemented in the system which is further controlled by skillful ANFIS controller and then the whole network is connected in with the grid and the load. The PV and PMSG Based Wind systems are connected in parallel with the load and the wind system output generated of 400 Vac and connected with the bus and then to the DVR of the system and then to grid.

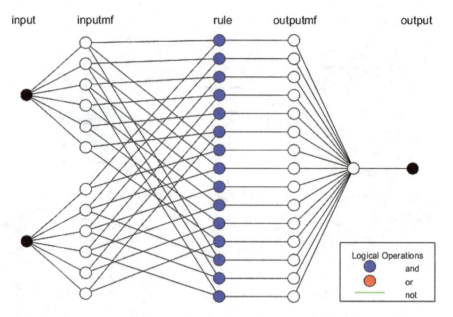

Fig. 25.15 Structure of the ANFIS obtained from the controller

Results Obtained from the Proposed Pv-Wind Hybrid System

PV System MPPT's Technique Outputs

In Figs. 25.16, 25.17 and 25.18, the proposed Neural Network-Based Model Predictive Controllers performance is evaluated based on the input and output given in the controller. As in Fig. 25.16 the whole training network performance is obtained with 200 epochs taken. And in Figs. 25.17 and 25.18 the performance plots generated by the controller. As Fig. 25.19 describes the performance of the PV system after training of the controller and in the output, the constant voltage is obtained. With this in Fig. 25.20 the plot of the active and reactive power are shown without the DVR's contribution in the network. Figure 25.21 shows the performance of valuable DVR of system when connected with it, constant output voltage is obtained.

PMSG Based Wind Turbine System Output

Figure 25.22 illustrates the training performance obtained by the proposed NN Based Model Reference Controller according to the data given in the controller. In Fig. 25.23 the training state is obtained by the controller after training of it is done. And

25 Enhancement of Hybrid PV-Wind System ...

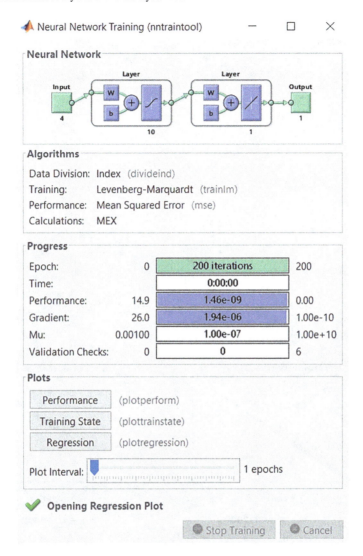

Fig. 25.16 NN predictive controllers performance after training of the data

also in Fig. 25.24 the regression plot is generated of the controller's performance for increasing the efficiency of the system. Figure 25.25 represents the Reference Controllers overall performance obtained after complete training of the network.

Figure 25.26 elucidates the performance obtained for the controlling of the pitch angle of the Wind Turbine of the system efficiently. The controllers used are Model Reference Controller with the PI controller in the system for making the operation of the turbine more consistent. As in Fig. 25.27 the consistency of the plants performance is presented by a constant electromagnetic torque (Te) for the system. Figure 25.28

Fig. 25.17 Testing data obtained after training of controller

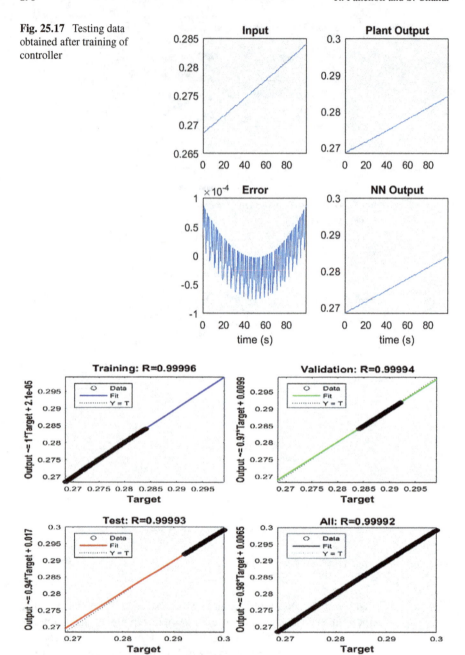

Fig. 25.18 Regression performance obtained after training the controller

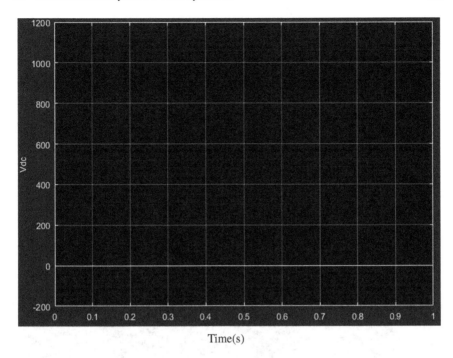

Fig. 25.19 Constant output voltage obtained in the PV panel after training of the NN Predictive Controller

determines the validated performance of the DVR implemented in the system, active and reactive powers are shown with constant reactive power is achieved by the system.

DVR's Performance

In Fig. 25.29 the constant output voltage achieved by the implementation of DVR in the system and the voltage is further supplied to the grid of the proposed PV-Wind Hybrid System. Figure 25.30 shows the performance obtained by the proposed ANFIS controller for controlling the DVR of the system systematically with 20 epochs achieved by the controller. And the continuous performance is obtained with the load data. Figure 25.31 demonstrates the performance of the Sugeno-type Fuzzy Inference System by the rules followed and then the surface viewer is generated by the system. The results generated from the DVR of the system is satisfactory as it reduces the THD (Total Harmonic Distortion in the system is (<5%) and since it is controlled by ANFIS controller that has also improved its performance further.

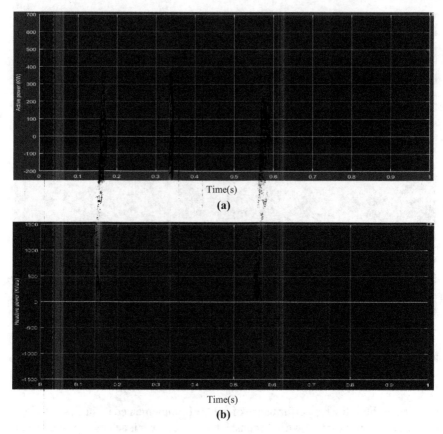

Fig. 25.20 Constant output **a**, **b** active and reactive power and obtained after training the controller without DVR'S contribution

PV-Wind Hybrid System Performance

As Fig. 25.32a, b displays the output generated results of the proposed PV-Wind Hybrid System in which the PV and Wind system are controlled by the efficient Neural Network techniques and also with the effective DVR connected in the Hybrid system and then to the grid which is obtained constant.

Conclusion

The prospective of the ingenious PV-Wind Hybrid System has validated its mandatory performance of the system by using of the virtuous Neural Networks Techniques in the system. The Neural Network-Based Model Predictive Controller technique is

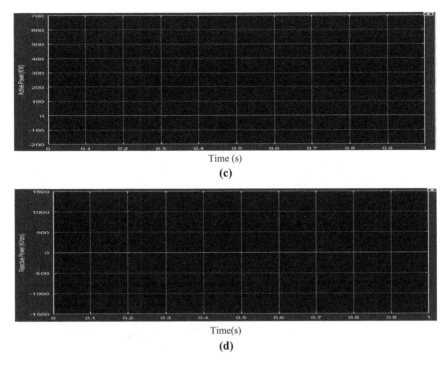

Fig. 25.21 Output power obtained **a, b** active and reactive power with the DVR's contribution in the grid

realized in the MPPT's of the PV system for extracting the best and maximal amount of power for the effective working of the PV system in any unrealistic and in agreeable conditions also. Then for an achieving a qualitative and significant amount of output the NN Based Model Reference Controllers Technique is implemented in the controlling of the pitch angle of PMSG Based Wind Turbine System for gain accuracy for imperative purposes for the system running efficiently. In addition, with it an important part of the proposed Hybrid Network as the DVR of the system. Its optimistic performance is worldwide known and that is paramount in the system for improving its profitability to the human beings of society. For increasing its restoration it is controlled by the proficient ANFIS controller with the Sugeno-type Fuzzy Inference System. And the performance of the DVR is well justified with its adequate responses. The more advanced version of Hybrid system can be further analyzed by adding more prolific techniques as neural network and artificial intelligence in the system. The results were procured by using the MATLAB and analyzed with the convenient and satisfactory systems approach.

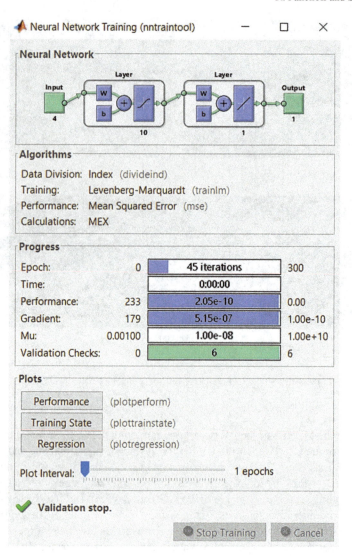

Fig. 25.22 Training of NN based model reference controller

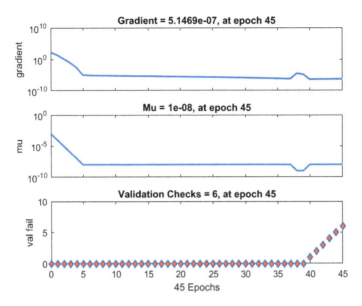

Fig. 25.23 The training state obtained after training of controller

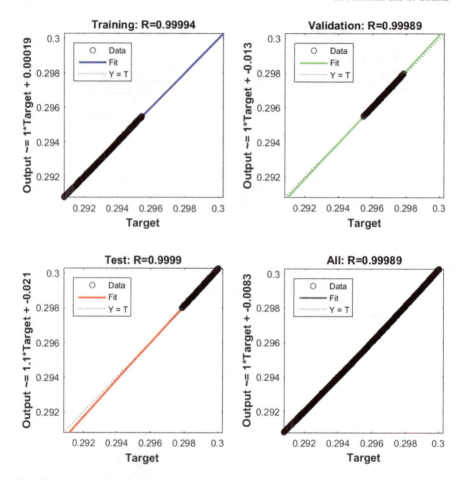

Fig. 25.24 Regression plot obtained after training of controller

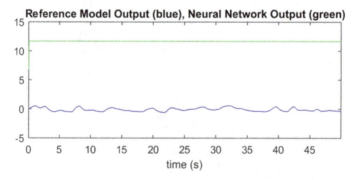

Fig. 25.25 The reference controller output is generated after training of whole controller

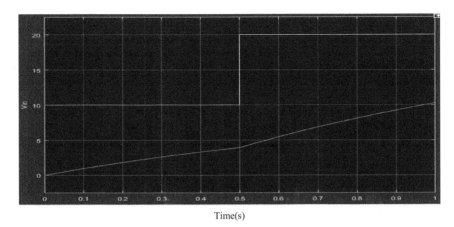

Fig. 25.26 The NN based reference controller with PI controllers performance

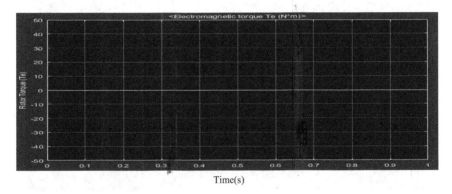

Fig. 25.27 Constant electromagnetic torque obtained with PMSG based wind turbine system controlled by PI and NN reference controller

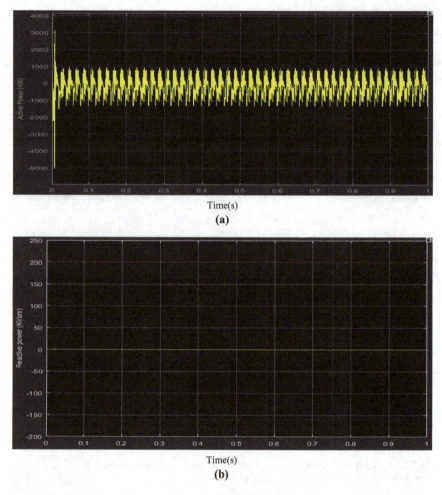

Fig. 25.28 a, b Active and reactive power of wind system with DVR's contribution

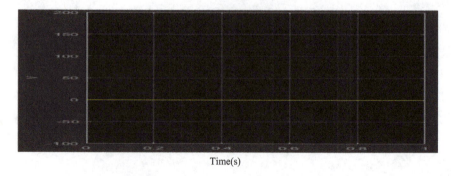

Fig. 25.29 Constant voltage obtained from DVR after supplying it to grid

25 Enhancement of Hybrid PV-Wind System ...

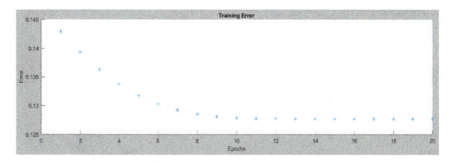

Fig. 25.30 The training performance obtained by the ANFIS controller

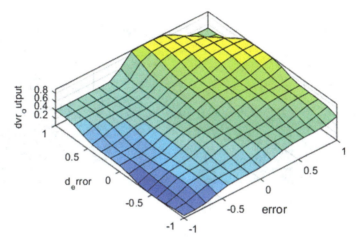

Fig. 25.31 The Sugeno-type fuzzy inference systems surface view

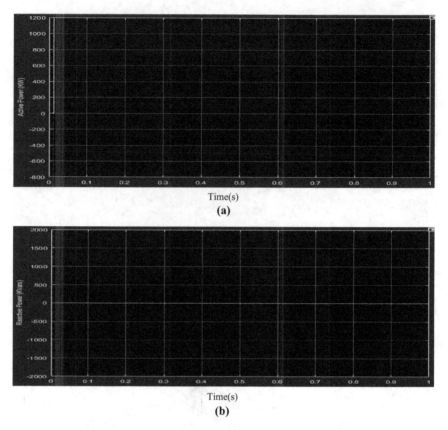

Fig. 25.17 Output power resulted from the proposed Hybrid System with DVR's Contribution as **a** total active power and **b** total reactive power of hybrid system injected to grid

References

1. Bendib B, Belmili H, Krim F (2015) A survey of the most used MPPT methods: conventional and advanced algorithms applied for photovoltaic systems. Renew Sustain Energy Rev 45:637–648
2. Rezk H, Eltamaly AM (2015) A comprehensive comparison of different MPPT techniques for photovoltaic systems. Sol Energy 112:1–11
3. Esram T, Chapman PL (2007) Comparison of photovoltaic array maximum power point tracking techniques. IEEE Trans Energy Conv 22(2):439–449
4. Bahgat ABG, Helwa NH, Ahmad GE, El Shenawy ET (2005) Maximum power point tracking controller for PV systems using neural networks. Renew Energy 30:1257–1268
5. Whei-Min L, Chih-Ming H, Chiung-Hsing C (2011) Neural-network based MPPT control of a stand-alone hybrid power generation system. IEEE Trans Power Electron 26:3571–3581
6. Rizzo SA, Scelba G (2015) ANN based MPPT method for rapidly variable shading conditions. Appl Energy 145:124–132
7. Tang R, Wu Z, Fang Y (2017) Configuration of marine photovoltaic system and its MPPT using model predictive control. Solar Energy 158:995–1005

8. Bouilouta A, Mellit A, Kalogirou SA (2013) New MPPT method for stand-alone photovoltaic systems operating under partially shaded conditions. Energy 55:1172–1185
9. Mosa M, Shadmand MB, Balog RS, Abu Rub H (2017) Efficient maximum power point tracking using model predictive control for photovoltaic systems under dynamic weather condition. IET Renew Power Gener 99:1–1
10. Lashab A, Sera D, Guerrero JM, Mathe L, Bouzid A (2018) Discrete model-predictive-control-based maximum power point tracking for PV systems: overview and evaluation. IEEE Trans Power Electron 33(8):7273–7287
11. Errami Y, Ouassaid M, Maaroufi M (2013) Control of a PMSG based wind energy generation system for power maximization and grid fault conditions, Mediterranean Green Energy Forum MGEF-13. Energy Procedia 42:220–229
12. Zhang S, Tseng K-J, Vilathgamuwa DM, Nguyen TD, Wang X-Y (2011) Design of a robust grid interface system for PMSG-based wind turbine generators. IEEE Trans Ind Electron 58(1):316–328
13. Uehara A, Pratap A, Goya T, Senjyu T, Yona A, Urasaki N, Funabashi T (2011) A coordinated control method to smooth wind power fluctuations of a PMSG-based WECS. IEEE Trans Energy Convers 26(2):550–558
14. Hwas AMS, Katebi R (2012) Wind turbine control using pi pitch angle controller. In: Proceedings of the IFAC conference on advances in PID Control PID'12, Brescia, Italy, pp 28–30
15. Dahbi A, Nait-Said N, Nait-Said MS (2016) A novel combined MPPT-pitch angle control for wide range variable speed wind turbine based on neural network. Int J Hydrog Energy 41:9427–9442
16. Lin WM, Hong CM, Ou TC, Chiu TM (2011) Hybrid intelligent control of PMSG wind generation system using pitch angle control with RBFN. Energy Convers Manag 52:1244–1251
17. Ghosh A, Ledwich G (2002) Compensation of distribution system voltage using DVR. IEEE Trans Power Del 17(4):1030–1036
18. Nielsen JG, Newman M, Nielsen H, Blaabjerg F (2004) Control and testing of a dynamic voltage restorer (dvr) at medium voltage level. IEEE Trans Power Electron 19(3):806–813
19. Banaei MR, Hosseini SH, Khanmohamadi S, Gharehpetian GB (2006) Verification of a new energy control strategy for dynamic voltage restorer by simulation. Simulat Model Pract Theor 14(2):112–125
20. Ferdi B, Dib S, Berbaoui B, Dehini R (2014) Design and simulation of dynamic voltage restorer based on fuzzy controller optimized by ANFIS. Int J Power Electron Drive Syst IJEPEDS 4(2):212–222
21. Benali A, Khiat M, Allaoui T, Denaï M (2018) Power quality improvement and low voltage ride through capability in hybrid wind-PV farms grid-connected using dynamic voltage restorer. IEEE Access 6:68634–68648
22. Karabacak K, Cetin N (2014) Artificial neural networks for controlling wind–PV power systems: a review. Renew Sustain Energy Rev 29:804–827
23. Rezvani A, Esmaeily A, Etaati H, Mohammadinodoushan M (2019) Intelligent hybrid power generation system using new hybrid fuzzyneural for photovoltaic system and RBFNSM for wind turbine in the grid connected mode. Front Energy 13(1):131–148
24. Whei-Min L, Chih-Ming H, Chiung-Hsing C (2011) Neural-network based MPPT control of a stand-alone hybrid power generation system. IEEE Trans Power Electron 26(12):3571–3581
25. Femia N, Petrone G, Spagnuolo G, Vitelli M (2013) Power electronics and control techniques for maximum energy harvesting in photovoltaic systems, 1st edn. CRC Press
26. Heidari M (2016) Improving efficiency of photovoltaic system by using neural network MPPT and predictive control of converter. Int J Renew Energy Res (IJRER) 6(4):1524–1529
27. Zangeneh Bighash E, Sadeghzadeh SM, Ebrahimzadeh E, Blaabjerg F (2018) Improving performance of LVRT capability in single phase grid-tied PV inverters by a model-predictive controller. Int J Electr Power Energy Syst 98:176–188
28. Al-Alawia A, Al-Alawib SM, Islam SM (2007) Predictive control of an integrated PV-diesel water and power supply system using an artificial neural network. Renew Energy J 32(8):1426–1439

29. Luo N, Vidal Y, Acho L (2014) Wind turbine control and monitoring. Springer, Berlin
30. Arani MFM, Mohamed YARI (2016) Assessment and enhancement of a full-scale PMSG based wind power generator performance under faults. IEEE Trans Energy Convers 31:728–73
31. Matayoshi H, Howlader AM, Datta M, Senjyu T (2016) Control strategy of PMSG based wind energy conversion system under strong wind conditions. Energy Sustain Dev 211–218
32. Ren Y, Li LY, Brindley J, Jiang L (2016) Nonlinear PI control for variable pitch wind turbine. Control Eng Pract 50:84–94
33. Narayana M, Sunderland KM, Putrus G, Conlon MF (2017) Adaptive linear prediction for optimal control of wind turbines. Renew Energy 113
34. Yilmaz AS, Ozer Z (2009) Pitch angle control in wind turbine above the rated wind speed by multilayer perception and radial basis function neural networks. IEEE Exp Syst Appl 36(6):9767–9775
35. Hwas AMS, Katebi R (2012) Wind turbine control using PI pitch angle controller. Proc IFAC Conf Adv PID Control, Brescia, Italy
36. Ro K, Choi H (2005) Application of neural network controller for maximum power extraction of a grid-connected wind turbine system. J Electr Eng 88:45–53
37. Ioannou P, Fidan B (2006) Adaptive control tutorial. SIAM, Philadelphia
38. Sastry S, Bodson M (2011) Adaptive control: stability, convergence and robustness. Dover Publications, New York
39. Fitzer C, Barnes M, Green P (2004) Voltage sag detection technique for a dynamic voltage restorer. IEEE Trans Indus Appl 40(1):203–212
40. Hingorani NG, Gyugyi L (2000) Understanding FACTS concepts and technology of flexible AC transmission systems, 1st edn. IEEE Press Inc
41. Bhavani R, Rathina Prabha N (2018) Simulation of reduced rating dynamic voltage restorer using srf anfis controller. Int J Fuzzy Syst 20(04)
42. Samira DIB, Brahim FERDI, Chellali BENACHAIBA (2011) Adaptive neuro-fuzzy inference system based DVR controller design. Leonardo Electr J Pract Technol 18. ISSN 1583–1078
43. Jang JSR (1993) ANFIS: adaptive-network-based fuzzy inference systems. IEEE Trans Syst Man Cybern 23(03):665–685
44. Hong C-M, Chen C-H (2014) Intelligent control of a grid-connected wind-photovoltaic hybrid power systems. Int J Electric Power Energy Syst 55:554–561
45. Lin W-M, Hong C-M, Chen C-H (2011) Neural-network based MPPT control of a stand-alone hybrid power generation system. IEEE Trans Power Electr 26(12):3571–3581
46. Rajkumar RK, Ramachandaramurthy VK, Yong BL, Chia DB (2011) Techno-economical optimization of hybrid pv/wind/battery system using Neuro-Fuzzy. Energy 36(8):5148–5153

Chapter 26
Modified Particle Swarm Optimization Technique for Dynamic Economic Dispatch Including Valve Point Effect

Gaurav Kumar Gupta and Mayank Goyal

Introduction

Energy cannot be retained in large quantities; hence the electricity produced must be consumed instantly. Among other terms, the need for energy will be exactly satisfied at any given instant. Any power imbalance of power can adversely influence the parameters of the control network. The produced power must therefore be equal to the amount of power used and the energy shortfall in the power grid. Each power company as a whole should also aim for higher profitability through the productive usage of the available capital, as well as power sector firms. In addition to increasing income, power producers need to uphold stringent technological restrictions, such as a balance of power. The Economic Dispatch (ED) system can be used by freight dispatch centers to achieve a technologically feasible and commercially efficient result [1–3]. Yet the ED issue has certain drawbacks, for example, if we perceive the load as continuous, but in real-life situations isn't true, since the load differs at every moment.

Dynamic Economic Dispatch (DED) was developed to resolve the limitations of the ED problem, that treats the load quasi static for one hour, which is the nearest logical solution to the issue. Generators need to be rescheduled every hour. Reassigning generators may eventually make an insane assumption that Generators would need to drastically change their power supply in a limited period, which might not be practicable. Within this paper we find generators' capacity to raise or decrease production power so as to take care of these abnormalities. This should be taken

---

G. K. Gupta (✉) · M. Goyal
Department of Electrical Engineering, GLA University, Mathura, India
e-mail: gaurav.gupta@gla.ac.in

M. Goyal
e-mail: mayank.goyal@gla.ac.in

© The Editor(s) (if applicable) and The Author(s), under exclusive license to Springer Nature Singapore Pte Ltd. 2021
A. K. Singh and M. Tripathy (eds.), *Control Applications in Modern Power System*, Lecture Notes in Electrical Engineering 710, https://doi.org/10.1007/978-981-15-8815-0_26

into consideration when calculating the level at which the output capacity may be modified as a limitation, regarded as the limits of ramp speed [6].

Additionally, the standard method of estimating the objective role of the ELD problem as a quadratic function of the actual power is not a fair approximation, because it does not take into account the impact of valves in steam turbines [3]. Whereas power production for the generator would increase, the turbine supply of steam would inevitably need to be expanded. There will be multiple valves in the steam path in a multi-stage turbine, because the turbine's output will be raised, it would therefore increasing the number of valves to be raised. If an exact, ideal solution to the problem is sought, the effect of opening and closing steam valves on production costs cannot be overlooked. The costs of opening and closing steam valves as a feature of costs can be modeled as a sinusoidal part, as they require additional operating costs and maintenance costs. Within this paper we take into account the impact of valves on the overall cost of generators.

Many approaches have been implemented in the recent past to address the ED problem for multiple test schemes [1–20]. Conventional approaches like Newton's method, the method of lambda-iteration, gradient approaches, etc. [1] are capable of solving the ELD issue effectively. While traditional analysis techniques have the drawback of modeling a linear process on cost curve [2]. Traditional cost-function modeling as quadratic does not reflect the impact of the valve on an objective functions. It help to makes the ELD issue non-linear, and it is more difficult to find a solution using the classical approach and might not be feasible with larger systems. Global optima methods based on the particle swarm optimization methods [17, 19] have recently become increasingly common in solving nonlinear multi-purpose optimization issues.

In this research paper, Modified Swarm of Particles Optimization (MPSO) solves the DED question during 24 h of the day in ten generator units system. In GLA University's simulation center, which is the open access counterpart to MATLAB®, applications are built for the DED problem solving methods. The approaches achieved was based on and implemented accurately on current structures.

Dynamic Economic Load Dispatch (DELD)

Economic load dispatch is clearly considered as an approach extending to electrical systems by identifying specific generation rates for various generating units in such a way that full load is transmitted from the device and the production and distribution cycle is completed. In the program which is more competitive in economic terms. Cost reductions are a principal characteristic of integrated networks.

When the valves are raised, even throttling losses under initial conditions are relatively small. The decrease gradually and are the lowest when valves are fully opened. The load is not continuous in real action, and varies instantaneously. The cost-function study cannot provide the running and operational costs of the steam valves as a quadratic. For the present function the generators' output curve is used,

Fig. 26.1 Running fuel cost characteristics including valve point effects

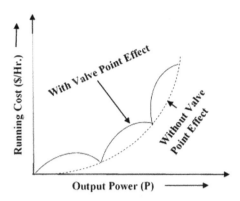

given the steam valve expenses. For each hour of the day the load is known as dynamic. Often, regarded is the generators total allowed capacity to increase or decrease generated strength, described as ramp down and ramp up the generator limits.

Effects of Valve Point

General ED problem that modern approaches can be addressed is to reduce the total expense of running of power system. Average running costs for generators are typically roughly quadratic. These cost function simulation takes little account of the complexity of a multi-stage turbine including valves. A multi-stage steam turbine's operating costs are nonlinear and include the sine wave components in the cost equation, as seen in Fig. 26.1.

Ramp Rate Limits

Thermal generation systems cannot change their processing power beyond the defined level because of mechanical inertia of machine systems of production units. Those limitations are called ramping up and running down the power of a generator. Such generator ramps must be included in the DED question to lift and drop limits in order to achieve a practical ED solution for time-varying loads. Changing load times are approximated as being relatively stagnant for one hour.

Mathematical Modelling for DELD

The challenge of solving the dynamic economic load dispatch (DELD) issue in the electric power system is to determine the generation rates for all operational units that minimized the overall cost of fuel in the system, thus fulfilling a set of constraints during the 24-h dispatch cycle.

$$\min F_T = \sum_{t=1}^{T=24} \sum_{k=1}^{N=10} F_{kt}(P_{kt}) \qquad (26.1)$$

where

F_T is the total running costs over the dispatch duration of 24 h.

$F_{kt}(P_{kt}) F_{kt}(P_{kt})$ is the total running cost of 'kth' alternator at time 't'.

P_k is the function of active power output of 'kth' alternator at time 't'.

Generally, the typical economic dispatch problem of determining the optimum power generation combination, which reduces the overall cost of fuel of a generating unit, is defined by a quadratic power output function P_k while meeting the overall demand as follows:

$$F_{kt}(P_{kt}) = a_k P_{kt}^2 + b_k P_{kt} + c_k \qquad (26.2)$$

In reality, the object of the DELD issue is not a differentiable point because of the valve point loading effects and the fuel adjustment. Development cost function is normally adopted, and a quadratic method is then used to address the DELD problem. This is often the case when the valve point effect is regarded. The fuel cost function of the kth generator units was modified as bellow in this regard:

$$F_{rt}(P_{rt}) = a_r P_{rt}^2 + b_r P_{rt} + c_r + |e_r \sin(f_r(P_{\min} - P_{rt}))|. \qquad (26.3)$$

where a_r, b_r and c_r are system coefficients; e_r and f_r are constants of valve point effect of kth generating unit.

The objective function is subject to the limitations (constraints) to be minimized.

Active Power Balanced Constraints

To order to maintain the system frequency constant and to keep the system stable the amount of actual power output will be equivalent to the total of power demand and actual power loss of the power system.

$$\sum_{k=1}^{N_G} P_{kt} = P_D + P_L \tag{26.4}$$

P can be determined using the equation Kron's Loss

$$P_L = \sum_{r=1}^{N_G}\sum_{k=1}^{N_G} P_r B_{rk} P_k + \sum_{r=1}^{N_G} P_r B_{0r} + B_{00} \tag{26.5}$$

Alternator Power Limit Constraints

For technological and economic reasons respectively, each generator cannot generate above a certain limit and beyond the defined limit.

$$P_k^{\min} < P_{kt} < P_k^{\max} \tag{26.6}$$

where $P_k^{\min}$, and $P_k^{\max}$ are the generator lower and upper bound limits of power outputs, respectively. The generator's overall output capacity is constrained by thermal calculation, and a boiler's flame instability limits its total power generation.

Alternator Ramp Rate Limit Constraints

The amount at which output power generated in the generating units cannot adjust above a certain amount influenced by mechanical inertia. The ramp rate limitations of the generator units are defined below:

If power production rising, $P_{kt} - P_k^{t-1} \leq UR_k$ f power production declines, $P_k^{t-1} - P_{kt} \leq DR_k$.

where P_k^{t-1} is the "kth" unit power output, and UR_k and DR_k are the ramp's upper and lower limits, respectively. The use of ramp restrictions shifts the generator's operational constraints (26.7) as seen below

$$\max\left(P_k^{\min}, UR_k - P_k\right) \leq P_r \leq \min\left(P_k^{\max}, P_k^{t-1} - DR_k\right) \tag{26.7}$$

Introduction to Particle Swarm Optimization (PSO)

Particle swarm optimization was discovered by James Kennedy and Russell Eberhart in 1995 as a form of stochastic optimization focused on population. PSO is built on the Swarms theory, its intelligence and its operation. PSO consists of a collection of creatures (particles) performing the same action while searching for space. This behavior focused on a flock of birds or feeding fish's social behavior. Use a set of random particles the PSO algorithm scans in parallel. Every particle in the swarm is the corresponding way to solve the problem. Particles in a swarm seek an ideal solution from its present size, past experience and neighbors' experience. Every particle in a swarm is modified to two best information in each generation. The first one is the best approach it has achieved to information. That information is known as P_{best}. A further best value recorded by the optimization of the Particle Swarm is the best information of any particle in the population so far achieved. That best information is the best in the search space, and it's called G_{best}.

Implement Optimization Technique for Solving DELD

Particle Swarm Optimization (PSO)

The position and velocity of the kth particle in the search space of n dimensions can be expressed as $X_k = [x_{k1}, x_{k2}, \ldots, x_{kn}]^T$ and $V_k = [v_{k1}, v_{k2}, \ldots, v_{kn}]^T$ respectively, particles have their own best (P_{best}) location $P_k(t) = [p_{k1}(t), p_{k2}(t), \ldots, p_{kn}(t)]^T$ refers to the highest personal intrinsic value obtained in the 't' generation so far. The best global particle (G_{best}) is labelled with $Pg(t) = [pg_{k1}(t), pg_{k2}(t), \ldots, pg_{kn}(t)]^T$, which is the strongest particle detected in the whole swarm at generation 't' so far. For each particle the new velocity is measured as follows:

$$v_k^{t+1} = w * v_k^t + c_a * \text{rand}_1 * \left(P_{best} - x_k^t\right) + c_b * \text{rand}_2 * \left(g_{best} - x_k^t\right) \quad (26.8)$$

In which c_a and c_b are constants called coefficients of acceleration related to conceptual and social behavior, ω is actually the factor of inertia, n is the population scale, rand$_1$ and rand$_2$ are two separate random numbers randomly assigned within [0, 1]. Therefore, each particle's location at each stage is modified according to the following equation:

$$x_k^{t+1} = v_k^{t+1} + x_k^t \quad (26.9)$$

Modified Particle Swarm Optimization (MPSO)

Equation (26.8) illustrates that the latest velocity is modified as per its past velocity and the difference between its present location and Its best ancient location and the best global swarm position. In general, the value of each and every variable in V_k can be attached to the [$V_{k\min}$, $V_{k\max}$] controlling range of unnecessary particle wandering beyond the search space [$X_{k\min}$, $X_{k\max}$]. According to Eq. (26.9), the particle flies into a new location. The cycle will be replicated until the user has achieved a given end criterion.

In the simple PSO cycle the inertia weight is held constant for all particles in one stage, but the most important element that moves the present location to the optimal position is the weight of inertia (ω). The algorithm should really be adopted to optimize the quest power, such that the swarm's motion cannot be controlled by the objective function. In our adaptive PSO, the direction of the particles is modified, so that the strongly equipped particle (best particle) progressively travels relative to the poorly fitted particle. This could be achieved by choosing different ω values for each particle, based on its size, between $\omega_{\min}$ and $\omega_{\max}$ as given below:

$$\omega_k = \omega_{\max} - \frac{\omega_{\max} - \omega_{\min}}{\text{itr}_{\text{Total}}} * \text{Rank}_k \quad (26.10)$$

Therefore, with Eq. (26.10), this could be found that the strongest particle would be in the first category and the weight of inertia for such a particle is meet to the optimal value, while the maximum inertia weight for the poorest fitted particle is taken, allowing the particle travel at a high velocity. Every particle's velocity is modified using Eq. (26.11).

$$\begin{aligned} v_k^{\prime t+1} &= \omega_k * v_k^t + c_1 * \text{rand}_1 * \left(P_{\text{best}} - x_k^t\right) \\ &+ c_2 * \text{rand}_2 * \left(g_{\text{best}} - x_k^t\right) \end{aligned} \quad (26.11)$$

The new position of the particles is observed with Eq. (26.12),

$$x_k^{\prime t+1} = v_k^{t+1} + x_k^t \quad (26.12)$$

After a certain number of generations the principle of re-initialization is implemented into the proposed MPSO algorithm if there is no change in the convergence of the algorithm. At the conclusion of the foregoing unique generation the population of the proposed APSO is re-initialized with newly random created individuals. The quantity of such new individuals is chosen from the initial population's poorest fitting population, which 'k' is the proportion of the overall population to shift. In a way, this result of initialization of population is close to that of the mutation operator in such like a genetic algorithm (GA). This result is beneficial if the algorithm meets unnecessarily to an optimal local and there is no change any more. This population initialization is achieved after testing the adjustments in the 'F_{best}' value throughout each unique number of alternators.

Algorithm of MPSO

Step 1   For each of the coefficients c_a, c_b, take input variables with range (min., max.) and assuming that the iteration counter is equivalent to zero, V_{max}, ω_{min} and ω_{max}.

Step 2   Randomly initialize the n number of particle population dimensions, such as positions and velocity.

Step 3   Increase the counter Iteration by that one.

Step 4   Measurement of the fitness function among all clusters within a group, evaluate the particles with each particle's best P_{best} position and change its objective cost. Secondly, finding the best global position (G_{best}) of the all the particles, and change their objective cost.

Step 5   If the criteria of stoppage is achieved go to move Eq. (26.1). Then carry on.

Step 6   Measure the inertia factor as per (26.10), such that every motion of the particles is specifically influenced by its fitness cost.

Step 7   Now updated the velocity through Eq. (26.11).

Step 8   Measure the inertia factor as per (26.10), such that every motion of the particles is specifically influenced by its fitness cost.

Step 9   To order to retain the finest particle discovered so far, the rulers are put into the first position of the new population.

Step 10 This F_{Best}, new value is contrasted with the F_{best}, old value for every 10 generations, when there's no significant change, then reset k percent of the population. Switch to step-3.

Step 11 The efficient output of the particle having in G_{best} and its now gets the optimized cost.

Case Study for DELD Problem

Such a research paper tends to have 10 generator units including valve point loading results, the device specifics and load demand trends shown in Table 26.1 and Fig. 26.2, respectively, should be taken from the reference paper [17].

Table 26.1 System characteristics data

Gen.	a_$/(Mw)^2$ h	b_$/(Mw)$h	c_$/hr	e_$/hr	f_1/Mw	P_{min} Mw	P_{max} Mw	UR_Mw/hr	DR_Mw/hr
1	0.000430	21.600	958.200	450.0	0.0410	150.0	470.0	80.0	80.0
2	0.000630	21.050	1313.60	600.0	0.0360	135.0	460.0	80.0	80.0
3	0.000390	20.810	604.970	320.0	0.0280	73.0	340.0	80.0	80.0
4	0.000700	23.900	471.600	260.0	0.0520	60.0	300.0	50.0	50.0
5	0.000790	21.620	480.290	280.0	0.0630	73.0	243.0	50.0	50.0
6	0.000560	17.870	601.750	310.0	0.0480	57.0	160.0	50.0	50.0
7	0.002110	16.510	502.700	300.0	0.0860	20.0	130.0	30.0	30.0
8	0.004800	23.230	639.400	340.0	0.0820	47.0	120.0	30.0	30.0
9	0.109080	19.580	455.600	270.0	0.0980	20.0	80.0	30.0	30.0
10	0.009510	22.540	692.400	380.0	0.0940	55.0	55.0	30.0	30.0

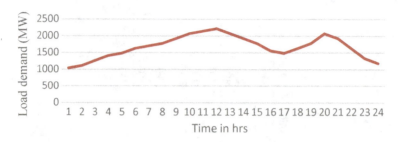

Fig. 26.2 Daily load demand patterns

Results and Discussions

The proposed algorithm was implemented in the MATLAB environment at research lab of GLA University Mathura. The proposed algorithm is introduced to 10 generating units including generator limitations, the ramp rate and also subjected to valve point effects. The simulation is conducted for 100 samples in order to find the right approach for best power scheduling of generation corresponding to optimal fuel cost. Table 26.2 indicates the overall fuel costs achieved from the new MPSO approach and comparisons with other approaches such as SQP, EP, HS, DE, MDE and Hybrid DE in terms of the consistency of the solution and the performance of the simulation shown in Table 26.3 as well as Fig. 26.3.

Conclusion

To find the optimum solution of DELD problem, multiple meta heuristic approaches are suggested by researchers. In real time (RTS) the system need to respond the parameter variation in quick manner in the presence of various constraints such as ramp rate limit, valve point effect and non-linear cost curve associated with power transmission. The MPSO mathematics is coded in MATLAB environment and it is found effective in handling of all of these constraints related to the dynamic load dispatch. The cost of power generation and improved efficiency can be achieved if and only if the optimization of DELD problem reach to its solution as soon as possible and must be highly convergent. As the demand for power changes frequently and in large quantities, the system needs to respond rapidly to any disequilibrium between demand and power supply in each iteration. The MPSO is capable to efficiently reach the optimum solution, simulation results support the approach and offers the most cost effective solution. The proposed MPSO algorithm is highly flexible to accommodate and remove the number of generators easily from the system without failing its reliability, stability and performance indices.

Table 26.2 Best scheduling of generation corresponding to optimal fuel cost by MPSO technique

Hr	PG1 (MW)	PG2 (MW)	PG3 (MW)	PG4 (MW)	PG5 (MW)	PG6 (MW)	PG7 (MW)	PG8 (MW)	PG9 (MW)	PG10 (MW)	Demand (MW)	Operating cost ($)
1	227.2956	135.1106	76.3695	64.3664	118.4773	160	130	49.0033	20.3773	55	1036	28,689
2	304.3299	135.4922	74.8043	61.3664	122.0839	160	129.7235	47.0606	20.1393	55	1110	30,128
3	382.2141	135.938	80.3764	111.3664	172.0839	123.1724	130	47.734	20.1149	55	1258	33,523
4	455.6735	135.2692	156.1924	62.7951	222.0839	121.971	129.309	47.3595	20.3463	55	1406	36,630
5	457.2741	215.2692	192.0511	61.3208	175.9258	124.6641	130	47.8977	20.5972	55	1480	38,198
6	457.898	225.4325	272.0511	60.8681	225.9258	133.4138	130	47.3685	20.0422	55	1628	41,493
7	455.47	305.4325	297.6448	61.1352	208.4646	121.1628	130	47.0527	20.6374	55	1702	42,938
8	457.1309	385.4325	299.7439	61.3278	174.5086	145.4403	130	47.0313	20.3847	55	1776	44,657
9	459.1616	397.9743	318.1054	111.3278	224.5086	160	130	47.5545	20.3678	55	1924	48,051
10	456.5048	460	340	118.4721	224.1008	160	130	77.5545	50.3678	55	2072	52,117
11	456.8337	460	340	168.4721	238.4914	160	130	85.7179	51.4849	55	2146	54,001
12	456.3925	460	340	218.4721	231.9692	160	129.9087	115.7179	52.5395	55	2220	56,003
13	457.356	395.693	301.0917	235.2316	226.0859	127.9596	130	120	23.5822	55	2072	51,631
14	380.1401	396.0151	289.8657	186.8107	221.6844	124.3208	130	120	20.1632	55	1924	48,118
15	302.799	397.044	296.7834	136.8653	221.3141	125.4852	130	90.6186	20.0903	55	1776	44,798
16	225.8161	317.2294	292.2094	86.9912	218.894	122.401	130	85.1845	20.2744	55	1554	40,000
17	226.548	309.7559	300.2111	60.563	169.0094	124.7158	129.4783	84.3406	20.378	55	1480	37,920
18	303.2578	310.2725	291.7304	60.6973	219.0094	123.9052	129.382	84.3675	50.378	55	1628	41,389
19	380.2226	390.2725	297.7063	62.8119	222.6824	130.2106	130	85.572	21.5217	55	1776	44,535
20	456.3015	460	337.5484	112.8119	222.9002	160	130	85.9163	51.5217	55	2072	51,966
21	461.3292	395.8845	302.1844	122.1574	225.7168	125.0627	129.7155	85.2014	21.7482	55	1924	47,884

(continued)

Table 26.2 (continued)

Hr	PG1 (MW)	PG2 (MW)	PG3 (MW)	PG4 (MW)	PG5 (MW)	PG6 (MW)	PG7 (MW)	PG8 (MW)	PG9 (MW)	PG10 (MW)	Demand (MW)	Operating cost ($)
22	381.6117	316.8803	295.722	73.5009	177.4141	94.6229	129.6126	83.5185	20.1169	55	1628	41,928
23	304.5173	249.3251	241.8936	61.9031	128.9821	58.9573	130	81.1055	20.316	55	1332	35,964
24	232.3229	223.8944	185.8698	61.2949	114.28	108.9573	130	52.0208	20.36	55	1184	32,056
Optimal fuel cost of system for 24 h load pattern ($/day) =												1,024,600

Table 26.3 Results comparison with other techniques

Method	Minimum total cost ($/day)
SQP	1,051,163.00
EP	1,048,638.00
HS	1,046,725.908
DE	1,033,958.00
MDE	1,031,612.00
Hybrid DE	1,031,077.00
MPSO	1,024,600.00

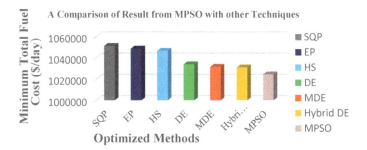

Fig. 26.3 Results comparison with other techniques

References

1. Wood AJ, Wollenberg BF (1996) Power generation, operation and control. Wiley. New York, pp 29–90
2. Sasaki Y, Yorino N, Zoka Y (2017) Stochastic dynamic load dispatch against uncertainties. IEEE Trans Smart Grid 9(6):5535–5542
3. Zhu J (2009) Optimization of power system operation. Willey, New York, pp 85–210
4. Jadoun VK, Pandey VC, Gupta N, Niazi KR (2018) Integration of renewable energy sources in dynamic economic load dispatch problem using an improved fireworks algorithm. IET J Renew Power Gener 12(9):1004–1011
5. Zhou M, Wang M, Li J, Li G (2017) Multi-area generation-reserve joint dispatch approach considering wind power cross-regional accommodation. CSEE J Power Energy Syst 3(1):74–83
6. Huang C, Yue D, Xie J, Li Y, Wang K (2016) Economic dispatch of power systems with virtual power plant based interval optimization method. CSEE J Power Energy Syst 2(1):74–80
7. Gautham S, Rajamohan J (2016) Economic load dispatch using novel bat algorithm. IEEE: 1–4
8. Bharathi S, Reddy AVS, Reddy MD (2017) Optimal placement of UPFC and SVC using moth-flame optimization algorithm. Int J Soft Comput Artif Intell 5(1):41–45
9. Chellappan R, Kavitha D (2017) Economic and emission load dispatch using cuckoo search algorithm. IEEE: 1–7
10. Wu Y, Zhao B, Liu L (2017) Glowworm solving economic load dispatch problem with valve point effect using mean guiding differential evolution swarm optimization algorithm for solving non-smooth and non-convex economic load dispatch problems. IEEE: 103–109
11. Reddy AVS, Reddy MD (2017) Network reconfiguration of distribution system for maximum loss reduction using sine cosine algorithm. Int J Eng Res Appl IJERA 7(10):34–39

12. Shahinzadeh H, Moazzami M, Fadaei D, Rafiee-Rad S (2017) Glowworm swarm optimization algorithm for solving non smooth and non-convex economic load dispatch problems. IEEE: 103–109
13. Ul Hassan HT, Asghar MU, Zamir MZ, Aamir Faiz HM (2018) Economic load dispatch using novel bat algorithm with quantum and mechanical behavior. IEEE: 01–06
14. Wood WG (1982) Spinning reserve constrained static and dynamic economic dispatch. IEEE Trans Power Apparatus Syst 101(2):381–388
15. Yuan X, Su A, Yuan Y, Nie H, Wang L (2008) An improved PSO for dynamic load dispatch of generators with valve-point effects. Sci Direct J Energy 34:67–74
16. Ravikumar Pandi V, Panigrahi BK (2011) Dynamic economic load dispatch using hybrid swarm intelligence based harmony search algorithm. Sci Direct Expert Syst Appl 38:8509–8514
17. Balamurugan R, Subramanian S (2008) Differential evolution-based dynamic economic dispatch of generating units with valve-point effects. Taylor Francis Electric Power Components Syst 36(8):828–843
18. Swaroop KS, Natarajan A (2005) Constrained optimisation using evolutionary programming for dynamic economic dispatch. Proc ICISIP: 314–319
19. Gaing ZL (2004) Constrained dynamic economic dispatch solution using particle swarm optimization. Power Eng Society General Meeting 1:153–158
20. Jadoun VK, Gupta N, Niazi KR, Swarnkar A (2014) Economic emission short-term hydrothermal scheduling using a dynamically controlled particle swarm optimization. Res J Appl Sci Eng Technol 8(13):1544–1557

Chapter 27
Electromagnetic Compatibility of Electric Energy Meters in the Presence of Directional Contactless Electromagnetic Interference

Illia Diahovchenko and Bystrík Dolník

Introduction

Electric energy meters (EEMs) are used in private households and industry to measure the electricity consumed. Consumers with high electric energy demand require accurate data to make energy management decisions, while utilities collect big data arrays to improve their services. Therefore, there is a pronounced tendency towards the use of electronic static and digital EEMs for registration of electric energy flows and for billing purposes [1]. Contemporary smart meters are built on integrated microcircuits, which allow to perform more value-added features, along with compactness of size. The multidimensionality of the requirements for the computational capabilities of the microprocessors stimulates the complication of the electric energy metering devices' architecture. State-of-the-art meters are equipped with options of remote sensing, power quality monitoring, recording of tampering events with time stamping, multi-tariff billing, Internet connectivity, peak demand, etc. [2].

The above-mentioned advances in technology have stimulated the topicality of the electromagnetic interference (EMI) issues. The circuitry/subassemblies of equipment in use can emit electromagnetic energy, which can corrupt the performance and accuracy of the sensitive elements of an EEM. Malfunction of EEMs when used in combination with photovoltaic (PV) inverters has been reported in [3–5]. The authors of the work [3] reported a gap in standardization of emissions and electromagnetic immunity in the 2–150 kHz range. Moreover, manipulation with strong magnetic

---

I. Diahovchenko (✉)
Sumy State University, Rymskogo-Korsakova st. 2, Sumy 40007, Ukraine
e-mail: i.diahovchenko@etech.sumdu.edu.ua

B. Dolník
Technical University of Košice, Letná st. 9, 04200 Košice, Slovakia

© The Editor(s) (if applicable) and The Author(s), under exclusive license to Springer Nature Singapore Pte Ltd. 2021
A. K. Singh and M. Tripathy (eds.), *Control Applications in Modern Power System*, Lecture Notes in Electrical Engineering 710,
https://doi.org/10.1007/978-981-15-8815-0_27

fields can be employed for tampering purposes [6]. Events such as switching of inductive loads lightning, electrostatic discharges, and the presence of telecom, radio and video broadcast signals cause some types of EMI. It is essential to produce EEMs that are electromagnetically compatible [2]. Consequently, the regulations associated with the high-frequency electromagnetic interference suppression are steadily tightened [7]. The purpose of this work is to verify the EMC performance of contemporary power meters and to emphasize the urgency of the EMI issues that already seemed resolved in the twenty-first century.

Applicable Standards

The common standards for EMC of power meters were established by the American Standards Institute (ANSI) or International Electrotechnical Commission (IEC). Some countries set up their domestic standards, but most of them can trace their origin to either ANSI or IEC standards. The largest markets for ANSI standards are in Canada, Mexico and the United States of America. Except that, they serve as a reference in some countries of Asia, Central America and South America. In Europe, the EN standards (which follow the IEC standards) are used [2].

There are two categories of EMC tests: (a) emission tests; (b) immunity tests. Emission trials are to verify that the EEM does not conduct or radiate EMI beyond tolerable limits, which are specified in standards. Immunity tests ensure the power meter operates properly in the presence of EMI. Depending upon how the EMI gets coupled, both the emission tests and immunity tests are categorized into conducted tests and radiated tests, as shown in Fig. 27.1.

In Europe, requirements for EMC are governed by [8–11]. The IEC 62132 standard determines electromagnetic immunity for integrated circuits (IC) of all types in the

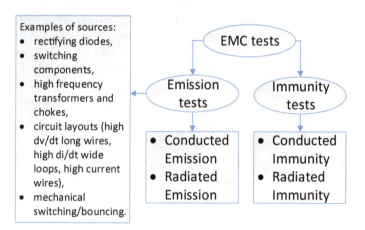

Fig. 27.1 EMC tests categories

range of 150 kHz–1 GHz. For the study presented in this paper is the IEC 61000-4-3 standard [12], was used as a reference. Additionally, an original testing method was used, which consists in applying an electromagnetic field of varying frequency in the range of 30 MHz–1 HHz during a single experiment. The latter is referred to as a Sweep Mode in this paper.

This document describes the procedure of the radiated immunity test, and the meter under test (MUT) should have voltage and auxiliary circuit energized with reference voltage. The standard covers frequency band 80–2000 MHz. The unmodulated field strength can be: 10 V/m with the basis current and 30 V/m without basis current.

Laboratory Tests

One single-phase (further referred as MUT 1) and two three-phase electric energy meters (further referred as MUT 2 and MUT 3) widely distributed in the Slovak electricity distribution system were chosen for the study. The 2.3 kW adjustable resistive load with power factor 0.9993 was connected to the EEMs to imitate an active energy consumer.

The MUT 1 is a digital electronic watt-hour meter, which is rated for 230 V and 0.25–5(60) A, accuracy class 2 (according to IEC 62053-21), the calibration factor is 5000 pulses/kWh. The MUT 2 is a three-phase four-wire direct switch digital power meter designed to measure consumed electrical energy (active and reactive, direct and reverse). Accuracy classes: active energy – class 1 (IEC 62053-21), reactive energy – class 2 (IEC 62053-23), four quadrants. It uses centralized collection of information about power consumption by means of GSM/GPRS, rated for 3 × 230/400 V and 0.25–5(100) A. The calibration factors: 1000 pulses/kWh for active energy and 1000 pulses/kVArh for reactive one. The MUT 3 records active and reactive energy consumption. Its accuracy complies with class 1 (IEC 62053-21) for active energy and class 1 (IEC 62053-23) for reactive energy. Rated voltage and current are 3 × 230/400 V and 5(100) A, respectively. The calibration factors are 500 pulses/kWh for active energy and 500 pulses/kVArh for reactive energy.

All the EEMs were preliminary tested under sinusoidal conditions, and then were subjected to EMC experiments. The radiated immunity test, also known as the electromagnetic high frequency (EMHF) field test, was conducted to ensure the proper functioning of the MUTs, when EMI is present in the surrounding environment.

The tested meters were installed in the EMC chamber Comtest EMC 1710-1. During the chosen non-contact testing method the MUT was installed in a certain distance from the radiation source, and the AC electromagnetic field of the certain frequencies was radiated towards the apparatus, penetrating through its housing to the elements of ICs, current and voltage transducers and influencing its operation [13]. The MXG Analog Signal Generator N5183A served as a source of electromagnetic impulses, and antenna BiConiLog 3142E was chosen as a radiating device. The hybrid antenna 3142E is applicable for operation in a frequency range between

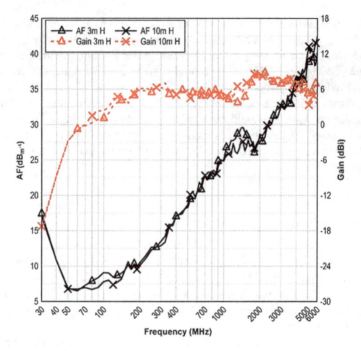

Fig. 27.2 The curves of AF and gain for 3142E

20 MHz and 6 GHz in a single sweep, refuting the necessity for multiple antennas and labor-intensive testbed setup. It can be employed for experiments with both immunity and emission methods. Core characteristics of the antenna (i.e. antenna factor (AF), power gain) are demonstrated in Fig. 27.2. The power generation features (i.e. average power required in vertical polarization) are depicted in Fig. 27.3.

The 3142E model increases the upper-frequency limit to accommodate the upper limit of 6 GHz included in the IEC 61000-4-3 standard [12]. Power gain is the product antenna's efficiency E_{ant} and directivity D [14]:

$$G = E_{ant}D \tag{27.1}$$

AF can be calculated as the ratio of the electric field E strength to the voltage V induced throughout the antenna's terminals [14]:

$$\text{AF} = \frac{E}{V} \tag{27.2}$$

The vertically polarized antenna was located in the Comtest EMC 1710-1 chamber at in 3.5 m from a tested device and radiated power in its maximum mode at the frequencies of 30 MHz, 120 MHz, 200 MHz, 1 GHz, Sweep Mode (the entire tuned range). The respective patterns of the antenna are given in Fig. 27.4. Five tests were

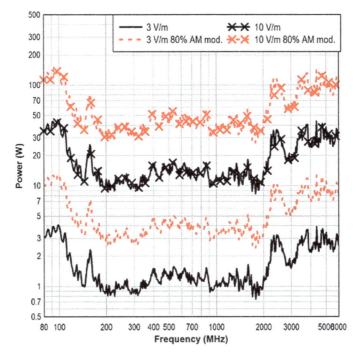

Fig. 27.3 3142E antenna average power required in vertical polarization

conducted in series for each of the three meters picked up for this study. The duration of each generating mode was set to 10 min. The experimental laboratory setup is demonstrated in Fig. 27.5.

Even though the MUT 2 and MUT 3 can measure both active and reactive energy, only active energy consumption was considered for this study. The results of estimated error of tested meters are summarized in Table 27.1.

During the experiments the values of electric E and magnetic H fields at the spot of the meter's placement were measured by means of the high-frequency spectrum analyzer SPECTRAN HF-60105. The highest values were supervised at 1 GHz frequency, namely $E = 42.11$ V/m, $H = 10.31 \cdot 10^{-3}$ A/m. The values of electric field and magnetic field for different experimental conditions are shown in Table 27.1. Positive errors correspond to the cases when an EEM overestimates electric energy and, on the contrary, negative errors indicate that a devise underestimates.

The investigated power meters demonstrated good immunity to radiated electromagnetic interferences. In general manufacturers ensured the immunity of EEMs to high-frequency EMC radiation in the considered range from 30 MHz to 1 GHz. However, some errors were observed for all tested samples.

For MUT 1 all the readings variations are within its accuracy class. It can be noted that the error constantly increases with the raise of frequency and electromagnetic field and reaches its highest value at 1 GHz frequency, where the AF ratio is the

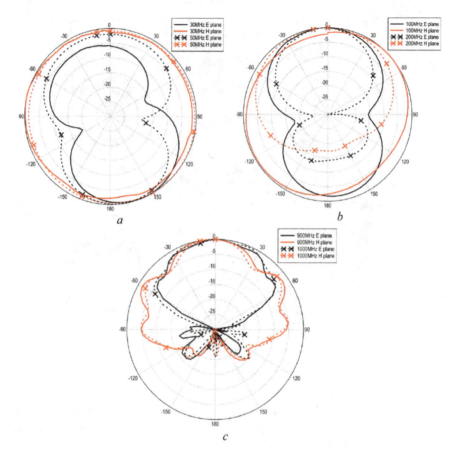

Fig. 27.4 Emission patterns of 3142E model for typical frequency ranges: **a** 30–50 MHz; **b** 100–200 MHz; **c** 900–1000 MHz

biggest. The MUT 2 and the MUT 3 went beyond their tolerance thresholds, but their reaction on the frequency increment was different. The readings of the MUT 2 were maximally compromised at 200 MHz frequency, where the error reached −1.55%, while MUT 3 went out of its accuracy class at 1 GHz and at Sweep Mode frequency (positive errors of 1.24% and 1.06% respectively), where EMI levels are expected to be the highest.

The performed tests have revealed that some of the contemporary power meters on the market are vulnerable to influence of powerful high-frequency electromagnetic fields, which can introduce distortions and compromise the accuracy of the electric energy meters. Therefore, to minimize measuring uncertainty it is essential to reinforce the protection of EEMs against electromagnetic fields of the MHz-HHz range. The obtained results complement the existing EMC solutions presented in the scientific literature and demonstrate that not only fixed frequency, but varying frequencies (like during the Sweep Mode) can be used for tampering purposes. This

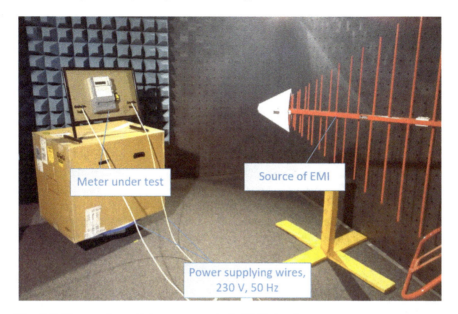

Fig. 27.5 The experimental laboratory setup in the EMC chamber

Table 27.1 Results of immunity tests for energy meters

Frequency		30 MHz	120 MHz	200 MHz	1 GHz	Sweep Mode
E, V/m		7×10^{-3}	1.173	2.23	42.11	–
H, mA/m		21×10^{-3}	3.45	4.21	10.31	–
Error, %	MUT 1	−0.13	−0.77	−0.77	−1.03	−0.26
	MUT 2	−0.77	−1.03	−1.55	−0.77	−0.51
	MUT 3	−0.63	−0.10	0.73	1.24	1.06

should be considered by manufacturers of electric energy meters while developing ways to combat the threats of EMI.

Before performing EMI immunity tests, it is important to make thorough preparation and to take preventive actions to avoid failure in these tests [15]. Common reasons for failure of the tests with electromagnetic high-frequency field radiation are openings, improper shielding and grounding [16]. If the experiment gave unsatisfactory results, it is worth to check whether the mating of top and bottom/side covers was proper during the test, whether the shielding and grounding were compliant with thresholds established in standards. If any of the named criteria was not satisfied during the experiment, it is recommended to fix the determined weak place and to repeat an EMI immunity test.

Conclusions

Accurate billing of the power consumed is of great importance to both utilities and consumers. Therefore, it is necessary to guarantee that EEMs operate properly, particularly in the event of any EMI occurrence. External impacts of strong electromagnetic fields can tamper the integrity of the electric energy-related information processing and, therefore, lead to breaches in the operation of metering equipment. This paper reviews the above topics, highlights EMC standards and testing procedures, and demonstrates laboratory results on testing immunity of modern electric energy meters to the EMI. Two out of three tested power meters went beyond their tolerance thresholds. Some meters overestimated power consumption, while others underestimated, which means that EMI can influence the readings of different meters in different ways. For the MUT 1 and the MUT 3 the highest errors correspond to the case, when the electromagnetic energy in the environment was the highest, which was observed for frequency 1 GHz.

References

1. Diahovchenko I, Volokhin V, Kurochkina V, Špes M, Kosterec M (2019) Effect of harmonic distortion on electric energy meters of different metrological principles. Frontiersn Energy 13(2):377–385
2. Clarion Energy: Testing Energy Meters for Electromagnetic Compatibility, https://www.smart-energy.com/top-stories/testing-energy-meters-for-electromagnetic-compatibility/, last accessed 2020/04/28
3. Kotsampopoulos P et al (2017) EMC issues in the interaction between smart meters and power-electronic interfaces. IEEE Trans Power Delivery 32(2):822–831
4. Volokhin VV, Diahovchenko IM, Derevyanko BV (2017) Electric energy accounting and power quality in electric networks with photovoltaic power stations Eds 2017. In: IEEE international young scientists forum on applied physics and engineering YSF 2017. Lviv, Ukraine, pp 36–39
5. Shevchenko SY, Volokhin VV, Diahovchenko IM (2017) Power quality issues in smart grids with photovoltaic power stations. Energetika 63(4):146–153
6. Diahovchenko IM, Olsen RG (2020) Electromagnetic compatibility and protection of electric energy meters from strong magnetic fields. Electr Power Syst Res 186:106400
7. Carlton RM (2004) An overview of standards in electromagnetic compatibility for integrated circuits. Microelectron J 6:487–495
8. IEC 62132-2 (2010) Integrated circuits—Measurement of electromagnetic immunity—Part 2: Measurement of radiated immunity—TEM cell and wideband TEM cell method
9. IEC 62132-4 (2006) Integrated circuits—Measurement of electromagnetic immunity 150 kHz to 1 GHz—Part 4: Direct RF power injection method
10. IEC 62132-6 (2009) Integrated circuits—measurement of electromagnetic immunity—part 6: Local Injection Horn Antenna (LIHA) method
11. IEC 62132-8 (2012) Integrated circuits—measurement of electromagnetic immunity—Part 8: Measurement of radiated immunity—IC stripline method
12. IEC 61000-4-3 (2006) Radiated, radio-frequency, electromagnetic field immunity test
13. Morva G et al (2017) Electromagnetic compatibility of digital electricity meters under influence of directional contactless electromagnetic fields. In: 33rd Kandó Conference. Budapest, Hungary pp 91–93

14. Electronic warfare and radar systems—engineering handbook, 4th ed. US Naval Air Warfare Center Weapons Division (2013)
15. Schwab AJ (1991) Elektromagnetische Verträglichkeit. Springer Verlag, Berlin
16. Zhezhelenko IV et al (2012) Electromagnetic compatibility of power consumers. Mashinostroenie, Moscow

Chapter 28
Improvement of Small-Signal Stability with the Incorporation of FACTS and PSS

Prasenjit Dey, Anulekha Saha, Sourav Mitra, Bishwajit Dey, Aniruddha Bhattacharya, and Boonruang Marungsri

Introduction

Expansion of power networks and ever increasing demands has compelled the engineers to develop newer techniques to optimize the cost–utility factor in power systems. Many devices related to power system have been developed since the past few decades in order to achieve reliable operation of power system in a more economical way [1].

Power systems can be split into transmission and distribution level. If factories employing highly sensitive and expensive equipment are not provided with quality power by the grid, they may suffer huge monetary losses. To avoid such a condition, custom power devices to improve voltage quality are installed at distribution levels. These devices are capable to reduce flicker, perform active filtering, and mitigate voltage dips and even interruptions by having additional sources of supply besides the main grid. Although DC is more efficient than AC for long-distance transmission of power, most of the existing lines transmit power in AC. Since total conversion

---

P. Dey (✉) · A. Saha
Department of Electrical and Electronics Engineering, NIT Sikkim, Sikkim, India
e-mail: deyprasenjit87@gmail.com

S. Mitra
Department of Electrical Engineering, IIT Kharagpur, Kharagpur, India

B. Dey
Department of Electrical Engineering, ISM Dhanbad, Dhanbad, India

A. Bhattacharya
Department of Electrical Engineering, NIT Durgapur, Durgapur, India

B. Marungsri
School of Electrical Engineering, Suranaree University of Technology, Nakhon Ratchasima, Thailand

© The Editor(s) (if applicable) and The Author(s), under exclusive license to Springer Nature Singapore Pte Ltd. 2021
A. K. Singh and M. Tripathy (eds.), *Control Applications in Modern Power System*, Lecture Notes in Electrical Engineering 710,
https://doi.org/10.1007/978-981-15-8815-0_28

from AC to DC is not economically feasible, engineers came up with devices called flexible AC transmission systems or, in short, FACTS [1], to maximize this AC power transmission capability.

FACTS can be broadly categorized into two subgroups: shunt and series. FACTS are a combination of dissimilar reactive elements required to regulate the excess reactive power that does not allow the active power to rise. They also help to improve power transmission by modifying the line reactance [1].

FACTS use thyristors to control the equivalent reactance. Many FACTS devices have already been developed but their applications are limited by high price and low efficiency. In the present work, only SVC (shunt type) and TCSC (series type) are considered since they are the most popular FACTS in their categories [2, 3].

Three evolutionary algorithms such as collective decision optimization algorithm (CDO) [4], grasshopper optimization algorithm (GOA) [5], and squirrel search algorithm (SSA) [6] have been applied to tune the controllers' parameters and their results compared. Observation of the results shows that SSA is the best-performing meta-heuristic among all three and TCSC performs better than SVC in maintaining small-signal stability of the system.

Small-Signal Model

Inclusion of the SVC and TCSC Together with PSS

Introducing an SVC in the system improves system stability [7–9]. The reactive power associated with the device is treated as an injection into the bus where it is located. The state equations to represent the small-signal model for SVC are given in [8]. TCSC has a significant impact on the system stability similar to that of SVC. The state equations to represent the small-signal model for TCSC are given in [8]. By having a clear idea of how the FACTS controller is modeled, it is possible to define specifically its inclusion in the system matrix [8]. Using the system of equations as a starting reference and equations for SVC [8] and for TCSC [8], one can modify the system in order to get the new DAE model including PSS [10]. The final system of equations takes the form as shown below:

$$\Delta \dot{x} = P_{1,\,\text{mod}}\, \Delta x + Q_1 \Delta I_g + Q_2 \Delta V_g + Q_{3,\,\text{mod}}\, \Delta V_l + E_1 \Delta u \qquad (28.1)$$

$$0 = R_1 \Delta x + S_1 \Delta I_g + S_2 \Delta V_g \qquad (28.2)$$

$$0 = R_2 \Delta x + S_3 \Delta I_g + S_4 \Delta V_g + S_5 \Delta V_l \qquad (28.3)$$

$$0 = R_{3,\,\text{mod}}\, \Delta x + S_6 \Delta V_g + S_{7,\,\text{mod}}\, \Delta V_l \qquad (28.4)$$

Using (28.1)–(28.4), the state-space matrix form becomes,

$$\begin{bmatrix} P_{1,\text{mod}} & Q_1 & Q_2 & Q_{3,\text{mod}} \\ R_1 & S_1 & S_2 & 0 \\ R_2 & S_3 & S_4 & S_5 \\ R_{3,\text{mod}} & 0 & S_6 & S_{7,\text{mod}} \end{bmatrix} \tag{28.5}$$

Simulations and Results

This section studies the effects of incorporating FACTS and PSS on the WSCC 3-machine 9-bus system [11], where the tuning of these controllers has been done using recently developed algorithms such as CDO, GOA, and SSA. The load disturbance has been created at bus 5 for system under study. The perturbation from $P_{(\text{Load})} = 1.25$ p.u to $P_{(\text{Load})} = 1.90$ pu has been given at bus 5. SSA is used to tune the controller parameters and is compared with GOA-based and CDO-based results. The objective function represented by (28.6) [10] is subjected to the constraints represented using Eqs. (28.7) and (28.8).

$$Z = Z_1 + Z_2 \text{ s.t. } Z_1 = \sum_{k=1}^{m} (\sigma_0 - \sigma_k)^2 \text{ and } Z_2 = \sum_{k=1}^{m} (\xi_0 - \xi_k)^2 \tag{28.6}$$

$$\left. \begin{array}{l} K_i^{\min} \leq K_i \leq K_i^{\max} \\ K_p^{\min} \leq K_p \leq K_p^{\max} \\ T_b^{\min} \leq T_b \leq T_b^{\max} \\ T_C^{\min} \leq T_C \leq T_C^{\max} \end{array} \right\} \text{ SVC parameters} \tag{28.7}$$

$$\left. \begin{array}{l} K_I^{\min} \leq K_I \leq K_I^{\max} \\ T_t^{\min} \leq T_t \leq T_t^{\max} \\ T_S^{\min} \leq T_S \leq T_S^{\max} \end{array} \right\} \text{ TCSC parameters} \tag{28.8}$$

The boundary values for T_b and T_c are set from 0.01 to 0.2 s in case of SVC. K is considered as 0.1. K_i varies in the boundary limit of 50–150 and K_p varies in the range of 0.01–0.5. For TCSC, T_t and T_S are in the boundary limit of 0.01–0.2 s and K_I from 0.02–20.

Case 1: SVC is Connected to the System

For investigating the impact of load variation on small-signal stability, the system eigenvalues (EVs) are evaluated after inclusion of FACTS. SVC is installed at bus 5 of the WSCC system as shown in Fig. 28.1a. This is because, bus 5 has the highest load and also the voltage is lowest at this bus [11]. SVC is installed to act as a voltage controller and provide additional damping through auxiliary control [8]. This also helps in getting a better insight into the impact of SVC on the small-signal stability. The main aim is to see its impact on small-signal stability of the system when it is equipped with SVC. Equations (28.1)–(28.4) have been used to construct the SVC-connected system matrix for obtaining the system EVs. Since electromechanical modes corresponding to the system EVs dictate small-signal stability, they have been obtained using SSA and compared with those obtained using GOA and CDO. The tuned parameters of SVC are given in Table 28.1 and the electromechanical modes are shown in Table 28.2.

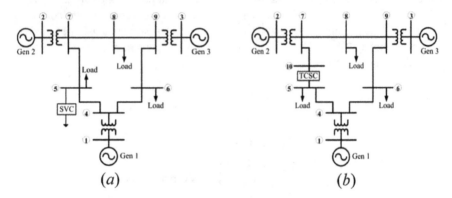

Fig. 28.1 a WSCC 3-machine 9-bus system equipped with SVC, b WSCC 3-machine 9-bus system equipped with TCSC

Table 28.1 Tuned parameters obtained for SVC

PSS parameters	CDO	GOA	SSA
K_i	84.104100	84.611900	79.242100
K_P	0.0415236	0.0306511	0.0110050
T_b (s)	0.0927996	0.0905143	0.1315680
T_C (s)	0.1277630	0.1303450	0.1044120

Table 28.2 EVs and damping ratio with SVC in WSCC system

	EVs and damping ratio	Dominant machine variables
Without SVC	$-0.7958 \pm j12.6386, 0.0628$	δ_3, ω_3
	$-0.2690 \pm j8.2179, 0.0327$	δ_2, ω_2
CDO-based SVC	$-0.82756 \pm j12.767, 0.0647$	δ_3, ω_3
	$-0.50434 \pm j8.2134, 0.0613$	δ_2, ω_2
GOA-based SVC	$-0.82758 \pm j12.767, 0.0647$	δ_3, ω_3
	$-0.50967 \pm j8.2204, 0.0619$	δ_2, ω_2
SSA-based SVC	**$-0.83669 \pm j12.767, 0.0654$**	δ_3, ω_3
	$-0.51102 \pm j8.2221, 0.06203$	δ_2, ω_2

Table 28.3 Tuned parameters obtained for TCSC

PSS parameters	CDO	GOA	SSA
K_1	0.315610	0.256520	0.543200
T_t (s)	0.039250	0.030032	0.028790
T_s (s)	0.095610	0.030028	0.154310

Case 2: TCSC Integration to the System

TCSC provides adequate slackening to the system along with maximizing the loading of transmission lines [12, 13]. Altering the firing angle of the thyristors makes the TCSC change its apparent reactance smoothly and rapidly. Due to its rapid and flexible regulation, it can improve the dynamic performance and is also capable of providing positive damping to the power system's electromechanical modes of oscillation. TCSC is installed between bus 5 and bus 7 of WSCC 9-bus system as shown in Fig. 28.1b, since bus 5 is having the highest loading. TCSC will have an impact on the system stability similar to that of SVC since they both act towards improving overall stability of the system. Equations (28.1)–(28.4) have been used to form the system matrix for the system having TCSC for obtaining the system EVs. SSA, GOA, and CDO optimization techniques have been used for tuning of TCSC parameters. The tuned parameters are shown in Table 28.3. Electromechanical modes that are obtained after tuning of TCSC parameters using the three algorithms and their comparison are shown in Table 28.4.

Case 3: PSSs are Connected in the System of Case 1 and Case 2

Power system stabilizer or PSS [10] is generally used for providing adequate slackening to the rotor oscillations. Idea behind the coordinated control by PSS together

Table 28.4 EVs and damping ratio with TCSC in WSCC system

	EVs and damping ratio	Dominant machine variables
Without TCSC	$-0.7958 \pm j12.6386, 0.0628$	δ_3, ω_3
	$-0.2690 \pm j8.2179, 0.0327$	δ_2, ω_2
CDO-based TCSC	$-0.881 \pm j12.728, 0.0690$	δ_3, ω_3
	$-0.505 \pm j8.2393, 0.0613$	δ_2, ω_2
GOA-based TCSC	$-0.9211 \pm j12.566, 0.07311$	δ_3, ω_3
	$-0.35094 \pm j8.329, 0.0421$	δ_2, ω_2
SSA-based TCSC	$\mathbf{-1.014 \pm j12.544, 0.08057}$	δ_3, ω_3
	$\mathbf{-0.7683 \pm j8.2865, 0.09232}$	δ_2, ω_2

with FACTS devices is to overcome the shortcomings of FACTS devices acting alone. Three optimization techniques which are mentioned in the previous section are considered here for tuning SVC and TCSC controller's parameters. Tuned parameters obtained using different optimization algorithms for the system having PSS together with SVC are shown in Table 28.5. Similarly Table 28.6 represents tuned

Table 28.5 Tuned parameters obtained for SVC and PSS (T in sec)

		CDO	GOA	SSA
Generator1	Kpss	37.3428900	30.528200	24.318200
	T_1	0.56301770	0.7218800	0.6123300
	T_2	0.04840599	0.0248835	0.0132033
	T_3	1.23912700	1.4553200	0.7054600
	T_4	0.01723039	0.0100000	0.0387581
Generator2	Kpss	3.28982200	3.1635600	2.3335600
	T	0.99425580	1.1912800	1.1349500
	T_2	0.01000000	0.0220860	0.0209160
	T_3	0.30518070	0.8663520	0.7583000
	T_4	0.01321604	0.0132453	0.0100000
Generator3	Kpss	1.09431200	1.1907400	6.0230000
	T_1	0.10099840	0.8735090	0.1000000
	T_2	0.12466170	0.0100000	0.0324000
	T_3	1.21203000	0.9712250	0.3278000
	T_4	0.10729460	0.0100000	0.1500000
SVC parameters	K_i	117.071400	76.682600	86.742000
	K_p	9.64856300	3.5571400	0.6114920
	T_b	0.16824390	0.0539470	0.2105470
	T_C	0.15784650	0.2000000	0.0154300

Table 28.6 Tuned parameters obtained for TCSC and PSS (T in sec)

		CDO	GOA	SSA
Generator1	Kpss	12.976500	43.597600	36.330000
	T_1	0.1273290	0.6940270	1.7861700
	T_2	0.1500000	0.0519461	0.0100000
	T_3	0.3568460	1.1496600	1.5000000
	T_4	0.0289991	0.0100000	0.0231800
Generator2	Kpss	2.5041900	4.5738400	7.3971000
	T_1	0.8822820	0.5657910	0.2381600
	T_2	0.0114734	0.0120770	0.0100000
	T_3	1.2144300	1.2298100	1.2044900
	T_4	0.0100000	0.0100000	0.0100000
Generator3	Kpss	1.7931900	4.0289300	3.1140000
	T_1	1.4655200	0.8564790	0.3672000
	T_2	0.0103135	0.0100000	0.0100000
	T_3	1.5000000	1.3868300	0.3274040
	T_4	0.0103509	0.0100000	0.1325000
TCSC parameters	K_1	0.4915540	2.6534700	1.4277600
	T_t	0.0185369	0.0635762	0.0100000
	T_s	0.1251820	0.0100000	0.0720567

parameters achieved using different optimization algorithms for the system equipped with PSS and TCSC.

It can be seen from Tables 28.7 and Table 28.8 that the EVs associated to electromechanical modes so obtained using SSA are used to evaluate stability of the system and a comparative analysis with those achieved by GOA and CDO. Results emphasize superiority of SSA over GOA and CDO in assessing small-signal stability of the system.

Table 28.7 EVs and damping ratio with SVC and PSS

	EVs and damping ratio	Dominant machine variables
Without TCSC	$-0.7958 \pm j12.6386, 0.0628$	δ_3, ω_3
	$-0.2690 \pm j8.2179, 0.0327$	δ_2, ω_2
CDO-based TCSC	$-0.82687 \pm j12.999, 0.06348$	δ_3, ω_3
	$-2.1585 \pm j8.1386, 0.25635$	δ_2, ω_2
GOA-based TCSC	$-1.1559 \pm j\,14.261, 0.08079$	δ_3, ω_3
	$-3.1826 \pm j\,6.9437, 0.41666$	δ_2, ω_2
SSA-based TCSC	**$-1.0526 \pm j13.431, 0.07813$**	δ_3, ω_3
	$-3.215 \pm j6.6814, 0.4336$	δ_2, ω_2

Table 28.8 EVs and damping ratio with TCSC and PSS

	EVs and damping ratio	Dominant machine variables
Without TCSC	$-0.7958 \pm j12.6386, 0.0628$	δ_3, ω_3
	$-0.2690 \pm j8.2179, 0.0327$	δ_2, ω_2
CDO-based TCSC	$-1.5914 \pm j15.678, 0.10099$	δ_3, ω_3
	$-2.3836 \pm j8.5973, 0.26717$	δ_2, ω_2
GOA-based TCSC	$-1.4373 \pm j16.264, 0.08803$	δ_3, ω_3
	$-3.2599 \pm j6.791, 0.43275$	δ_2, ω_2
SSA-based TCSC	**$-1.5140 \pm j13.588, 0.11074$**	δ_3, ω_3
	$-3.712 \pm j6.503, 0.49574$	δ_2, ω_2

SVC versus TCSC: A Comparative Analysis

Table 28.9 demonstrates the comparison between the system with SVC and the system with TCSC using best values obtained by SSA. It can be observed that damping ratio of the electromechanical modes have been improvised from 0.0628 to 0.065395 and 0.0327 to 0.062032 in case of SVC and 0.0628 to 0.08057 and 0.0327 to 0.062032 in case of TCSC incorporated system. Further improvement can be observed when PSSs are installed in the system. Damping ratio further improved from 0.065395 to 0.07813 and 0.062032 to 0.4336 for the system with SVC coordinated PSS. For TCSC, coordinated PSS system provides improved damping ratio from 0.08057 to 0.11074 and 0.09232 to 0.49574. Therefore, it can be concluded from Table 28.9 that TCSC is providing better damping ratio as compared to SVC. Also, the inclusion of PSS with TCSC is improving the damping ratio further as compared to the inclusion of PSS with SVC.

Figure 28.2 demonstrates the comparison between electromechanical modes for different cases discussed above. Figure 28.2a addresses the comparison between best

Table 28.9 EVs and damping ratio comparison between SVC and TCSC

	EVs and damping ratio	Dominant machine variables
Without TCSC	$-0.7958 \pm j12.6386, 0.0628$	δ_3, ω_3
	$-0.2690 \pm j8.2179, 0.0327$	δ_2, ω_2
SSA-based SVC	$-0.83669 \pm j12.767, 0.065395$	δ_3, ω_3
	$-0.51102 \pm j8.2221, 0.062032$	δ_2, ω_2
SSA-based TCSC	**$-1.014 \pm j12.544, 0.08057$**	δ_3, ω_3
	$-0.7683 \pm j8.2865, 0.09232$	δ_2, ω_2
SSA-based PSS and SVC	$-1.0526 \pm j13.431, 0.07813$	δ_3, ω_3
	$-3.215 \pm j6.6814, 0.4336$	δ_2, ω_2
SSA-based PSS and TCSC	**$-1.5140 \pm j13.588, 0.11074$**	δ_3, ω_3
	$-3.712 \pm j6.503, 0.49574$	δ_2, ω_2

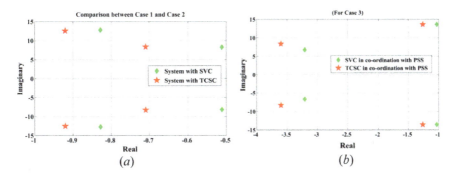

Fig. 28.2 Comparative analysis for the eigenvalues of different scenario (*Case 1* to *Case 3*)

EVs obtained using SSA for both SVC-based and TCSC-based controllers. From the figure, it is observed that the system with SVC gives proper damping to the system but further improvement can be obtained when TCSC is connected. Also, it is observed that the real parts of EM mode are shifted more toward left half of S plane in case of TCSC compared to SVC. The performance is further enhanced for TCSC and PSS coordinated system as indicated by Fig. 28.2b.

Again to embellish the superiority of TCSC over SVC, a 3ø fault is created near bus 7 at 0.1 s, which is mitigated at 0.2 s without tripping any line. Though there are various parameters that can be shown in time domain response, change in speed deviation is enough for arriving at a conclusion regarding system stability. Therefore, only the change in rotor speed deviations obtained after time domain simulation is demonstrated in Fig. 28.3a, b for **Case 3** using best values obtained by SSA. It is observed that TCSC included system is more stabilized and also requires lesser settling time to mitigate the system oscillations as compared to SVC incorporated system.

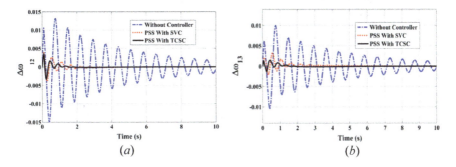

Fig. 28.3 Change in $\Delta\omega_{12}$ and $\Delta\omega_{13}$ for *Case 3*

Conclusion

This paper considers the tuning of FACTS controllers individually as well as in coordination with PSS for reducing the LFOs. Controller parameters have been tuned using SSA optimization technique and its performance compared with those obtained using GOA and CDO algorithms. Eigenvalue analysis demonstrates effective damping achieved by FACTS controllers for reduction of LFO problem when coordinated with PSS. It is also observed that TCSC performs better than SVC. TCSC has strong effect on rotor angle stability as it regulates the power flow and power is tightly coupled with angle. SVC has a strong effect on voltage stability and does reactive power compensation by controlling the voltage magnitude. Incorporation of SVC can also improve small-signal stability to a certain extent.

References

1. Acha E, Fuerte-Esquivel CR, Ambriz-Perez H, Angeles-Camacho C (2004) FACTS: modelling and simulation in power networks. Wiley
2. Hingorani NG, Gyugyi L (2000) Understanding FACTS: concepts and technology of flexible AC transmission systems. IEEE Press
3. Barve G (2014) Application study of FACTS devices in indian power system. Int J Comput Technol IJCAT 1(1):57–59
4. Zhang Q, Wang R, Yang J, Ding K, Li Y, Hu J (2017) Collective decision optimization algorithm: a new heuristic optimization method. Neurocomputing 221:123–137
5. Mirjalili SZ, Mirjalili S, Saremi S, Faris H, Aljarah I (2018) Grasshopper optimization algorithm for multi-objective optimization problems. Appl Intell 48(4):805–820
6. Saha A, Dey P, Bhattacharya A, Marungsri B (2019) A new meta-heuristic algorithm for solving transient stability constrained optimal power flow. SSA 1(2):1–11
7. Hassan MO, Cheng SJ (2009) Steady-state modeling of SVC and TCSC for powerflow analysis. Int MultiConf Eng Comput Sci 2
8. Pai MA, Gupta DS, Padiyar KR (2004) Small signal analysis of power systems. Alpha Science Int'l Ltd.
9. Anjos C (2011) Optimal location of facts for the enhancement of power system security. In: Master'sthesis, Instituto Superior Tecnico, Technical University of Lisbon
10. Dey P, Mitra S, Bhattacharya A, Das P (2019) Comparative study of the effects of SVC and TCSC on the small signal stability of a power system with renewables. J Renew Sustain Energy 11(3):033305
11. Sauer PW, Pai MA (1998) Power system dynamics and stability. Urbana
12. Trindade M (2013) Optimal location of FACTS to optimize power system security. In: Master's thesis, Instituto Superior Tecnico, Technical University of Lisbon
13. Pai MA et al. (2016) Small signal analysis of integrated power systems. Narosa Publishing House Pvt. Ltd.

Chapter 29
Optimal Threshold Identification of Fault Detector Using Teaching and Learning-Based Optimization Algorithm

Ch. Durga Prasad and Monalisa Biswal

Introduction

Ideal values of thresholds of FD's based on the normal operating conditions of the power system may not produce correct decisions during disturbances. Hence, investigations of all types of faults by varying the fault parameters in their boundaries are needed to set a proper threshold for FD yields reliable and fast decisions during faults [1–5]. This task is difficult particularly for transmission line fault detector since the differences of the boundaries (limits) of the fault parameters like line length and fault resistance are large in magnitude. Several fault detectors implemented with simple mathematical approaches were available in literature with either ideal threshold settings or extensive simulation-based threshold setting [3–5]. However, magnitudes of the fault detection indices vary with fault parameters and these parameters such as fault type, fault location, fault inception, and fault resistance are generated randomly in their upper and lower limits for investigating extensive case studies. Therefore, population-based techniques may provide an acceptable solution for this threshold setting problem. Earlier, swarm optimization algorithm is applied for threshold setting in different ways [6–8]. PSO is introduced first time for mean error-based statistical fault detector employed on instantaneous current signal in transmission line protection with two primary fault parameters known as fault location and fault inception angle [6]. Later, this swarm intelligence assistance is provided to FD implemented with the help of time frequency transformation of current signals. Along with aforementioned fault parameters, fault resistance is also

---

Ch. Durga Prasad (✉) · M. Biswal
Department of Electrical Engineering, NIT Raipur, Raipur, Chattishgarh, India
e-mail: dpchinta@srkrec.edu.in

M. Biswal
e-mail: mbiswal.ele@nitrr.ac.in

considered in the threshold setting process [7]. Recently, superimposed-based power differential scheme is enhanced with this PSO threshold setting mechanism [8]. This PSO assistance provides better threshold than existing mechanisms.

However, several advanced population search-based optimization algorithms are originated after PSO. Since PSO needs selection of control parameters which influence the process of identification of overall global best values and also suffer with prematurity condition. Therefore, PSO is replaced in this paper by TLBO and the mean error estimation technique equipped with TLBO-based fault detector is implemented and tested its performance on two bus power system with wide variety of cases. Compared to PSO, TLBO produces better optimal threshold and ensured more reliable results.

Proposed Method

In this paper, mean error-based fault detector algorithm is considered for setting TLBO-assisted threshold. This method is implemented on instantaneous current signal since the disturbances/faults influences it in greater extent. The concept is based on the error value calculated between actual current sample and estimated current sample. For estimation of current sample, three consecutive samples are considered in each phase of the line currents where the arithmetic mean of the first and last samples is the middle sample. Suppose, $i_{p(n-1)}, i_{p(n)}$ and $i_{p(n+1)}$ are the three consequetive current samples of phase-p at instant 'n'. Then, the error E_{pn} is expressed as (29.1)

$$E_{pn} = i_{p(n)\text{actual}} - i_{p(n)\text{estimated}} \qquad (29.1)$$

In Eq. (29.1), $i_{p(n)\text{estimated}}$ is the current sample estimated from the preceding and succeeding samples given by

$$i_{p(n)\text{estimated}} = \frac{i_{p(n-1)} + i_{p(n+1)}}{2} \qquad (29.2)$$

This error is evaluated for all three phases and final detection index is designed from the absolute sum of all three phase errors given by

$$\beta_n = E_{an} + E_{bn} + E_{cn} \qquad (29.3)$$

The value of the fault detection index β_n exceeds the maximum error value attained during normal operating conditions and hence detects the fault, if

$$\beta_n > \varepsilon \qquad (29.4)$$

Since maximum error value attained for normal operating condition may not ensure reliability and at the same time, PSO assistance struggles with prematurity and needs multitrials creating a gap for fine threshold setting. Hence, TLBO algorithm is opted in this paper to overcome both drawbacks. TLBO is a simple population search based optimization algorithm implemented by R.V. Rao in the year 2011 based on the concept of learning of a student in the classroom environment. In this algorithm, searching and updating mechanisms are done in two stages known as teacher phase and learner phase. As this algorithm is familiar in engineering research, complete details are available in [9, 10]. To identify optimal value for the threshold of the mean error estimation technique using TLBO, a fitness function is framed given by

$$\varepsilon = \min\{\max(\beta(x, t))\} \quad (29.5)$$

In Eq. (29.5), β is the fault index function whose value depends on the fault location x and fault inception time t included in the problem with upper and lower limits given by

$$0 \leq x \leq L \quad (29.6)$$

$$t_{min} \leq t \leq t_{max} \quad (29.7)$$

where L is the total length of the transmission line and t_{min} and t_{max} are the minimum and maximum limits of fault initiation time in seconds.

System Studied

A power system model shown in Fig. 29.1 is considered for implementation proposed TLBO threshold setting mechanism. This test system specifications are 400 kV, 50 Hz with positive and zero-sequence impedance (per km base) of the transmission line are $(0.03 + j\,0.34)\,\Omega$ and $(0.28 + j\,1.04)\,\Omega$, respectively. The FD utilized instantaneous current signal with1 kHz sampling rate for verification of TLBO threshold efficacy.

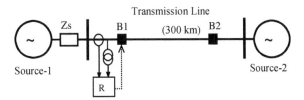

Fig. 29.1 Single-line representation of test system

Simulation Results

For the fault detector, initially, intelligent threshold is set by using TLBO and later, several case studies are investigated to show/check the effectiveness of the proposed fault detector with TLBO threshold.

Threshold Setting Using TLBO

While setting optimal threshold for fault detector using TLBO, process is initiated with 300 populations with random solutions and final optimal value is achieved at the end of final iteration 50. These random solutions are generated within the limits of the variables as mentioned in Eqs. (29.6) and (29.7). The lower limits of variables fault location x, fault inception time t are 1, 0.02 and upper limits are 299, 0.04, respectively. There is no additional burden regarding selection of control parameters since TLBO learning and updating equations are free from such control parameters. As suggested in [6], only line-to-ground faults are simulated by TLBO in threshold setting process. Initially, AG fault is taken for TLBO and at the end of optimization process, the optimal threshold value converges at 0.8358. This solution is achieved when the fault location is 151.0179 km and fault inception time is 0.0397 s. Similar process is carried out for other line-to-ground faults and corresponding results are presented in Table 29.1.

Out of 3-line-to-ground faults, minimum fault detection index is obtained for BG fault and hence the optimal threshold value of the fault detector is set at 0.7347 and this value obtained is at the end of final iteration of TLBO. In this process, initial iteration and final iteration of TLBO solutions are plotted in Fig. 29.2 to show the convergence of TLBO in the optimal threshold setting mechanism. Figure 29.3 shows the entire surface of fault detection index from initial iteration to final iteration. This complex surface shows the importance of the TLBO in this threshold setting problem and the reliability is tested in reverse engineering manner with few more random case studies presented in followed cases.

Table 29.1 Optimal indices identified using TLBO for various LG faults

Type of fault	Fault location (km)	Fault inception (s)	Optimal threshold
AG	151.0179	0.0397	0.8358
BG	177.1939	0.0363	0.7347
CG	266.9326	0.0329	0.8431

Fig. 29.2 Data pattern generated by TLBO of all particles for initial and final iterations in case of BG fault

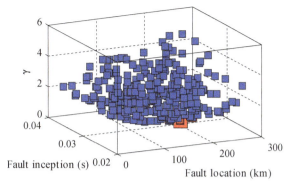

Fig. 29.3 Fault detection index surface generated by TLBO for BG fault

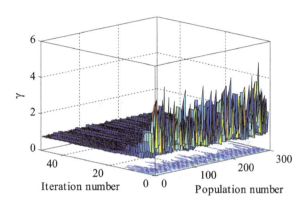

Performance of Proposed Method During Remote End Faults

Fault detection indices have less magnitude for remote-end faults in comparison with other locations. But the interesting key observation from the TLBO results is that the optimal threshold corresponding fault location points. These fault locations are nearer to middle of the line. Hence, to strengthen the applicability of TLBO in this aspect, remot-end faults need to be verified. When AG fault with fault resistance of 20Ω is located at 270 km (90% of section of line), the response of fault detector with proposed TLBO threshold is shown in Fig. 29.4. From the inception of fault (0.042 s), FD detected it in 3 ms and generated trip command.

Performance of Proposed Method During High Resistive Faults

Fault resistance is also one parameter which influences magnitude of the fault but it is not included in the objective function while applying TLBO and hence a high resistive

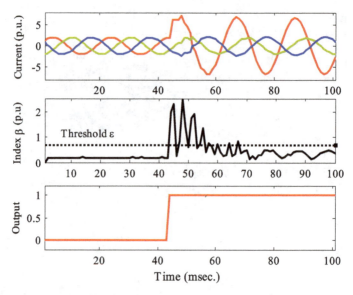

Fig. 29.4 Performance of proposed method during remote end fault

fault case is considered to test its effect on FD. For this purpose, BG fault is created at 90 km from R with a fault resistance of 100Ω initiated at 0.05 s. Figure 29.5 shows the corresponding results and even the fault resistance is not taken into account; in the

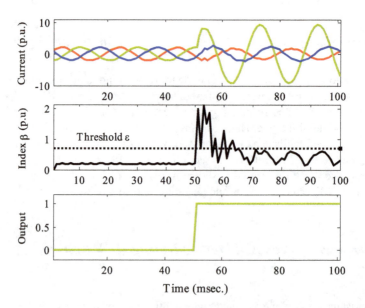

Fig. 29.5 Response of proposed method during high resistance fault

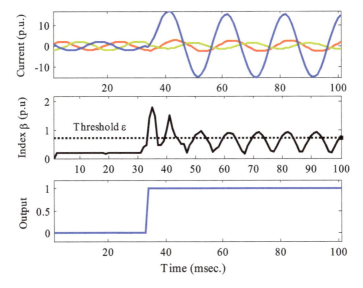

Fig. 29.6 Performance of proposed method during zero inception fault

fitness function formation, faults with high resistance are detected by the proposed method.

Performance of Proposed Method for Faults at Zero Inception Angles

When fault occurs at zero inception angle, the index is low compared to other inception angle cases. Of course, it is included in the objective function of TLBO. However, to show the reliability of the proposed method, CG fault is created at 150 km from with fault resistance of 30 Ω initiated at 0.0314 s. The corresponding detection plots of proposed FDU are presented in Fig. 29.6. From Fig. 29.6 it is clearly observed that the fault is detected within 5 ms from its inception since TLBO simulated these cases in the process of optimal threshold setting.

Improvement in the Threshold Value Compared to PSO

TLBO provides better solutions in less time with high convergence rate than PSO [9]. In this paper also, the results obtained by TLBO are compared with PSO [6] and reported in Table 29.2. From Table 29.2, it is concluded that the TLBO provides comparatively minimum threshold value than PSO. Hence, TLBO is suggested in

Table 29.2 Comparison of PSO and TLBO for BG fault

Algorithm	Fault location	Fault inception	Optimal threshold
PSO [6]	177.1283	0.0264	0.7349
TLBO	177.1939	0.0363	0.7347

Table 29.3 Performance of proposed method with and without source impedance

Fault	Location (km)	Inception (s)	Resistance (Ω)	Without Z_s		With Z_s	
				Index	Detection time (s)	Index	Detection time (S)
AG	120	0.06	30	0.923	2	0.813	7
BG	180	0.07	50	3.104	2	3.001	3
CG	240	0.08	20	6.611	5	3.816	3

place of PSO for better results. Of course, the results seem to be closer since several runs were executed in case of PSO.

Impact of Source Impedance on Proposed Method

The proposed threshold setting mechanism is adopted when the test system offers source impedance. In this case, different faults are verified with and without source impedance and reliable results are achieved in both cases with proposed intelligent threshold setting. Table 29.3 shows the response of proposed method for various line-to-ground faults with and without source impedance.

Conclusions

TLBO-assisted threshold setting mechanism is presented in this paper for enhancing the performance of mean error-based fault detector in transmission line protection. An objective function is framed in this paper for fault detector with fault parameters and the minimum value of the function is achieved by TLBO. The method is tested on a standard power network and results for different fault cases prove the efficacy of the technique. Finally, comparisons are made with PSO to show the improvements in the threshold setting mechanism by using TLBO.

References

1. Phadke AG, Thorp JS (2009) Computer relaying for power systems. Wiley, Chichester, West Sussex, England
2. Biswal M (2016) Faulty phase selection for transmission line using integrated moving sum approach. IET Sci Meas Technol 10(7):761–767
3. Mohanty SR, Pradhan AK, Routray A (2007) A cumulative sum-based fault detector for power system relaying application. IEEE Trans Power Delivery 23(1):79–86
4. Prasad CD, Nayak PK (2019) A DFT-ED based approach for detection and classification of faults in electric power transmission networks. Ain Shams Eng J 10(1):171–178
5. Prasad CD, Nayak PK (2018) A mixed strategy approach for fault detection during power swing in transmission lines. Artif Intell Evol Comput Eng Syst: 597–607
6. Prasad CD, Nayak PK (2018) Performance assessment of swarm-assisted mean error estimation-based fault detection technique for transmission line protection. Comput Electr Eng 71:115–128
7. Prasad CD, Biswal M, Nayak PK (2020) Wavelet operated single index based fault detection scheme for transmission line protection with swarm intelligent support. Energy systems Springer. https://doi.org/10.1007/s12667-019-00373-9
8. Prasad CD, Biswal M (2020) Application of particle swarm optimization for threshold setting in fault detection unit. In: Proceedings of iCASIC conducted VIT Vellore
9. Rao RV, Savsani VJ, Vakharia DP (2012) Teaching–learning-based optimization: an optimization method for continuous non-linear large scale problems. Inf Sci 183(1):1–15
10. Boudjefdjouf H, Mehasni R, Orlandi A, Bouchekara H, De Paulis F, Smail MK (2015) Diagnosis of multiple wiring faults using time-domain reflectometry and teaching–learning-based optimization. Electromagnetics 35(1):10–24

Chapter 30
A Novel MTCMOS Stacking Approach to Reduce Mode Transition Energy and Leakage Current in CMOS Full Adder Circuit

Anjan Kumar and Sangeeta Singh

Introduction

Because of the developing requirement for high performance, low voltage and small static power consuming systems, several static power reduction techniques have been proposed. MTCMOS is widely used technique to limt static power consumption [4]. In this technique, transistors with higher threshold voltage and lower threshold voltages are used. Low V_{th} threshold transistors are applied to meet the speed requirement and high V_{th} transistors are applied to achieve the power requirement [2, 3, 9]. In non operatory mode, low threshold circuits are isolated from ground or power rail using high threshold transistors. Energy consumption during mode transition is a major limitation of the MTCMOS technique. Several techniques are proposed by the various research group to minimize leakage current and mode transition energy [8, 10].

Liu et al. proposed an innovative MTCMOS technique based on charge recycling to reduce mode transition energy [7]. The major limitation of these techniques is the requirement of a complex timing control unit to excite the action of capacitor charge recycling in sleep to active mode transition. Abdollahi et al. proposed another alternative approach to reduce the mode transition energy [1, 6]. In this technique, the short circuit current produced during mode transition is reduced by connecting separate sleep transistor for each stage. Each High V_{th} transistors are on at different time using different control signal during mode change. The main limitations of this technique are high wakeup delay and area overhead because of more than one sleep transistor.

A. Kumar (✉)
GLA University, Mathura, India
e-mail: anjan.kumar@gla.ac.in

A. Kumar · S. Singh
National Institute of Technology Patna, Patna, India

© The Editor(s) (if applicable) and The Author(s), under exclusive license to Springer Nature Singapore Pte Ltd. 2021
A. K. Singh and M. Tripathy (eds.), *Control Applications in Modern Power System*, Lecture Notes in Electrical Engineering 710,
https://doi.org/10.1007/978-981-15-8815-0_30

Jiao et al. gave a new idea based on the multiple-phase high V_{th} sleep signal modulation technique to reduce the mode transition energy [5]. When the ground line (virtual) reaches close to the ground value of sleep signal increases quickly using the additional circuit in phase 3. The overall process decreases the reactivation time. The major limitation of this technique is the need of complex analog circuit to generate a three-phase sleep transistor control circuit.

Most of the technique proposed so far needs a complex circuit to control the mode transition. Here an attempt is being made to reduce the mode transition energy while keeping overall circuit simple. In this technique concept of stacking is used to reduce mode transition energy. In general stacked MTCMOS circuit only NMOS is stacked but in our proposed technique combination of NMOS and PMOS is used. Mathematical analysis of complete stacked NMOS and PMOS is performed in sec. The result is very much encouraging as discussed in Section 'Simulation Result'.

Proposed Design

A novel low energy stacking MTCMOS technique is proposed. Here a PMOS transistor is connected above the NMOS transistor. In sleep mode, both series-connected PMOS and NMOS are turned off. In active mode, the extra series-connected

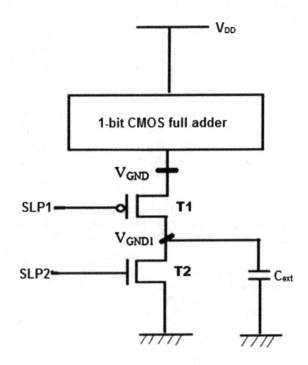

Fig. 30.1 Proposed stacked MTCMOS structure

transistors (NMOS and PMOS) are turned on. In sleep mode, both NMOS and PMOS is turned off. This additional modification significantly reduces leakage power dissipation in non operatory mode in contast with a single off device. Minimize in leakage current is achieved by stacking effect of off-state NMOS and PMOS transistors as depicted in Fig. 30.1. When stacked PMOS and NMOS is turned off, the voltage of VG1 is increased to positive value because of little drain current. This reduces the drain to the source voltage (Vds) of transistor T1.

Methodology

Strategy to Minimize Energy

In proposed stacked based design, the VGND is retained at a small voltage close to VGND in active mode. In sleep mode, both PMOS and NMOS are turned off. The potential at node VGND is now steadily charged to a voltage power line voltage close to V_{DD}. In sleep mode, the small amount of the total consumed energy is stored at capacitor C_{int}. When both NMOS and PMOS is switched on, the energy stored at capacitor C_{int} is consumed by the ON state PMOS transistor and charge the intermediate capacitance C_{ext}. This charged capacitor doesn't discharge too quickly because of the inverted sleep signal, which is fed to the NMOS sleep transistor. When the sleep transistor T2 is active, the charge stored at capacitor C_{ext} gets discharged. Because of the stacked structure, the energy is resumed on the circuitry and a small amount of energy is wasted on the ground.

$$E_{CONV} = E_{dyn} + E_{stat} + E_{transition} \tag{30.1}$$

$$E_{dyn} = 0.5 \times C_L \times V_{DD}^2 \tag{30.2}$$

$$E_{stat} = I_{LEAK} \times V_{DD} \times Period \tag{30.3}$$

$$E_{TRANSITION} \cong C_{V_{GND}} \times V_{DD}^2 \tag{30.4}$$

Equivalent RC circuit in different modes of operation is as follows.

Operatory Mode(Active State):

In this mode, Transistors connected at footer are in ON state. Both sleep transistors show small value of resistance and potential at virtual ground node is equal to zero. So the net voltage across the combinational circuit is equal to V_{DD} (Fig. 30.2).

When T1 and T2 transistors are turned on, the equivalent resistance of T1 and T2 are $R_{p,ON}$ and $R_{n,ON}$ respectively. The values of these resistances are very small as

Fig. 30.2 Equivalent RC circuit in active mode

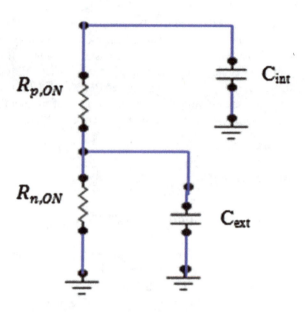

both the transistors are in ON state. Capacitor (C_{int}) is an internal parasitic capacitance of first transistor T1. An additional external capacitor (C_{ext}) is placed between the transistors T1 and T2 to limit the potential of VGND1. Internal capacitance at the drain of T2 is neglected, as external capacitance at node VGND1 has a high value. In the normal active mode of operation both the capacitors (C_{int} and C_{ext}) are discharged, because of the low resistance path. Potential across internal capacitor $C_{int} \approx$

$$VC_{int\,ACTIVE} \approx V\left(R_{p,ON}\right) + V\left(R_{n,ON}\right) \approx 0V \qquad (30.5)$$

Potential across external capacitor $C_{ext} \approx$

$$VC_{ext\,ACTIVE} \approx V\left(R_{n,ON}\right) \approx 0V \qquad (30.6)$$

Non-Operatory Mode(Sleep State):

Both PMOS and NMOS are turned off by connecting NMOS to ground and PMOS to power line (Fig. 30.3).

In this mode resistance offered by both the transistors are very high as both the transistor (NMOS and PMOS) are turned off. Because of high resistance, both the capacitors are not finding any low resistance path to discharge. As sub-threshold leakage current flows from low threshold full adder to the capacitor C_{int}, first capacitor C_{int} charges close to the. External capacitor C_{ext} also charges to a small value, because

Fig. 30.3 RC equivalent circuit in standby mode

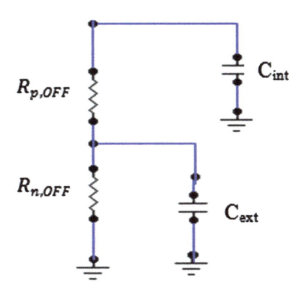

most of the charge is acquired by the capacitor C_{int}, providing no path for capacitor C_{ext} to charge.

Potential across internal capacitor $C_{int} \approx$

$$VC_{int\,SLEEP} \approx V(R_{p,ON}) + V(R_{n,ON}) \approx 0V \tag{30.7}$$

Potential across external capacitor $C_{ext} \approx$

$$VC_{ext\,SLEEP} \approx V(R_{n,ON}) \approx 0V \tag{30.8}$$

Transition Mode (Sleep to Active):

During mode transition, the first PMOS sleep transistor is turned on by connecting SLP1 to the ground whereas the second NMOS transistor is switched on after some delay. By doing this, fluctuation at a virtual ground node can be reduced.

Mathematical Modeling of Proposed Stacked Based MTCMOS Circuit

It is a known fact that capacitor voltage changes slowly, the potential across external capacitors C_{ext} and internal capacitor C_{int} remain equal by turning ON the sleep transistor T1. Figure 30.4 shows the equivalent circuit of stacked PMOS and NMOS at $t = 0^+$ (Fig. 30.5).

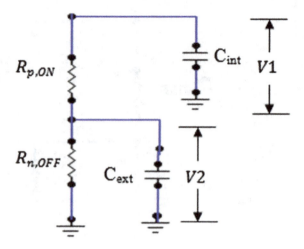

Fig. 30.4 Equivalent RC diagram of NMOS and PMOS transistor in 0<t<t1

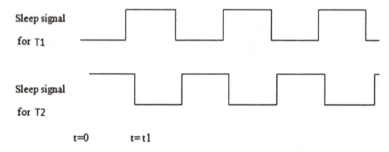

Fig. 30.5 Wave form applied on SLP and SLP1 of Transistor T1 and T2

Potential across internal capacitor $C_{int} \approx V1$, and
Potential across external capacitor $C_{ext} \approx V2$.

Since NMOS is off, so it shows high resistance $R_{nmos,OFF}$ value and it can be treated as open switch as shown in Fig. 30.6.
The current equation calculated using Fig. 30.7a:

$$i = \frac{V1 - V2}{R_{p,ON}} \qquad (30.9)$$

At t=0$^+$, the above circuit becomes,
As we know that $V_C = \frac{1}{c} \int i \, dt$, Applying KVL in the circuit shown in Fig. 30.7b

$$\frac{1}{C_{int}} \int i \, dt + \frac{1}{C_{ext}} \int i \, dt + i R_{p,ON} = 0 \qquad (30.10)$$

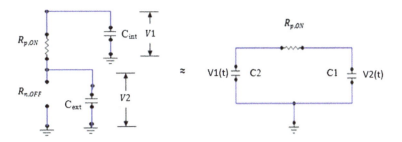

Fig. 30.6 Equivalent circuit of stacked PMOS and NMOS at $t = 0^+$

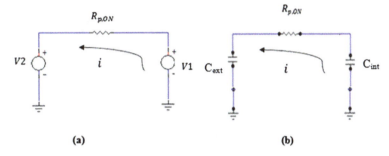

Fig. 30.7 **a** Circuit configuration at $t = 0^+$, **b** circuit configuration at $t > 0$

Differentiating both sides

$$\frac{i}{C_{int}} + \frac{i}{C_{ext}} + R_{p,ON}\frac{di}{dt} = 0 \qquad (30.11)$$

After solving

$$i = ke^{-t/R_{p,ON}C_{eq}} \qquad (30.12)$$

At Time $t = 0^+$

$$i(0^+) = k = \frac{V1 - V2}{R_{p,ON}} \qquad (30.13)$$

Putting this value in the above equation, we get

$$i(t) = \frac{V1 - V2}{R_{p,ON}} e^{-\left(\frac{t}{R_{p,ON}C_{eq}}\right)} \qquad (30.14)$$

During 0<t<t1, internal capacitor C_{int} starts discharging and external capacitor C_{ext} starts charging. This charging discharging process continues till both C_{int} and C_{ext} have the equal potential. Voltage across capacitor can be calculated using

$$V2(t) = \frac{1}{C_{ext}} \int_{-\infty}^{t} i\, dt \qquad (30.15)$$

$$V2(t) = \frac{1}{C_{ext}} \int_{-\infty}^{0} i\, dt + \frac{1}{C_{ext}} \int_{0}^{t} i\, dt \qquad (30.16)$$

$$V2(t) = V2\left(0^{-}\right)\left[= V2\left(0^{+}\right]\right. + \frac{1}{C_{ext}} \int_{0}^{t} i\, dt \qquad (30.17)$$

Above equation gives the voltage across the external capacitor C_{ext}

$$V2(t) = V2 + \frac{1}{C_{ext}} \int_{0}^{t} \frac{V1 - V2}{R_{p,ON}} e^{-\left(\frac{t}{R_{p,ON} C_{eq}}\right)} dt \qquad (30.18)$$

Similarly, the voltage across the capacitor C1 will be calculated using

$$-V1(t) = \frac{1}{C_{int}} \int_{-\infty}^{t} i\, dt \qquad (30.19)$$

$$-V1(t) = -V1 + \frac{1}{C_{int}} \int_{0}^{t} \frac{V1 - V2}{R_{p,ON}} e^{-\left(\frac{t}{R_{p,ON} C_{eq}}\right)} dt \qquad (30.20)$$

Solution of above equation gives

$$V_{C_{int}}(t) = \frac{V2 C_{ext} + V1 C_{int}}{C_{int} + C_{ext}} + \frac{C_{ext}}{C_{int} + C_{ext}} (V1 - V2) e^{-\frac{t}{\tau}} \qquad (30.21)$$

where

$$\tau = R_{p,ON} \left(\frac{C_{int} C_{ext}}{C_{int} + C_{ext}} \right) \qquad (30.22)$$

Equation 30.21 shows the net voltage drawn from power supply for charging both the capacitors (C_{int} and C_{ext}) during complete reactivation. When making transition from sleep to active mode, intermediate node capacitors get discharged due to on current I_{ON} flowing from low threshold full adder to ON sleep transistors. Because of the stacked PMOS and NMOS, considerable amount of charge is saved rather than disposing on the ground. Hence the net energy consumed from power supply during sleep to active mode ($E_{SLEEP-TO-ACTIVE}$) transition for charging intermediate virtual ground node (VGND and VGND1) is reduced.

Simulation Result

Leakage Current in Sleep Mode

Leakage power in sleep mode is measured when the circuit is not operatory. High threshold PMOS and NMOS is turned off by applying an input of 1 and 0 V on SLP and SLP1.

Figure 30.8 shows a reduction in sleep mode leakage current as compared to the conventional Technique in 1-bit full adder circuit.

Table 30.1 shows the comparison of leakage current of 1-bit full adder circuit with conventional MTCMOS Technique for different input combinations. Results clearly reflect that proposed technique reduces leakage current by significant amount.

Static Energy

For each operation energy of circuit is calculated by summing both static energy and dynamic energy. Integration of leakage current during complete sleep time period gives static energy.

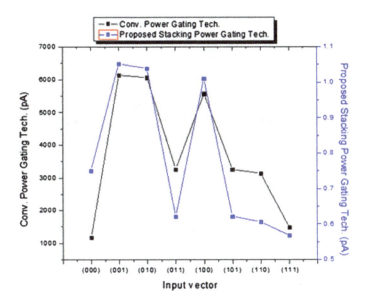

Fig. 30.8 Comparison of leakage current with different input combination in conventional and proposed technique

Table 30.1 Comparison of Leakage current in conventional and proposed MTCMOS technique

Input vector C A B	Conventional MTCMOS (nA)*	Proposed MTCMOS (PA)*
000	1.1795	0.7503
001	6.1382	1.0524
010	6.0658	1.0397
011	3.2568	0.6218
100	5.5682	1.0110
101	3.2568	0.6218
110	3.1325	0.6060
111	1.4824	0.5686

Table 30.2 Comparison of static energy in conventional and proposed MTCMOS technique

Input vector C A B	Conventional MTCMOS (nA)*	Proposed MTCMOS (PA)*
000	58.975	37.517
001	306.91	52.622
010	303.29	51.989
011	162.84	31.090
100	278.41	50.550
101	162.84	31.090
110	156.62	30.303
111	74.123	28.432

As shown in the Table 30.2 the static energy dissipation is greatly reduced as compared to the conventional technique by a good amount. When the entire input vector is same the energy dissipation is minimum.

Mode Transition Energy

When making the transition from sleep to active vice versa, a considerable amount of mode transition energy is consumed. Consumption in energy is attributed to the charging and discharging of a capacitor connected at a virtual ground node. To satisfy performance requisite, the size of the sleep transistor is generally taken larger in comparison to other transistors connected in the main logical block. A large-sized sleep transistor further increases the internal capacitance value of virtual lines (Fig. 30.9).

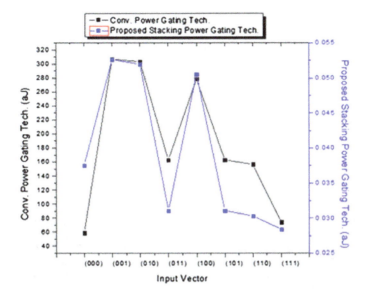

Fig. 30.9 Comparison mode transition energy dissipation with different input combination in conventional and proposed technique

Conclusion

A novel stacking MTCMOS circuit is proposed and their mathematical analysis is performed. The proposed circuit is simulated and performance parameters like leakage current, static energy, and mode transition energy are calculated. Results show that the proposed technique minimizes the leakage current and static energy dissipation by more than 80% as compared to the conventional MTCMOS circuit. Simulation is also performed to calculate mode transition energy dissipation. Results show that, when the same input is applied to all the inputs of full adder, the average reduction in mode transition energy is around 27%.

References

1. Abdollahi A, Fallah F, Pedram M (2007) A robust power gating structure and power mode transition strategy for mtcmos design. IEEE Trans Very Large Scale Integr (VLSI) Syst 15(1):80–89
2. Bhanuprakash R, Pattanaik M, Rajput S, Mazumdar K (2009) Analysis and reduction of ground bounce noise and leakage current during mode transition of stacking power gating logic circuits. In: TENCON 2009-2009 IEEE region 10 conference. IEEE, pp 1–6
3. Chowdhury MH, Gjanci J, Khaled P (2008) Controlling ground bounce noise in power gating scheme for system-on-a-chip. In: 2008 IEEE computer society annual symposium on VLSI. IEEE, pp 437–440

4. Jiao H, Kursun V (2010) Tri-mode operation for noise reduction and data preservation in low-leakage multi-threshold cmos circuits. In: IFIP/IEEE international conference on very large scale integration-system on a chip. Springer, pp 258–290
5. Jiao H, Kursun V (2012) Multi-phase sleep signal modulation for mode transition noise mitigation in mtcmos circuits. In: 2012 international SoC design conference (ISOCC). IEEE, pp 466–469
6. Kim S, Kosonocky SV, Knebel DR, Stawiasz K, Papaefthymiou MC (2007) A multi-mode power gating structure for low-voltage deep-submicron cmos ics. IEEE Trans Circuits Syst II Express Briefs 54(7):586–590
7. Liu Z, Kursun V (2008) Characterization of wake-up delay versus sleep mode power consumption and sleep/active mode transition energy overhead tradeoffs in mtcmos circuits. In: 2008 51st midwest symposium on circuits and systems. IEEE, pp 362–365
8. Pattanaik M, Raj B, Sharma S, Kumar A (2012) Diode based trimode multi-threshold cmos technique for ground bounce noise reduction in static cmos adders. In: Advanced materials research, vol. 548. Trans Tech Publ, pp 885–889
9. Solanki S, Kumar A, Dubey R (2016) Stacked transistor based multimode power efficient mtcmos full adder design in 90nm cmos technology. In: 2016 international conference on communication and signal processing (ICCSP). IEEE, pp 0663–0667
10. Thomas SP, Jose A (2019) Transistor full adder: a comparative performance analysis. Recent Trends Electron Commun Syst 5(3):22–31

Chapter 31
A Backward/Forward Method for Solving Load Flows in Droop-Controlled Microgrids

Rahul Raj and P. Suresh Babu

Introduction

In the operation, analysis, control, and optimization of a power system, the power flow studies are important. The system reconstruction and reconfiguration use the power system analysis tools [1]. These are employed in the operation and planning of distribution networks [2]. The classical strategy like fast decoupled, Newton Raphson, and Gauss-Seidel are also used in the load flow analysis. But there are situations when they are inefficient and not easy to implement, mostly in the case of a poorly conditioned system. Therefore, a method which also has been used in the analysis of power flow in radial type of distribution networks [3] that is backward/forward sweep. The method of BFS is based on Kirchhoff's voltage law and Kirchhoff's current law. The improvement of the distributed generators has put together an experienced dimension to the standard distribution network after using bidirectional power flow. A microgrid can be formed by the combination of several DGs. The power can be supplied by them to the loads without any reinforcement of the master grid. Microgrids are the parts of a smart grid, which can guarantee high reliability. Microgrids can be handled in two ways, that is, grid-connected mode and islanded mode [4]. The frequency of the whole system is controlled by the grid and it is designed as a slack bus and all DG buses are designed as PQ bus in the grid-connected mode. For the islanded mode the DG bus are not designed as PQ, PV, or slack bus. The reactive and active power of DGs depends on the frequency and the voltage of the system which is a droop-based case. The powers of the generators are inversely dependent on their respective droop coefficients. The classical methods taken earlier do not work for change in the frequency which makes them insufficient for the study of the islanded microgrids.

R. Raj (✉) · P. S. Babu
National Institute of Technology, Warangal, Telangana, India
e-mail: r.r.r.12895@gmail.com

The solution of the load flow can be established if the main grid is attached to the microgrid. It is divided into two parts: (1) one which requires derivatives and (2) one which do not require derivatives. A method called Newton trust region [5] has been taken into account considering the several operating modes of DG containing droop to establish a number of equations for the solution of power flow. These equations are nonlinear which increases with the DGs. This method disregarded the effect of the droops. The methods which are derivative based may not be proper to solve practical problems. The BFS is the most efficient load flow methods for the distribution systems. It is a derivative-free method and it does not require calculation of the Jacobian matrix. A direct method [6] has been given to solve the load flows in islanded microgrids. The results of the algorithm have been compared with that of [6]. BFS is a productive method for the radial as well as the distribution networks which are weakly meshed. The mathematics required is just the calculation of flow variables such as apparent powers and the complex current.

The contributions of this paper are following

(1) A method which is free of derivative to solve the power flow problem in the islanded microgrids.
(2) The given method succeed in dealing with the restriction of the conventional BFS methods.

    In the proposed method, a microgrid with DGs which is controlled according to their droops has been considered. The droops taken into consideration are the P-f and Q-V droops.

    In what follows, models of the system in Sect. 2 are provided followed by the proposed method in section
(3) Section 4 is the results part and Sect. 5 is the conclusions of this paper.

Models of the System

There are two models to be considered. The DG model and the load model
Load Model
This static load model is used for modeling loads. The load power absorbed depends upon the voltage magnitude and the frequency. The equations representing the static load model are

$$P_{iL} = P_{ioL} \left(\frac{|U_i|}{|U_o|}\right)^a (1 + S_{pf}(\omega - \omega_o)) \quad (31.1)$$

$$Q_{iL} = Q_{ioL} \left(\frac{|U_i|}{|U_o|}\right)^b (1 + S_{qf}(\omega - \omega_o)) \quad (31.2)$$

given $|\omega_o|$ and U_o are nominal frequency and nominal voltage, respectively. $|U_i|$ is the voltage magnitude at ith bus and the system frequency is ω. P_{io} and Q_{ioL}, at nominal working voltage of system, are representing active and reactive power of ith bus. a

and b are the active power and the reactive power exponents [5]. S_{pf} varies from 0 to 3 and S_{qf} varies from -2 to 0 are the frequency sensitivity factors [7].

DG Model

The DGs which are implemented here are droop controlled. It means that each individual DG supplies to the net system power depending upon the coefficients of their droops. The DG with larger droop coefficients will contribute lesser power. The islanded DGs droop operation is represented by the following equation

$$\omega = \omega_0 + x_p(P_G - P_{oG}) \quad (31.3)$$

$$|U| = |U_0| + x_q(Q_G - Q_{oG}) \quad (31.4)$$

where P_G and Q_G are the DG generated active and reactive powers; PoG and QoG are the nominal active and reactive power generated, x_q and x_p are the coefficients of voltage droop and frequency droop, respectively [8].

The suggested method

The suggested method is divided into 4 stages

Stage 1 (Before BFS)

In this stage, all the procedures which are done before the starting of the backward sweep takes place. In this, a virtual bus (VB) is added to a DG bus which will act as a fake-grid. A correction variable of frequency is put to zero and all the voltages of the buses are put to one. The DGs supply electric power which is computed with

$$P_{iG} = P_{ioG} + \Delta P_{iG}$$
$$\Delta P_{iG} = \frac{\Delta f}{x_{ip}} \quad (31.5)$$

where P_i represents active power produced at ith bus, P_{ioG}, at nominal frequency, f_o, is the active power(nominal) produced at ith bus. ΔP_i the active power produced with respect to the deviation in the frequency, Δf, and x_{ip} is coefficient of the frequency droop at ith bus.

The voltage deviation, ΔU, is used to determine the reactive power produced. The DGs supply reactive power as

$$Q_i = Q_{ioG} + \Delta Q_G \quad, \Delta Q_{iG} = \frac{\Delta U_i}{x_{iq}} \quad (31.6)$$

where Q_{iG}, the reactive power produced at ith bus, Q_{ioG}, at nominal voltage, U_i,. ΔQ_{iG} is the reactive power produced with respect to the deviation in voltage, ΔU_i and x_{iq} is coefficient of the voltage droop at ith bus.

Stage 2 (Backward Sweep)

The voltages at entire buses are presumed to be well known in this stage. Each bus current is evaluated by the help of the voltages at the bus and the apparent power. The apparent power is determined at each bus as

$$A_i = P_i + jQ_i = (P_{iG} + Q_{iG}) - (P_{il} + jQ_{il}) \tag{31.7}$$

After this, the current is computed as

$$I_i = \frac{P_i - jQ_i}{U_i^*} \tag{31.8}$$

The branch currents regarding the main junction that is the virtual bus can be given by summing the nodal currents backwards. This procedure can be represented as

$$I_{ij} = I_j + \sum_{k \in b_j} I_{jk} \tag{31.9}$$

where is the bus set which are next to jth in the direction of the virtual bus. The next equation establishes the relation of the bus nodal currents with the branch current

$$[\mathbf{B}] = [\mathbf{BIBC}][\mathbf{I}] \tag{31.10}$$

where the darked letters represent the matrix or vector variables, the current in branch is **B**, **I** represents current in bus, **BIBC** represents the evolution vector [9]. This vector working creates the branch currents from the nodal currents by the help of the evolution vector. The evolution vector contains ones and zeros. Also, an n bus radial system will have $n - 1$ number of branch currents. Hence, the evolution vector is an upper triangular matrix whose dimension is $(n - 1 \times n - 1)$. Then, stage 2 is completed and the algorithm is moved forward to stage 3 as soon as the solution of the branch currents is achieved.

Stage 3 (Forward Sweep)

The stage 3 includes the determination of the voltages beginning from the connected virtual bus and moving toward the lower bus. The branch currents, the substation voltage, and the line parameters can be used to represent the bus voltages [8]. The dependency of the bus voltages on the virtual bus can be given as

$$[\mathbf{U}_1] - [\mathbf{U}] = [\mathbf{\Delta U}] \tag{31.11}$$

where U_1 is the virtual bus voltage, **U** is the bus voltages at the buses from two to n, $\mathbf{\Delta U}$ is the difference of voltages between the bus 1 voltage and the remaining bus voltage. Equation (31.11) can be rewritten as

$$[U] = [U_1] - [\Delta U] = [U_1] - [BCBV][B] \quad (31.12)$$

Given **BCBV** vector is a evolution vector which uses the branch currents to get the bus voltages as explained in [8]. From Eqs. (31.10) and (31.12), we can get the following relation

$$[U] = [U_1] - [BCBV][BIBC][I] \quad (31.13)$$

Stage 4 (After-BFS Update)

It can be the frequency and the voltage rectification stage. Generally, the system frequency and the voltage of the virtual bus are not permanent in a droop-controlled islanded microgrid. Therefore, these are adjusted in this stage. It takes place after a number of iterations in a backward/forward method. A virtual bus concept is equivalent to connecting a pseudo-grid to the islanded microgrid. Therefore, at the start of the algorithm, through the pseudo grid, there will be power flow which implies that the steady-state solution has not been reached by the system. According to the power flowing away from or towards the pseudo-grid, the voltage of the virtual bus and the frequency of the system are modified. On the voltages of the overall system, there will be a ripple effect due to the modification of the virtual bus voltage. The total active power variation on the virtual bus, ΔP, is found after subtracting the active power coming out of the virtual bus from the inserted active power on the virtual bus denoted as P_{1G}, which is put to zero. The active power coming out of the virtual bus is apparent power's real part produced by the virtual bus denoted as $(U_1 B)$. The relation between frequency update
Δf and ΔP is

$$\Delta P = \sum_{i=1}^{z} \frac{1}{x_{ip}} \Delta f \quad (31.14)$$

where $\Delta f = f - f_o$ and z represents the number of DGs taken into the system. Equation (31.14) can also be written as

$$\Delta f = x_{peq} \left[P_{1G} - R(U_{1B}^*) \right] \quad (31.15)$$

where

$$x_{peq} = \frac{1}{\sum_{i=1}^{z} x_{ip}} \quad (31.16)$$

P_1 is the virtual bus active power generation, U_1 is the virtual bus voltage, and B_{j-1} is the branch to jth bus from the virtual bus approaching the identical method in [8]. Consequently, there will be an upgradation in the line impedances as well, due to the frequency update as

$$Z_{ij} = R_{ij} + jX_j \frac{f_{k+1}}{f_k} \quad (31.17)$$

where Z_{ij} is the impedance between the bus i and bus j, R_{ij} and X_{ij} are the resistance and reactance between the bus i

and bus j, respectively, f_{k+1} and f_k represent the frequency at the $(k+1)_{th}$ and the kth iterations, respectively.

The variation, at the virtual bus, of the total reactive, ΔQ, is found by subtracting the reactive power flowing out of the virtual bus from the inserted virtual bus reactive power designated as Q_{1G}. Q_{1G} is the reactive power produced by the virtual bus that is put to zero. Therefore, the variation of reactive power a signal that the steady state has not been reached by the solution yet. Similarly, to the computation of the frequency variation, the variation of reactive power provides a root to the renovation of the deviations in the voltage, ΔU, as

$$\Delta Q = \sum (1/x_{iq}) \Delta U_i, \quad \Delta U_i = |U_i| - 1 \quad (31.18)$$

The update of virtual bus, $\Delta U1$, is given as,

$$\Delta U_1 = x_{\text{qeq}} \left[Q_{1G} - I(U_1 B_j - 1^*) \right] \quad (31.19)$$

where

$$x_{qeq} = \frac{1}{\sum_{i=1}^{z} 1/x_{iq}} \quad (31.20)$$

Q_{1G} is the virtual bus reactive power. For the virtual bus, the reactive and the active power both will come to zero showing the convergence of the frequency and the voltages of the system. The convergence condition can be the inclination of the reactive and the active power flows from the virtual bus to zero and ΔU_1 to zero. The algorithm is stopped when ΔU_1 is not more than 10^{-4}. The flowchart in Fig. 31.1 represents the whole algorithm.

Results

The system taken under study is test system of 33-bus [9]. The system shown in Fig. 31.2 contains 5 DG buses at 1, 6, 13, 25, and 33. All of these DGs have their own droop coefficient

The nominal powers of all the DGs has been taken as,

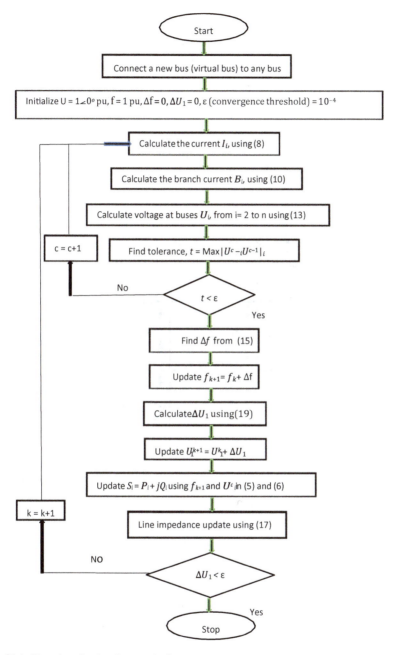

Fig. 31.1 Flowchart for the given method

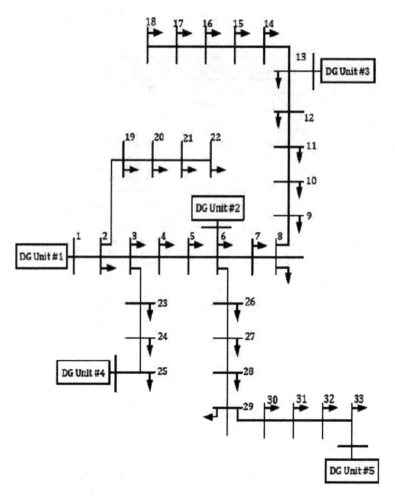

Fig. 31.2 Test system of 33-bus [10]

$P_{ioG} + jQ_{ioG} = 0.9 + j0.9$. The droop set taken into consideration is shown in Table 31.1.

Table 31.1 DG droops

DG	Bus location	x_p	x_q
1	1	−0.04	−0.04
2	6	−1.00	−1.00
3	13	−0.20	−0.10
4	25	−0.50	−0.30
5	33	−0.20	−0.20

The values of the constants a, b, Spf and Sqf has been taken as $a = 2$, $b = 2$, $Spf = 1$ and $Sqf = 1$. The power flow results using the algorithm have been shown in Table 31.2.

Table 31.2 Power flow results

Bus number	Voltage (p.u.)	Voltage angle (degree)	Load, P_L (p.u.)	Load, Q_L (p.u.)
1	0.9989	0	0	0
2	0.9959	−0.0070	−0.19	−0.11
3	0.9943	−0.0462	−0.17	−0.07
4	0.9927	−0.0486	−0.23	−0.15
5	0.9933	−0.0551	−0.11	−0.05
6	0.9920	−0.0959	−0.11	−0.03
7	0.9912	−0.0960	−0.39	−0.20
8	0.9895	−0.1832	−0.39	−0.20
9	0.9900	−0.2173	−0.11	−0.03
10	0.9899	−0.2858	−0.11	−0.03
11	0.9899	−0.3477	−0.08	−0.05
12	0.9917	−0.3603	−0.11	−0.06
13	0.9945	−0.3861	−0.11	−0.06
14	0.9932	−0.4530	−0.23	−0.15
15	0.9919	−0.5125	−0.11	−0.02
16	0.9906	−0.5411	−0.11	−0.02
17	0.9891	−0.5575	−0.11	−0.03
18	0.9878	−0.6159	−0.17	−0.07
19	0.9979	−0.6229	−0.17	−0.07
20	0.9950	−0.0165	−0.17	−0.07
21	0.9944	−0.0727	−0.17	−0.07
22	0.9938	−0.0898	−0.17	−0.10
23	0.9941	−0.1081	−0.17	−0.40
24	0.9930	−0.1819	−0.83	−0.40
25	0.9942	−0.2410	−0.83	−0.04
26	0.9909	−0.0701	−0.11	−0.05
27	0.9889	−0.0335	−0.11	−0.05
28	0.9875	0.0833	−0.11	−0.04
29	0.9845	0.1815	−0.23	−0.15
30	0.9840	0.2471	−0.39	−1.20
31	0.9871	0.2269	−0.29	−0.15
32	0.9882	0.2345	−0.41	−0.20
33	0.9911	0.2848	−0.11	−0.07

Generation (P_C)	Generation (Q_C)
2.8510	0.9079
0.9779	0.9085
1.2904	0.9462
1.0560	0.9085
1.2902	0.9460

	P_G	Q_G	P_L	Q_L	P_{Loss}	Q_{Loss}
Total	7.4655	4.628	7.42	4.5	0.354	0.0268

Conclusion

A method of backward/forward sweep is proposed in the paper for solving the power flow for an islanded AC microgrid. This algorithm is a divergence of the BFS approach generally preferred in the study of power system networks. This method is taken such that one of the unknown variables is the system frequency that is standard for an islanded system. There is no slack bus considered in the system. The algorithm was tested on 33-bus distribution system. The results were contrasted with the direct method and the proposed algorithm was more accurate. The instant implementation of the proposed algorithm is to examine the equilibrium of the islanded microgrid in the specified state of droop and load positions. In respect to the conventional or classical methods, this method is straightforward and uncomplicated to implement which makes it an interesting apparatus to solve the power flows of islanded microgrids. The results were more accurate under various droop gains.

References

1. Lisboa C, Guedes LSM, Vieira DAG, Saldanha RR (2014) A fast power flow method for radial networks with linear storage and no matrix inversions. Int J Elect Power Energy Syst 63:901–907
2. Cheng S, Shirmohammadi D (1995) A three- phase power flow method for real-time distribution system analysis. IEEE Trans Power Syst 10(2):671–679
3. Shirmohammadi D, Hong HW, Semlyen A, Luo GX (1988) A compensation-based power flow method for weakly meshed distribution and transmission networks. IEEE Trans Power Syst 3(2):753–762
4. Pogaku N, Prodanovic M, Green TC (2007) Modeling, analysis and testing of autonomous operation of an inverter-based microgrid. IEEE Trans Power Electron 22(2):613–625
5. Abdelaziz MMA, Farag HE, El-Saadany EF, Mohamed YARI (2013) A novel and generalized three-phase power flow algorithm for islanded microgrids using a Newton trust region method. IEEE Trans Power Syst 28(1):190–201

6. Diaz G, Gomez-Aleixandre J, Coto J (2016) Direct backward/forward sweep algorithm for solving load power flows in AC droop-regulated microgrids. IEEE Trans Smart Grid 7(5):2208–2217
7. Kundur P, Balu NJ, Lauby MG (1994) Power system stability and control. McGraw-Hill, New York NY, USA
8. De Brabandere K et al (2007) A voltage and frequency droop control method for parallel inverters. IEEE Trans Power Electron 22(4):1107–1115
9. Teng JH (2012) A direct approach for distribution system load flow solutions. IEEE Trans Power Del 18(3):303–314
10. Baran ME, WSu FF (1989) Network reconfiguration in distribution systems for loss reduction and load balancing. IEEE Trans Power Del 4(2):1401–1407

Chapter 32
Monte Carlo Simulation Application in Composite Power System Reliability Analysis

Atul Kumar Yadav, Soumya Mudgal, and Vasundhara Mahajan

Introduction

In the modern society, as the renewable energy sources (RES) are fastly integrated with existing power generation, maintaining the reliability of system without any load interruption and customer satisfaction is a major concern to focus on. RES can be modelled in different states to analysis the reliable power supply. As different RES have state model, MCS technique is one of the leading approaches to accumulate the RES with existing thermal and hydropower project [1, 2]. Markov chain process is the key factor for determination of transition of power system components from one state to another. The transition rate of each state can be estimated through random number selection in MCS technique. Several points have been discussed below which may lead to improve the system reliability performance. Failure of any components of the power system will lead to several effects on security performance. The maintenance of the system under fault situation will cause to make the system more secure and according to load demand in the system, under faulty condition how to improve the system performance in terms of reliability is a major concern of the MCS modelling [3, 4]. Computational efficiency gets improved by using MCS in reliability study of complex system. A slop-based stability approach is discussed to make system stable and reliable through its state of position. The equilibrium states based reliability is studied through direct MCS that also called the adaptive MCS [5, 6].

A. K. Yadav · S. Mudgal (✉) · V. Mahajan
Department of Electrical Engineering, SVNIT, Surat, Gujarat, India
e-mail: mudgalsoumya@gmail.com

A. K. Yadav
e-mail: yatul76@gmail.com

V. Mahajan
e-mail: vasu.daygood@gmail.com

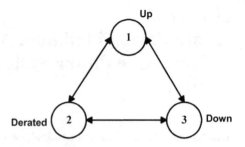

Fig. 32.1 Three state model of generating unit

- Consider interior transmission restrictions in the generation system reliability assessment.
- Minimize comparative investments in composite systems.
- Improved trends in the distribution of resource.
- A better demonstration of generation effects in transmission system reliability study.
- Dispersed generation: battery storage, co-generation, etc.

Modelling of Components

The following components can be used in the reliability analysis of composite system generating units, transformers, transmission branches, bus bars, and circuit breaker. Each components model can be studied in several state operations.

State model of generating units: Generators are modelled in three states followed by UP, DOWN, and derated states. Transition diagram for different states of generating units is shown in Fig. 32.1.

Transmission branches: Transmission branches either are supposed to stay in the up or down state. The rate of failure and repair is supposed to be reliant on environmental condition. Transition diagram of transmission system for different states is represented in Fig. 32.2 below. Environmental scenarios are supposed to occur either in the usual or adverse state. A significant problem regarding the environment is the degree of coverage. When the system consists of a huge space zone, at any specified period, every zone may have dissimilar states of climate. A meticulous treatment related to this consequence is not conceivable and some modifications are required. One method is to patrician the entire zone into some different sections. The climate in every section is considered with a mean period of normal (usual) and adverse states [7–9]. The climate variations in each section are supposed to be autonomous. Each branch is consigned to a certain section, which actually represents that branch is mainly effected through the climate in the respective section.

Transformers and buses: These components are also considered as a two-state model like transmission branches. Failure and repair rate of these components are supposed autonomous behaviour with climate [7].

Circuit breakers: It has numerous failure ways as defined in several points below.

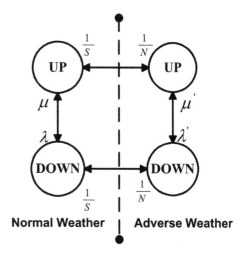

Fig. 32.2 Model of transmission line

Ground faults: It shows to the fault occurs in the breaker itself. The breaker is considered in a similar way as a bus or transformer for this type of related fault.

Operation failure: The main aim of a breaker is to separate the faulted section. Due to buried faults in the breaker, sometimes, breaker will not operate accordingly that means it will not open under the fault situation [10, 11].

Undesirable tripping: Sometimes breaker will be open without any external command. This is generally considered not to have a significant effect on the related outages.

The DC load flow is normally sensed to be a sensible conciliation among computational price and correctness related to forecasting analysis. It is frequently in an area where the system is powerfully interlaced and no under or over voltage situation occurs. The AC load flow model grips both features of real and reactive power. In the AC load flow model, there is a restriction over real and reactive power generation. It generates reactive power according to bus voltage constraints.

Backgrounds

Reliability parameter analysis of a real physical system could be predictable by assembling data on the existence of failures and times of restoration. Monte carlo simulation (MCS) technique elaborates on the failure and restoration antiquity of components states. Statistics are composed and parameters assessed by statistical implication. There exist two leading methodologies in MCS, sampling or non-sequential and sequential simulation. This section describes the basic idea of MCS and its application to composite power system reliability [12, 13]. MCS used random numbers for prediction of reliability indices and then describe the ideas of sampling and sequential simulation [14]. The necessity for numbers to be random is that each number should have an equal probability of captivating on any one of the probable values and it must be statistically autonomous of the other numbers in the system. Therefore, random numbers of a specified range monitor a uniform probability density functions (pdfs) [15, 16].

μ Normal climate repair rate
μ' Adverse climate repair rate
λ Normal climate failure rate
λ' Adverse climate failure rate
N, S Mean time of normal and adverse climate

Multiplicative congruential method: In this method, a random number is generated by the following mathematical relationship shown in Eq. (32.1).

$$R_{n+1} = (kR_n)(\mod m) \tag{32.1}$$

where R_n is nth random number and k, m are positive integers such that $k < m$. R_{n+1} is the remainder of kR_n after its division with m.

Random sampling or non-sequential simulation: This process can be more simply carried out by using a cumulative probability or pdfs. If the system contains n autonomous components, then n random numbers will be required to illustrate the state of every component.

Random Sampling of States Basic Ideas
Let $x = (x_1, x_2, \ldots, x_i)$ be the state of system components where
x_i ith component state
X Set of states
$P(X)$ Probability associated with state x
$F(x) =$ Test to be conducted to check the availability of required load.
The expected value of $F(x)$

$$E(F) = \sum_{x \in X} F(x)P(x) \tag{32.2}$$

For $E(F)$ to be LOLP
$F(x)$ is 1 if load curtailment in state x; otherwise, 0.

In random sampling, $x \in X$ are sampled from their joint distributions. Then estimate of $E(F)$.

$$E(F) = \frac{1}{NS} \sum_{i=1}^{NS} F(x^i) \tag{32.3}$$

where
$NS =$ total number of sample
$x^i =$ ith sampled value
$F(x^i) =$ test result associated with x^i

$$V(\hat{E}(F)) = \frac{V(F)}{NS} \tag{32.4}$$

Since $V(F)$ is unknown and can be estimated through relationship,

$$\hat{V}(F) = \frac{1}{NS} \sum_{i=1}^{NS} (F(x^i) - \hat{E}(F))^2 \tag{32.5}$$

Equation (32.2–32.5) shows the relationship of estimation and its variance based on the sampled values.

Algorithm

1. Initiate $NS = 0$
2. $NS = NS + 1$, state selection $x^i \in X$.
3. Calculate test value associated with, i.e. $F(x^i)$.
4. Estimate the value of $\hat{E}(F)$
5. Evaluate the mismatch of the estimate.
6. If the mismatch is satisfactory, stop; otherwise, return back to step 2 and continue the process.

Sequential Simulation: It can be accomplished either by advancing period in fixed steps or by advancing to the next event.

Fixed period interval method: This method is convenient in Markov chains process when transition probabilities over a period step are defined.

Next event method: This method is applicable when the times in system states are defined using continuous variables with pdfs. Discrete random variables are only a special case of continuous random variables. Equations (32.6–32.8) shows the relationship of probability distributions of random selection.

Let

$$z = F(X) \tag{32.6}$$

Let φ is inverse of F; then

$$x = \varphi(z) \tag{32.7}$$

Probability distribution determined as follows

$$\begin{aligned} P(x \leq X) &= P(F(x) \leq F(X)) \\ &= P(z \leq F(X)) \\ &= P(X) \end{aligned} \tag{32.8}$$

- Sequential simulation is usually sluggish to converge than non-sequential.
- Random sampling used lesser data for estimation of reliability parameter in comparison with sequential simulation.

- For time interrelated actions, sequential simulation is very appropriate.

Reliability parameter can be studied in two sections deterministic and probabilistic.

Deterministic reliability parameter: This approach of reliability study require fewer data to predict the parameter associated with this method. Following parameter is used to estimate the system reliability:

- *Per cent Reserve Margin*: This can be defined as the excess generation over the installed capacity of generating units with respect to annual peak load demand.

Probabilistic reliability parameter: Probabilistic method is the main focus of modern power system reliability evaluation. It has the capability to determine the several reliability parameters based on the individual and cumulative probability. Some parameters related to reliability are explained below.

- *Loss of Load Expectations (LOLE)*: LOLE is defined as the un-fulfilment of annual peak load demand in day per year or hour per year.
- *Expected Unserved Energy (EUE)*: EUE defined as the expected energy fail to supply per year due to the shortage of available capacity.

Conclusion

This paper discusses about MCS method to analyse the power system reliability with the state model of components. A random number can be generated through several methods with equal probability of each number to be chosen. Random sampling is used to estimate the reliability parameter within an acceptable range. Deterministic and probabilistic reliability parameters are explained in this paper. As the power system consists of several components which causes an unreliable situation in most of the unhealthy operation of components. MCS gives a state model of individual components and with a combination of components connected through bidirectional transition rates. The probability associated with each transition is used to estimate and predict the system reliability performance. This paper explains a review of the basic idea of MCS used in power system reliability evaluation and model of different states of the existing components of the system. The outcome of this review paper is that MCS makes the system more efficient in terms of computational approach of reliability parameters.

References

1. Appasani B, Mohanta DK (2020) Monte-Carlo Simulation models for reliability analysis of low-cost IoT communication networks in smart grid. In: Real-time data analytics for large scale sensor data, vol. 6. Elsevier, pp 73–96

2. Kumar S, Saket RK, Dheer DK, Holm-Nielsen J, Sanjeevikumar P (2020) Reliability enhancement of electrical power system including impacts of renewable energy sources: a comprehensive review. IET Gener Transm Distrib 14:1799–1815
3. Singh C, Jirutitijaroen P, Mitra J (2019) Reliability evaluation of composite power systems. pp 247–272
4. Yadav AK, Mahajan V (2019) Reliability improvement of power system network with optimal transmission switching. In: IEEE 1st international conference on energy, systems and information processing (ICESIP). Kancheepuram, pp 1–6
5. Browska EDA (2020) Monte Carlo simulation approach to reliability analysis of complex systems. J. KONBiN 50:155–170
6. Liu X, Li D-Q, Cao Z-J, Wang Y (2020) Adaptive Monte Carlo simulation method for system reliability analysis of slope stability based on limit equilibrium methods. Eng Geol. 264
7. Yadav AK, Mahajan V (2019) Transmission line switching for loss reduction and reliability improvement. In: International conference on information and communications technology (ICOIACT). Yogyakarta, Indonesia, pp 794–799
8. Zhao Y, Tang Y, Li W, Yu J (2019) Composite power system reliability evaluation based on enhanced sequential cross-entropy Monte Carlo simulation. IEEE Trans Power Syst 34:3891–3901
9. Fisher EB, O'Neill RP, Ferris MC (2008) Optimal transmission switching. IEEE Trans Power Syst 23:1346–1355
10. Yadav AK, Mahajan V (2019) Transmission system reliability evaluation by incorporating STATCOM in the system network. In: IEEE student conference on research and development (SCOReD). Perak, Malaysia, pp 198–203
11. Kumar N, Mahajan V (2018) Reconfiguration of distribution network for power loss minimization & reliability improvement using binary particle swarm optimization. In: IEEE 8th power India international conference (PIICON). Kurukshetra, India, pp 1–6
12. Zhaohong B, Xifan W (2002) Studies on variance reduction technique of Monte Carlo simulation in composite system reliability evaluation. Electr Power Syst Res 63:59–64
13. Mudgal S., Mahajan V (2019) Reliability and active power loss assessment of power system network with wind energy. In: IEEE student conference on research and development (SCOReD). Bandar Seri Iskandar, Malaysia, pp 186–191
14. Kovalev GF, Lebedeva LM (2019) Reliability of power systems. Springer
15. Mudgal SM, Yadav AK, Mahajan V (2019) Reliability evaluation of power system network with solar energy. In: 8th International conference on power systems (ICPS). Jaipur, pp 1–6
16. Yadav AK, Mudgal S, Mahajan V (2019) Reliability test of restructured power system with capacity expansion and transmission switching. In: 8th international conference on power systems (ICPS). Jaipur, pp 1–6

Chapter 33
Enhancement of Static Voltage Stability Margin Using STATCOM in Grid-Connected Solar Farms

S. Venkateswarlu and T. S. Kishore

Introduction

In the era of ever increasing load growth due to improvement of global economic rise, industrial infrastructure development, social living status of people, the power system faces new challenges every single day. This pressure is even increased on account of deregulation which is a consequence of reforms in electricity policies in accordance with new trends in electricity industry and its stakeholders. Under such situation, generating and transmission of power uninterruptedly are of significant importance and even forms the major indices of power system operation in many nations. In order to achieve this, the entire power system must be operated stably or in other words, the power system should maintain stability. Of all power system stabilities, voltage stability is of serious concern because of its inherent nature of gradually endangering the power system network operation leading to system blackouts. It is because, most of the power system protective devices will not consider this voltage stability phenomenon as faults and will not provide any isolation mechanism, even though to some extent, the damage can be limited to brownouts. Therefore, the power system should be supervised for an extensive range of system circumstances in order to assess voltage stability [1, 2].

In general, static voltage stability analysis is used to examine the system security against an operating point, assess the stability margin, and define the voltage collapse limits. In a deregulated system, due to increased loading, limited expansion of generation and transmission capacity and stochastic market scenarios cause the system to

---

S. Venkateswarlu (✉) · T. S. Kishore
GMR Institute of Technology, Rajam, Andhra Pradesh 532127, India
e-mail: venkatchamala135@gmail.com

T. S. Kishore
e-mail: kishore.ts@gmrit.edu.in

© The Editor(s) (if applicable) and The Author(s), under exclusive license to Springer Nature Singapore Pte Ltd. 2021
A. K. Singh and M. Tripathy (eds.), *Control Applications in Modern Power System*, Lecture Notes in Electrical Engineering 710,
https://doi.org/10.1007/978-981-15-8815-0_33

operate at highly stressed conditions leading to voltage instability. This can lead to partial or complete blackout of the system. The generator reactive power limits also contribute significantly to this phenomenon [3]. The development of FACTS devices has opened new opportunities for enhanced efficient operation of power system along with fast control on system parameters such as phase, voltage, and impedance. Hence, better control of power flow and improved voltage regulation can be achieved, which further tries to maintain the system at prescribed voltage stability or thermal limits. In view of the present scenario, this paper presents a study on the impact of integrating a renewable energy (SPV) source in power system network on the static voltage stability of the system. The IEEE 14 bus test system was chosen to carry out the present research and is modified to incorporate the renewable energy source. The voltage stability analysis has been carried out by continuation power flow and the results are analyzed. Results indicate that FACTS devices have significant potential in enhancing the system voltage stability by increasing their limits.

SPV Impact on Grid

In recent years, in view of the reforms taking place in energy markets, new ventures are mushrooming to inject power produced from renewable energies especially wind and solar, into electricity grid at a higher percentage. At the same time, due to economic growth, energy demand is ever rising and power grids are forced to operate at maximum threshold level. In this scenario, maintaining voltage stability is an important concern regarding grid operation and planning to safeguard reliable power supply. Solar photovoltaic (SPV) power generation is considered the most prominent green energy source because of its inherent benefits such as the free and affordable alternative to conventional power generation. Apart from its advantages, it also has negative effects on the efficiency of electrical grid voltages, particularly in the case of high penetration rates. The issues that need to be addressed before injecting SPV energy into the power grids are: voltage regulation, high levels of grid-connected SPV result in problems like relay desensitization, nuisance tripping, interference with automatic reclosers, and ferroresonance and power quality issues also. Most of these issues are related to voltage stability and ways to mitigate this is highly essential to completely utilize the benefits of renewable energy sources [4].

System Modeling

The proposed work in this paper is intended to be implemented on IEEE 14 bus test system illustrated in Fig. 33.1. The system under consideration entails of five synchronous generators with IEEE type-1 exciters, three of which are synchronous condensers used only for reactive power support. There are nineteen buses, seventeen transmission lines, eight transformers, and eleven constant impedance loads. The total

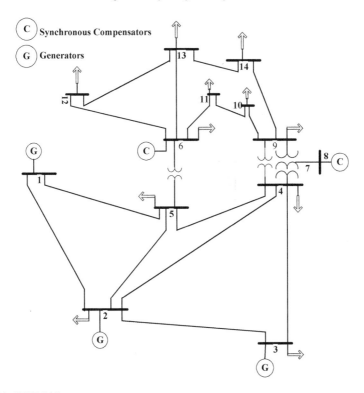

Fig. 33.1 IEEE 14 bus test system

load demand is 259 MW and 73.5 MVAr. Static voltage stability analysis has been carried out using continuation power flow method on test system with and without FACTS device (STATCOM). Further, a minor modification has been done to the test system by way of replacing a PV generator at a particular busbar and incorporating a bulk SPV power generating source of equivalent capacity and the similar analysis procedure has been carried out.

STATCOM Model

The compensation of real and reactive power can be effectively done by a shunt device, STATCOM (Voltage Source Inverter—VSI), which accepts a DC voltage as input and delivers AC voltage as output. STATCOM adjusts voltage and angle of internal source by controlling the voltage at the bus to which it is connected wr.r.t. a reference value. It displays continual current characteristics when the voltage is low/high under/over the limit which is responsible for delivering constant reactive power at the limits [5]. The STATCOM model and its typical characteristics are illustrated in Fig. 33.2.

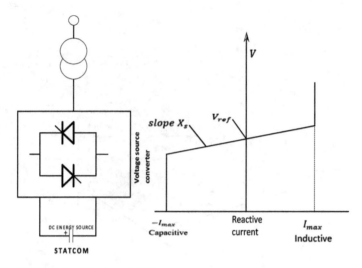

Fig. 33.2 STATCOM and its characteristics

The capacitor voltage V_{DC} is given by,

$$V_{dc} = \frac{P}{CV_{dc}} - \frac{V_{dc}}{R_C C} - \frac{R(P^2 + Q^2)}{CV^2 V_{dc}} \qquad (33.1)$$

The power injection at the AC bus as from:

$$P = V^2 G - kV_{dc}\cos(\theta - \alpha) - kV_{dc} V B \sin(\theta - \alpha) \qquad (33.2)$$

$$Q = -V^2 B - kV_{dc}\cos(\theta - \alpha) - kV_{dc} V G \sin(\theta - \alpha) \qquad (33.3)$$

SPV Model

SPV generator used for supplying the equivalent power generally consists of PV system, boost converter, inverter, and a transformer which steps up voltage to make the system grid interconnected. A PV array is composed of many modules joined in parallel and a module is composed of many cells joined in series. Since the generated voltage of PV system is very low, a boost converter is used to step it up so that it can be given as input to the inverter. This is done by implementing maximum power point tracking to optimize the PV system working. Inverter controls the active and reactive power pumped into grid. SPV generators, being enormous in size, are considered as PV models with limited VAr capacity, for supplying reactive power [6]. The single line diagram and constant PV mathematical model used are presented in Fig. 33.3.

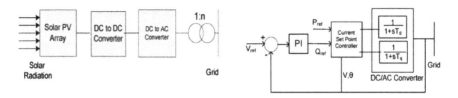

Fig. 33.3 SPV single line diagram and constant PV model

Simulation and Results

Continuation power flow analysis is a powerful tool to assess the static voltage stability analysis of a power system network [7–9]. It basically uses a modified power flow equations and then employs a predictor–corrector method to find the point of collapse or voltage stability margin as the loading parameter is increased [10, 11]. The loading parameter is usually the reactive power. This technique is adopted in the present study [12, 13]. Voltage stability analysis is first carried out on IEEE 14 bus test system and the voltage magnitudes corresponding to static voltage collapse point or limit at all buses are identified. Later, STATCOM is place at each of the 9 buses excluding the generator buses and the same analysis is repeated. The results of both these simulations are presented in Table 33.1. The voltage stability limit or voltage collapse point can be identified pictorially by plotting the PV curves illustrated in Fig. 33.4. Further, a ranking table has been constituted to assess the improvement in bus voltage stability limit w.r.t the STATCOM placement at a particular bus. From this assessment, the optimal location for placement of STATCOM is identified. The results of these are presented in Table 33.2. The entire process discussed above is repeated by incorporating a SPV farm at bus 2 for the test system considered and thus is designated as modified IEEE 14 bus test system. The results of the simulations carried out on modified IEEE 14 bus test system are presented in Tables 33.3 and 33.4. The QV curves illustrating the results of modified IEEE 14 bus test system are illustrated in Fig. 33.5.

Conclusions

In this paper, keeping in view the current stressful operating conditions of power systems posed by multifaceted constraints and the importance of maintaining stability, an attempt has been made to incorporate SPV generation with high level of penetration into a conventional power grid and determine the voltage stability margins for a wide variation of reactive power limits. The simulations were successfully implemented by modifying IEEE 14 bus test system so as to incorporate a SPV farm in MATLAB. The results can be interpreted as follows.

Table 33.1 Voltage magnitudes of IEEE 14 bus test system

Bus number	Base case	IEEE 14 Bus test system with STATCOM placed at bus number										
		14	13	12	11	10	9	7	5	4
14	0.55	0.73	1.04	1.04	1.04	1.04	1.04	1.04	1.04	1.04
9	0.6	0.7	0.812	0.851	0.825	0.84	0.84	0.84	0.825	0.825
10	0.6	0.73	0.82	0.82	0.84	0.85	0.84	0.84	0.833	0.832
4	0.65	0.7	0.685	0.7	0.7	0.7	0.7	0.7	0.7	0.75
5	0.65	0.67	0.663	0.68	0.667	0.68	0.68	0.68	0.72	0.7
7	0.7	0.8	0.83	0.85	0.85	0.85	0.85	0.86	0.86	0.87
11	0.76	0.88	0.926	0.926	0.955	0.94	0.94	0.93	0.93	0.935
13	0.80	0.935	1.0405	1.015	1.005	1.005	1.005	1.005	1.002	1.003
12	0.86	0.98	1.037	1.05	1.017	1.017	1.017	1.017	1.016	1.016
3	0.93	1.01	1	1.01	1.01	1.02	1.01	1.01	1.01	1.01
6	0.95	1.07	1.07	1.07	1.07	1.07	1.07	1.07	1.07	1.07
8	0.98	1.09	1.09	1.09	1.09	1.09	1.09	1.09	1.09	1.09
2	1	1.04	1.1	1.05	1.045	1.045	1.045	1.045	1.045	1.045
1	1.06	1.06	1.1	1.06	1.06	1.06	1.06	1.06	1.06	1.06

IEEE 14 Bus test system results without STATCOM

IEEE 14 Bus test system results with STATCOM

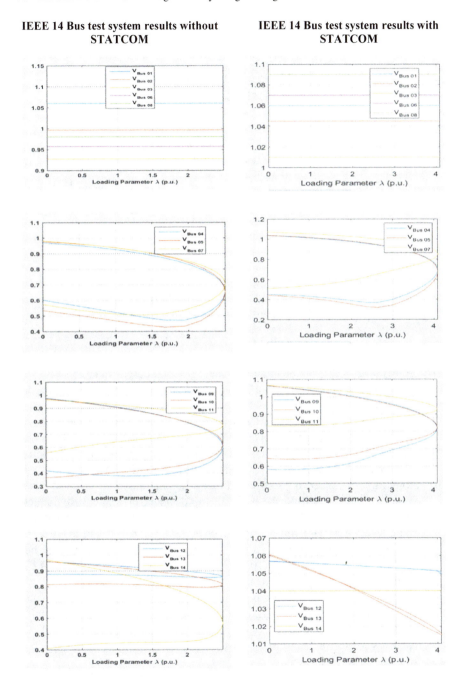

Fig. 33.4 PV curves for IEEE 14 bus test system

Table 33.2 Ranking table for IEEE 14 bus test system

Bus number	Without STATCOM	Voltage magnitude rank-wise with STATCOM	Optimal location
14	0.55	13, **12**, 11, 10, 9, 7, 5, 4(**1.04**), 14(0.73)	12
9	0.6	**12(0.851)**, 10, 9, 7(0.84), 5, 4(0.825), 13(0.812), 14(0.7)	
10	0.6	10(0.85), 9, 7, 11(0.84), 5, 4 (0.8332), **12**, 13(**0.82**), 14(0.73)	
4	0.65	4(0.75), **12**, 11, 10, 9, 7, 5, 14(**0.7**), 13(0.685)	
5	0.65	5(0.72), 4(0.7), **12**, 10, 9, 7(**0.68**), 14(0.67), 11(0.667), 13(0663)	
7	0.7	4(0.87), 7, 5(0.86) **12**, 11, 10, 9(**0.85**), 13(0.83), 14(0.8)	
11	0.76	11(0.955) 10,9 (0.94) **12**, 13 (**0.926**), 4(0.935), 7, 5(0.93), 14(0.88)	
13	0.807	13(1.045) **12(1.015)** 11, 10, 9(1.0050), 7(1.005), 4(1.0035), 5(1.0025),14(0.935)	
12	0.86	**12(1.05)** 13(1.037) 11, 10, 9, 7(1.017) 5, 4(1.016)	

(i) From Table 33.1, it can be observed that in the base case simulation, 14th bus is identified as weak bus with a voltage collapse point corresponding to 0.55 magnitude.

(ii) To improve this, a STATCOM has been placed at all buses except generator buses. When the STATCOM is placed at 14th bus, the magnitude was improved to 0.74 and at all remaining buses, it was improved to 1.04. A similar improvement has been found at all buses by using STATCOM.

(iii) From Table 33.2, it can be observed that a ranking has been given to buses employing STATCOM on the basis of improvement in voltage magnitudes. This is done to select the optimal bus location for STATCOM placement. For example, the base case voltage magnitude at bus 9 is 0.6. It has been observed that maximum improvement can be found by placing a STATCOM at bus 12 (0.851) followed by buses 10,9,7 (0.84), etc. This indicates that bus 12 is optimal location for improvement of voltage magnitude at bus 9.

(iv) A similar ranking analysis has been done to identify the optimal location and the best bus location exhibiting significant voltage magnitude improvement for all the buses has been considered and is found to be bus 12 for the IEEE 14 bus test system. An error of 0.05 was considered in identifying a unique bus for STATCOM placement.

(v) A similar simulation and ranking analysis has been performed for modified IEEE 14 bus test system represented in Tables 33.3 and 33.4. The optimal location for STATCOM placement has been identified as bus 9 for the modified IEEE 14 bus test system.

Table 33.3 Voltage magnitudes of modified IEEE 14 bus test system

Bus number	Base case	Modified IEEE 14 buses test system With STATCOM placed at											
		2	4	5	7	9	10	11	12	13	14
1	1.06	1.06	1.06	1.06	1.06	1.06	1.06	1.06	1.06	1.06	1.06
2	0.9555	1.01	1.01	1.02	1.01	1.01	1.01	1.01	1.012	1.01	1.01
3	0.8837	1.01	1.01	1.01	1.01	0.98	1.01	1.01	1.01	1.01	1.01
4	0.6208	0.76	0.8	0.79	0.76	0.82	0.76	0.75	0.75	0.75	0.75
5	0.6213	0.74	0.75	0.78	0.72	0.75	0.7	0.7	0.7	0.7	0.7
6	0.9141	1.06	1.07	1.07	1.07	1.08	1.07	1.07	1.07	1.07	1.07
7	0.6630	0.9	0.92	0.9	0.92	0.95	0.92	0.9	0.9	0.9	0.9
8	0.9392	1.08	1.09	1.09	1.09	1.07	1.09	1.09	1.09	1.09	1.09
9	0.5594	0.86	0.87	0.84	0.87	0.91	0.9	0.86	0.85	0.86	0.88
10	0.579	0.9	0.89	0.88	0.88	0.87	0.95	0.89	0.87	0.87	0.89
11	0.7266	0.95	0.96	0.96	0.96	1.01	1	1	0.95	0.96	0.97
12	0.8214	1.01	1.016	1.015	1.017	1.017	1.02	1.018	1.0414	1.035	1.028
13	0.7707	0.98	0.99	0.988	0.99	1.02	0.99	0.99	0.998	1.02	1.01
14	0.6748	0.85	0.86	0.85	0.86	0.89	0.8	0.87	0.85	0.87	0.95

Table 33.4 Ranking table for modified IEEE 14 bus test system

Bus number	Without STATCOM	Voltage magnitude rank-wise with STATCOM	Optimal location
2	0.9555	2(1.013), 4, 7, 10, 11, 12, 13, 14(1.012), 5(1.02), **9(1.011)**	9
4	0.62086	**9(0.82)**, 4(0.8), 5(0.79), 2, 7, 10(0.76), 11, 12, 13, 14(0.75)	
5	0.62133	5(0.78), 4,**9(0.75)**, 2(0.745), 7(0.72), 10, 11, 12, 13, 14(0.7)	
7	0.66304	**9(0.95)**, 4, 7, 10(0.92), 2, 5, 11, 12, 13, 14(0.9)	
9	0.55941	**9(0.91)**, 10(0.9), 14(0.88), 4, 7(0.87), 2, 11, 13(0.86), 12(0.85), 5(0.84)	
10	0.579	10(0.95), 2(0.9), 11, 14(0.895), 4(0.89), 5, 7(0.88), **9(0.877)**, 12, 13(0.87)	
11	0.7266	**9(1.01)**, 10, 11(1), 14(0.97), 4, 5, 7, 13(0.96), 2, 12(0.95)	
12	0.82146	12(1.0414), 13(1.035), 14(1.028), 10(1.02), 11(1.018), 7, **9(1.017)**, 10(1.02), 4(1.016), 5(1.015), 2(1.011)	
13	0.7707	13(1.025), **9(1.021)**, 14(1.012), 12(0.998), 10(0.997), 11, 7(0.99), 5(0.988), 2(0.98)	
14	0.6748	14(0.95), **9, 10(0.89)**, 11, 13(0.87), 7(0.868), 4(0.86)2, 5, 12(0.85)	

Hence, it is observed that voltage stability margin can be significantly improved using STATCOM for increased reactive power loadings and also incorporating renewables in the conventional power grids. Continuation power flow proves to be a powerful tool to carry out static voltage stability analysis and the presented methodology proves to be efficient and simple in identifying the buses prone to voltage collapse and improve their voltage stability margins.

Modified IEEE 14 bus test system results without STATCOM **Modified IEEE 14 bus test system results with STATCOM**

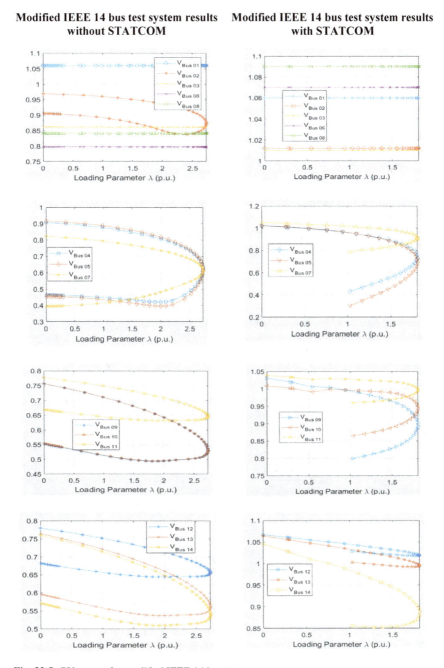

Fig. 33.5 PV curves for modified IEEE 14 bus test

References

1. Kishore TS, Singal SK (2015a) Economic analysis of power transmission lines using interval mathematics. J Electr Eng Tech 10(4):1472–1480
2. Kishore TS, Singal SK (2015b) Design considerations and performance evaluation of EHV transmission lines in India. J Sci Ind Res 74:117–122
3. IEEE/CIGRE (2004) Joint task force on stability terms and definitions. Definition and classification of power system stability. IEEE Trans Power Sys 19(2):1387–1401
4. Kishore TS, Kaushik SD, Venu Madhavi Y (2019) Modelling, simulation and analysis of PI and FL controlled microgrid system. In: 2019 IEEE international conference on electrical computer and communication technologies (ICECCT). Coimbatore, India, 1–8
5. Tien DV, Hawliczek P, Gono R, Leonowicz Z (2017) Analysis and modeling of STATCOM for regulate the voltage in power systems. In: 2017 18th international scientific conference on electrical power engineering (EPE) Kouty and Desnou, 1–4
6. Sohan C, Sumit P (2015) Influence of grid connected solar and wind energy on small signal stability. In: 2015 IEEE international conference on technological advance in power & energy, 1–6
7. Canizares CA, Alvarado FL (1993) Point of collapse and continuation methods for large AC/DC systems. IEEE Trans Power Sys. 8(1):1–8
8. H.K Chappa, T.Thakur: Identification of Weak Nodes in Power System Using Conditional Number of Power Flow Jacobian Matrix. ICUE Conf. on Green Energ. for Sustainable Development, 2018.
9. Mehrdad AK, Mostfa A, Hamid L, Nemat T (2018) Comparison of SVC, STATCOM, TCSC and UPFC controllers for static voltge stability evaluated by continuation power flow method. IEEE Electr Power Energ Conf
10. Hemanthkumar C, Tripta T (2019) A fast online voltage instability detection in power transmission system using wide-area measurements. Iranian J Sci. Tech Trans Electr Eng 43:427–438
11. Satish K, Ashwani K, Sharma NK (2016) A novel method to investigate voltage stability of IEEE-14 bus wind integrated system using PSAT. Front Energy 1–9
12. Telang AS, Bedekar PP (2016) Application of voltage stability indices for proper placement of STATCOM under load increase scenario. Int J Energ. Power Eng 10(7):998–1003
13. Hemanthkumar C, Tripta T (2020) Voltage instability detection using synchrophasor measurements: a review. Int Trans Electr Energ Syst 1–34

Chapter 34
Energy and Economic Analysis of Grid-Type Roof-Top Photovoltaic (GRPV) System

Abhinav Kumar Babul, Saurabh Kumar Rajput, Himmat Singh, and Ramesh C. Yadaw

Introduction

Renewable energy resources are producer of green energy which makes the world clean. These sources also improve the quality index of air globally. Among all other renewable energy resources, the solar energy source has great potential of generating electrical energy [1]. For GRPV with the balance-of-system (BOS), the embodied energy payback times (EinPBT) is evaluated. It is clear that EinPBT depends on the sun radiation intensity and with increase in sun radiation intensity, the EinPBT reduces. It was further pointed out that sun radiation 2-hour hike lowers 2–4 years EinPBT. Similarly, EinPBT with the roof-top PV is reduced from 2 to 6 years. Also by increasing the battery life, EinPBT reduces [2]. EinPBT of a greenhouse dryer and hybrid photovoltaic/thermal (PVT) was analyzed. The overall thermal output of PVT system has been calculated from a combination of thermal energy and yearly electrical generation; the overall yearly thermal output is 1056.74 kWh. EEin of PVT greenhouse dryer has calculated on different energy density of aluminum. For the cases of (Aluminum) with different energy density of PV modules, the EE in of hybrid PVT (HPVT) greenhouse dryer is calculated and E in PBT is determined. For case A (Aluminum 32.39 kWh/kg), the EE in 3726.77 kWh and E in PBT is 3.52 years, respectively, for case B (Aluminum 55.28 kWh/kg) of EE in is 1056.74 kWh and E in PBT are 5.25 years. For the energy density of a 249 kWh solar module, the EE in of Case C (aluminum 32.39 kWh/kg) is 3236.77 kWh and the E in PBT is 3.06 years, respectively, for Case D (Aluminum is 55.28 kWh/kg), EE in and E in

---

A. K. Babul · S. K. Rajput (✉) · H. Singh
Department of Electrical Engineering, MITS, Gwalior, Gwalior, MP, India
e-mail: saurabh9march@mitsgwalior.in

R. C. Yadaw
Engineering Division, NITRA, Ghaziabad, UP, India

© The Editor(s) (if applicable) and The Author(s), under exclusive license to Springer Nature Singapore Pte Ltd. 2021
A. K. Singh and M. Tripathy (eds.), *Control Applications in Modern Power System*, Lecture Notes in Electrical Engineering 710,
https://doi.org/10.1007/978-981-15-8815-0_34

PBT are 5065.11 kWh, 4.79 years. E in PBT is approximately 3-5 years compared to the 30-year life develop of HPVT greenhouse dryer [3]. The E in PBT with annual electrical energy output with one fan is 147.42 kWh and 12.55 years. E in PBT with annual thermal energy output with one fan is 187.05 kWh and 9.89 years. The total yearly thermal energy output is 575.00 kWh and the E in PBT is 3.22 years [4]. In another study, the E in PBT of HPVT system with standalone PV power supply was evaluated. Total EE in calculated as 1,98,166.89 kWh (annual), energy based on energy and e-x energy are 7826.00 kWh, 11,512 kWh [5]. E in PBT is calculated for PVT under variation in climatic conditions. The result includes E in PBT of 21.70 years for Srinagar station and 16.70 years for Jodhpur station [6]. An attempt to understand the trend of the difference between ideal efficiency and actual efficiency of solar modules is undertaken. Actual capacity of PV modules is calculated at voltage for open circuit and current for short circuit. It is observed that in the initial years, the percentage difference increase is more and in later years, it slows down [7]. In case of HIT-type PV modules, energy saving is high because of high module efficiency. The E in PBT counts on an external basis is high than on a thermal energy base [8]. The energy efficiency of Sodha BAG Energy Complex (Varanasi, UP), India, was analyzed and E in PBT of the room was determined to be 5.50 years [9]. In a study E in PBT for one-side slope and two-side slope-type passive solar stills is analyzed. The total EE in of solar stills is 1483.90 kWh, whereas the annual ex-energy of one-side slope passive solar still is 108.48 kWh and for two-side slope passive solar stills is 89.24 kWh, respectively. One-side and two-side slope solar still are having annual energy of 1159.43 kWh and 1037.00 kWh, respectively. E in PBT concluded based on energy and ex-energy is 1.42 years, 16.62 years [10]. The E in PBT for solar module integrated with roof top of Sodha BAG Energy Complex (Varanasi, UP), India, is further evaluated. The E in PBT decreases for the average per day sun radiation of 450.00 W/m^2 from 37.49% with an increase in temperature from 4 to 8 °C (due to greater thermal heat) [11]. The E in PBT with evacuated tubular-type collector which is integrated with compound parabolic concentrator is 1.20 years [12]. E in PBT has been evaluated for three different cases of partially, fully, and conventional N-CPVT collector. The E in PBT for these three cases is 8.22 years, 17.87 years, and 5.58 years, respectively [13]. E in PBT for passive-type two-slope solar still has been analyzed. E in PBT for Al_2O_3-water-based nanofluid is higher and based on base fluid is lower [14]. E in PBT for roof-top PV system was calculated as 8.75 years [15]. The E in PBT for the GRPV system is 6.00 years. It takes at least 5.31 years for GRPV system to pay off its EE in due to its high energy density compared to other systems [16].

The GRPV system itself take a huge amount of energy for its construction, manufacturing & installations. Also the initial investment cost of the system is very high, so it becomes very important to understand and know about the energy efficiency and economic viability of the GRPV system. This complete techno-economic analysis of the GRPV system is essential at system forecasting part. This paper presents a methodology for energy efficiency and economic analysis of GRPV system. The complete study is performed on a 100 kW GRPV system, which is located at the roof top of NITRA Ghaziabad, India.

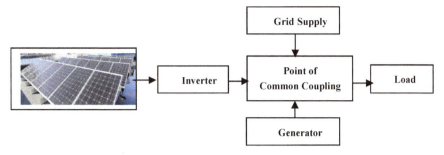

Fig. 34.1 Block diagram of roof-top PV system

The structure of paper:
Section 2 is the description of the GRPV system located at NITRA Ghaziabad, India.
Section 3 is the detailed energy-based analysis if GRPV system.
Section 4 includes the methodology for economic analysis of the system.
The conclusion part is covered in Sect. 5 of the paper

GRPV Setup

100 kW GRPV system has been installed at NITRA, Ghaziabad. An effective area of each module is 1.63 m² and the total effective area of the associated 318 number of modules in series/parallel string combination is 520.08 m². The three-phase commercial inverters are used in the system. The inverter output terminals are connected to the common coupling point, and the common coupling point is connected to the load. Due to fault or cloudy weather conditions, if the roof-top PV system is not able to supply the load, a change is made to the grid by a switch located at the common coupling point. This complete system is shown in Fig. 34.1.

The GRPV system consists of 318 number of PV modules. Each module is rated for peak power–320 W, open circuit voltage–46 V, rated voltage–37.7 V, short circuit current–9.03 Amp, and rated current–8.50 Amp.

Energy Analysis of Grid-Type Roof-Top Photovoltaic (GRPV) System

The energy analysis GRPV system is done in two parts: first is the measurement of annual energy (kWh) generation by the plant and second is the calculation of EE in of the plant. Then the time for E in PBT is calculated and analyzed.

(a) **Annual energy (kWh) generation of plant.**

Table 34.1 Solar electrical energy generated (Monthly)

Month	Generation (kWh)	CUF (%)
January	8767.67	12.04
February	8283.45	12.57
March	12,730.66	17.84
April	12,514.24	17.98
May	13,020.45	18.09
June	11,497.05	16.72
July	9268.34	12.48
August	9975.14	13.85
September	6821.79	9.817
October	10,281.59	14.08
November	6114.74	10.19
December	4510.33	7.301
	113,785.46 (Total)	13.58 (Average)

For the energy analysis of the plant, the data is taken by software installed along with the system. Table 34.1 shows monthly energy generated by GRPV in kWh with CUF in % for previous one year.

The total generation of electrical energy by the GRPV system is calculated by summing the monthly energy generation for previous one year. $E_{per\ year} = 113{,}785.46$ kWh.

If a system continuously distributes full rated power, its CUF will be unity 100%. CUF, PV depends on the system of location. High efficiency factor, better PV system. The highest CUF (18.09%) is observed in May 2019 due to the highest energy generation of 1309.45 kWh while the lowest CUF (7.30%) for lowest energy production 4510.33 kWh in December 2019. CUF varies from January to May, are 12.04–18.09%, which was agreed to limit the solar PV system for most roof in India is 16–17%. Figure 34.2 shows the monthly kWh energy generated by GRPV system.

(b) **Embodied energy** (EEin) **of GRPV system.**

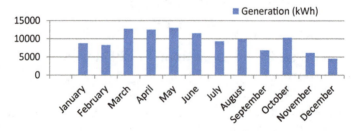

Fig. 34.2 Solar energy generated monthly (kWh)

The EE in consists of the energy, which is required for the making (construction, installation, etc.) of GRPV system. It is the necessary energy associated with the material production energy, PV system installation energy, energy used in maintenance, and energy used in administration of systems for the generation [15].

The EE in used in the part of material production energy (E_{mpe}) of GRPV system is given in Table 34.2.

Total material production energy (Empe) = 509,680.29 kWh.

The embodied energies and weights of materials used in the supporting structure (Table 34.3, 34.4 and 34.5).

Total material production energy (Einst) = 29972.63 kWh.

Total manufacturing energy (Emfg) = Empe + Emain = 514,797.87 kWh.

Total material production energy (Einst) = 29,972.63 kWh.

Total energy used in administration (Eadmin) = 27,824.28 kWh.

EEin = 514,797.87 + 29,972.63 + 27,824.28 = 572,594.78 kWh.

Table 34.2 Material production energy (E_{mpe})

Solar module material	EEin (kWh/m^2)	Total area (m^2)	Total EEin (kWh)
Processing and purification of silicon material	670.00	520.08	3,48,455.49
Solar cell production	120.00	520.08	62,409.60
Assembly and lamination of GRPV system module	190.00	520.08	98,815.20

Table 34.3 PV system installation energy ($E_{inst.}$)

Item	EEin	Total weight	Total EEin (kWh)
Support structure: Iron stand screw	7.70 (kWh/kg) 8.63 (kWh/kg)	1060.00 kg 26.5 kg	8162.00 (kWh) 228.69 (kWh)
Inverter	210.00 (kWh/kW)	100.00 kW	21000.00 (kWh)
Wires	3.00 (kWh/m^2)	193.98 m^2	581.94 (kWh)

Table 34.4 Energy used in maintenance (E_{main})

Item	EEin (kWh/m^2)	Total area (m^2)	Total EEin (kWh)
Human labor	9.84	520.08	5117.58

Table 34.5 Energy used in administration (E_{admin})

Item	EEin (kWh/m^2)	Total area (m^2)	Total EEin (kWh)
Transportation	53.50	520.08	27,824.28

Economic (Benefit–Cost) Analysis of GRPV System

For the complete analysis of GRPV system, the benefit–cost-based economic analysis of the GRPV system is also done. This analysis of the GRPV system is done by considering the total investments and tangible benefits of the GRPV system. The total cost of GRPV system includes installation cost, maintenance cost, running cost, and taxes, etc.; while the benefit of GRPV system mainly includes the savings in electricity bills.

If the GRPV system provides the output for 330 days in a year (for the climate condition under consideration) and the average capacity utilization factor (CUF) of 13.58%. Then, the unit (kWh) generation is calculated by formula:

$$\text{Unit generation (kWh)} = \text{kW output} \times \text{CUF} \times 24\,\text{h} \times 330\,\text{days} \quad (34.1)$$

For benefit–cost analysis, amount saved by generating electricity (kWh/units) by photovoltaic system is considered as benefit and cost is "Operation, maintenance and insurance cost". In the present case study, cost is considered as 2.50% of initial investment on system.

The annual monitory saving by the photovoltaic system is obtained by subtracting the total annual cost of GRPV system from the annual benefit of GRPV system. This is calculated by the formula:

$$\text{Annual saving} = \text{Annual Benefit} - \text{Annual cost}$$
$$\text{(Operation, Maintenance Insurance)} \quad (34.2)$$

As shown in the Fig. 34.3, the benefit–cost analysis of the solar system is explained by a pictorial representation of cash flow diagram [15].

In this study, the cost of the system is considered as installation, operation & maintenance (O&M) cost, whereas the benefit is considered as the savings in the electricity bills. The detailed (yearly) calculation of benefit–cost–saving is shown in Tables 34.6 and 34.7. This benefit–cost–saving analysis is done for output of one inverter rated for 50 kW.

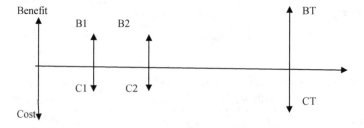

Fig. 34.3 Pictorial representation of cash flow diagram

Table 34.6 Yearly generation by GRPV system

Year	System % output	KW output	kWh(Unit) generation
Year–1	100.00	50.00	62,640.00
Year–2	97.00	48.50	60,760.80
Year–3	96.30	48.16	60,335.40
Year–4	95.60	47.82	59,913.20
Year–5	94.90	47.48	59,493.60
Year–6	94.30	47.15	59,077.00
Year–7	93.60	46.82	58,663.60
Year–8	92.90	46.49	58,253.40
Year–9	92.30	46.17	57,845.60
Year–10	91.60	45.85	57,440.20
Year–11	91.00	45.53	57,038.80
Year–12	90.40	45.21	56,639.00
Year–13	89.70	44.89	56,242.60
Year–14	89.10	44.58	55,849.20
Year–15	88.50	44.26	55,458.40
Year–16	87.90	43.95	55,070.00
Year–17	87.20	43.65	54,684.00
Year–18	86.60	43.34	54,301.40
Year–19	86.00	43.04	53,921.80
Year–20	85.40	42.74	53,544.00
Year–21	84.80	42.44	53,169.40
Year–22	84.20	42.14	52,796.80
Year–23	83.60	41.84	52,427.20
Year–24	83.10	41.55	52,060.80
Year–25	82.50	41.26	51,696.20

In order to calculate and properly analyze the Simple payback time of the GRPV system, a detailed calculation of monitory savings (on yearly basis) is done in Tables 34.8 and 34.9. As per this analysis, the total monitory saving at the end of ninth year will be the Rs. 2,019,800.00/-. This monitory saving is due to the output generated by one inverter. So the total savings after completion of ninth year will be Rs. 4,039,600.00/-.

Table 34.7 Life time benefit–cost–savings analysis of GRPV system

Year	Unit cost (Rs)	Benefit (Rs)	(O&M + Insurance) Cost	Savings (Rs)
Year–1	8.40	526,176.00	309,961.00	216,215.00
Year–2	8.50	520,598.50	307,173.30	213,425.20
Year–3	8.70	527,293.40	310,517.70	216,775.70
Year–4	8.90	534,075.80	313,909.90	220,165.90
Year–5	9.00	540,941.30	317,345.70	223,595.60
Year–6	9.20	5,47,896.90	320,821.50	227,075.40
Year–7	9.40	554,944.00	324,348.10	230,595.90
Year–8	9.60	562,083.90	327,918.00	234,165.90
Year–9	9.80	569,312.20	331,527.10	237,785.10
Year–10	10.00	576,629.90	335,194.00	241,435.90
Year–11	10.20	584,051.10	338,895.60	245,155.50
Year–12	10.40	591,558.10	342,653.10	248,905.00
Year–13	10.60	599,165.20	346,459.70	252,705.50
Year–14	10.80	606,873.90	350,308.00	256,565.90
Year–15	11.00	614,679.10	354,213.60	260,465.50
Year–16	11.30	622,582.00	358,167.00	264,415.00
Year–17	11.50	630,584.20	362,169.20	268,415.00
Year–18	11.70	638,694.20	366,219.20	272,475.00
Year–19	11.90	646,913.90	70,328.00	276,585.90
Year–20	12.20	655,229.90	374,494.00	80,735.90
Year–21	12.40	663,659.00	378,703.60	284,955.40
Year–22	12.70	672,187.00	382,971.60	289,215.40
Year–23	12.90	680,831.30	387,285.70	293,545.60
Year–24	13.20	689,594.00	391,669.00	297,925.00
Year–25	13.50	698,460.30	396,105.20	302,355.10

Table 34.8 Yearly monitory savings (Rs.) for first five years

At the end of first year	At the end of second year	At the end of third year	At the end of fourth year	At the end of fifth year
216,215.00	216,215.00 + 213,425.20	216,215.00 + 213,425.20 + 216,775.70	216,215.00 + 213,425.20 + 216,775.70 + 220,165.90	216,215.00 + 13,425.20 + 216,775.70 + 220,165.90 + 223,595.60
216,215.00	429,640.20	646,415.90	866,581.80	1,090,177.00

Table 34.9 Yearly monitory saving (Rs.) for next four years

At the end of sixth year	At the end of seventh year	At the end of eighth year	At the end of ninth year
216,215.00 + 213,425.20 + 216,775.70 + 220,165.90 + 223,595.60 + 227,075.40	216,215.00 + 213,425.20 + 216,775.70 + 220,165.90 + 223,595.60 + 227,075.40 + 230,595.90	216,215.00 + 213,425.20 + 216,775.70 + 220,165.90 + 223,595.60 + 227,075.40 + 230,595.90 + 234,165.90	216,215.00 + 213,425.20 + 216,775.70 + 220,165.90 + 223,595.60 + 227,075.40 + 230,595.90 + 234,165.90 + 237,785.10
Rs. 1,317,253.00	Rs. 1,547,849.00	Rs. 1,782,015.00	Rs. 2,019,800.00

Conclusion

After the energy and economic analysis of GRPV system, it is observed that the system itself takes a huge amount of energy for its construction, manufacturing & installations. Also, the initial investment cost of the GRPV system is very high. So it become necessary to know about the energy efficiency and "cost-benefits" analysis of the GRPV system. In this study, the GRPV system is installed with capital investment of Rs 40.00Lakh; for which the annual generation of electrical energy is 113,785.46 kWh and the EE in of GRPV system is 572,594.78 kWh. Considering this, the E in PBT of the system is 5.03 years and simple payback time is 9 years.

Although the system under the study is an energy as well as economically efficient system but the GRPV system performance could be further upgrade by proper maintenance, washing of panels, avoiding shading effect and by maintaining the standard test conditions, etc. In this study, the escalation in the price of the electricity pricing is not considered, so it is also recommended to consider it for more accurate forecasting of the payback time of the system.

References

1. Twidell J, Weir A (2006) Renewable Energy. Resources. https://doi.org/10.4324/9780203478721
2. Nawaz I, Tiwari G (2006) Embodied energy analysis of photovoltaic (PV) system based on macro- and micro-level. Energy Policy 34:3144–3152. https://doi.org/10.1016/j.enpol.2005.06.018
3. Barnwal P, Tiwari G (2008) Life cycle energy metrics and CO^2 credit analysis of a hybrid photovoltaic/thermal greenhouse dryer. Int J Low-Carbon Technol 3:203–220. https://doi.org/10.1093/ijlct/3.3.203
4. Tiwari A, Barnwal P, Sandhu G, Sodha M (2009) Energy metrics analysis of hybrid—photo voltaic (PV) modules. Appl Energy 86:2615–2625. https://doi.org/10.1016/j.apenergy.2009.04.020

5. Prabhakant Tiwari G (2009) Energy payback time and life-cycle conversion efficiency of solar energy park in Indian conditions. Int J Low-Carbon Technol 4:182–186. https://doi.org/10.1093/ijlct/ctp020
6. Nayak S, Tiwari G (2010) Energy metrics of photovoltaic/thermal and earth air heat ex changer integrated greenhouse for different climatic conditions of India. Appl Energy 87:2984–2993. https://doi.org/10.1016/j.apenergy.2010.04.010
7. Rajput SK, Singh O (2017) Reduction in CO_2 emission through photovoltaic system: a case study. In: 3rd IEEE international conference on nanotechnology for instrumentation and measurement. GBU, Greater Noida, India
8. Kamthania D, Tiwari G (2014) Energy metrics analysis of semi-transparent hybrid PVT double pass facade considering various silicon and non-silicon based PV module Hyphen is accepted. Sol Energy 100:124–140. https://doi.org/10.1016/j.solener.2013.11.015
9. Sudan M, Tiwari G (2014) Energy matrices of the building by incorporating daylight concept for composite climate—An experimental study. J Renew Sustain Energy 6:053122. https://doi.org/10.1063/1.4898364
10. Singh D, Tiwari G, Al-Helal I et al (2016) Effect of energy matrices on life cycle cost analysis of passive solar stills. Sol Energy 134:9–22. https://doi.org/10.1016/j.solener.2016.04.039
11. Gupta N, Tiwari G (2017) Energy matrices of building integrated photovoltaic thermal systems: case study. J Arch Eng 23:05017006. https://doi.org/10.1061/(asce)ae.1943-5568.0000270
12. Mishra R, Garg V, Tiwari G (2017) Energy matrices of U-shaped evacuated tubular collector (ETC) integrated with compound parabolic concentrator (CPC). Sol Energy 153:531–539. https://doi.org/10.1016/j.solener.2017.06.004
13. Tripathi R, Tiwari G, Dwivedi V (2017) Energy matrices evaluation and exergoe conomic analysis of series connected N partially covered (glass to glass PV module) concentrated-photovoltaic thermal collector: At constant flow rate mode. Energy Convers Manag 145:353–370. https://doi.org/10.1016/j.enconman.2017.05.012
14. Sahota L, Shyam Tiwari G (2017) Energy matrices, enviroeconomic and exergoeconomic analysis of passive double slope solar still with water based nanofluids. Desalination 409:66–79. https://doi.org/10.1016/j.desal.2017.01.012
15. Rajput SK (2017) Solar energy: fundamental, economic & energy analysis, 1st ed. NITRA Publication. ISBN: 978-93-81125-23-6 16
16. Yadav S, Bajpai U (2019) Energy, economic and environmental performance of a solar rooftop photovoltaic system in India. In J Sustain Energy 39:51–66. https://doi.org/10.1080/14786451.2019.1641499

Chapter 35
Automation of Public Transportation (Bus Stands)

Gaurav Yadav, Archit, Parth Dutt, and Sankalp Sharma

Introduction

Buses are a very important form of public transport in India, the significance of the same can be easily observed by taking into consideration the fact that in the National Capital Region alone, the total daily footfall of commuters in 2016 was estimated to be around 3 million, but the issue lies in the fact that 3 million lacks about 1.7 million commuters as compared to the figure of 4.7 million of the same calculated for the duration of 2012–13 which directly depicts the falling trend of reliability on the respective public transportation system [1].

The chosen strategy to tackle the issue at hand aims to firstly, reduce direct obstruction of traffic caused by buses and secondly, make bus transportation more reliable by addition of various utilities and features to make bus travel more reliable hence, attracting more commuters to travel by buses rather than personal transport, both of which will lead to reduced traffic congestions and decreased by road travel times thus fulfilling the main objectives.

For the attainment of the formerly mentioned objectives, this application utilizes RFID tags and intercommunication of specialized hardware to relay bus timings across informative visual surfaces, relay live bus tracking updates, provide charging options to commuters and permit buses to only stop very close to bus stops to ensure that public transportation, specifically buses, don't obstruct the flow of traffic near bus stops as is generally observed thus resulting in a significant reduction in traffic

---

G. Yadav · P. Dutt (✉) · S. Sharma
Department of Electrical and Electronic Engineering, Amity School of Engineering and Technology, Noida, Uttar Pradesh, India
e-mail: dutt.parth123@gmail.com

Archit
Amity University, Noida, Uttar Pradesh, India

© The Editor(s) (if applicable) and The Author(s), under exclusive license to Springer Nature Singapore Pte Ltd. 2021
A. K. Singh and M. Tripathy (eds.), *Control Applications in Modern Power System*, Lecture Notes in Electrical Engineering 710,
https://doi.org/10.1007/978-981-15-8815-0_35

congestion caused by buses leading to reduction in overall commute time and safer and smoother flow of traffic supplemented by the addition of charging docks, live bus tracking, integrated ticketing kiosk and interactive video surfaces to aid in making bus transportation more organized, reliable and practical which will in turn restore public faith in the bus transportation network in turn attracting more daily commuters thereby reducing the number of private or hired vehicles on the road, denting the traffic density significantly and, as well, consequentially reducing the area's carbon footprint in the process [2].

System Description

The system to be employed consists of various components which work simultaneously in cooperation to fulfill various needs to be met by the bus transportation network and the commuters at a specific bus stand for maximum efficiency in improving commuter experience and reliability of bus stops. To begin with, the system works with the help of a microcontroller which functions as the main control terminal tasked with sending and receiving information from a central server about the arrival of the scheduled bus along with the expected wait time before its arrival and its entire route along with scheduled timings and, in case of a delay or a breakdown, presenting the information of the same on the interactive luminous surfaces.

The microcontroller is also tasked with maintaining a log of the actual bus arrival timings as compared to the scheduled timing along with a log of the defaulters for the day (defaulter in this case being bus drivers who do not stop in proximity to the RFID tags thus obstructing traffic and posing a public safety violation). This microcontroller will also have the tasks of operating the advertising surfaces fixed in and around the bus stand, receiving updates on when to change/update to the next set of adverts as provided by the main server and directing power to the ticketing kiosk while simultaneously maintaining a log of the ticket sales. The system in question will be powered by solar panels mounted preferably at the top of the bus stand's structure providing enough power to run the ticketing kiosks, interactive informative surfaces, advertising surfaces, and RFID tags. Information of the components used for demonstrating the basic functionality of the mentioned system will be mentioned hereafter.

Arduino Microcontroller

Arduino is one of the fastest growing community and vastly used microcontroller in the world for beginners in Do It Yourself electronics, these boards are based on the ATMEGA AVR series microcontrollers from ATMEL. Different features of Arduino UNO are given in Table 35.1 [3] (Fig. 35.1).

Table 35.1 Features of Arduino UNO [4, 8]

S. No.	Feature	Range
1	Micro\controller	ATmega328
2	Operating voltage	5 V
3	Input voltage	7–24 V
4	Analog input pins	8
5	Flash memory	32 KB
6	EEPROM	1 KB
7	Digital I/O pins	14
8	SRAM	2 KB
9	Clock speed	16 MHz
10	DC current per I/O pin	40 A

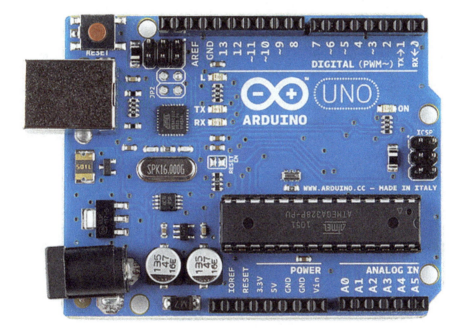

Fig. 35.1 Arduino Uno microcontroller

RFID Tag

RFID tags are a radio frequency identification system which use low-power radio frequency waves to communicate to and from the tags and readers due to which information can be scanned from a short wireless range depending upon the application required. RFID tags could come in active, passive, and semi-passive forms with the latest same technology being capable of reaching distances up to 20 feet (6

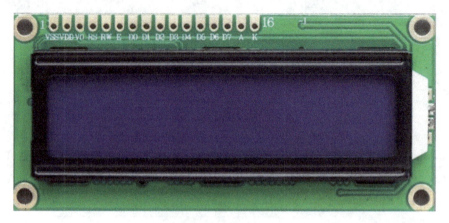

Fig. 35.2 16 × 2 LCD screen

meters) at relatively lower costs [9]. In the respect of the currently being worked on project, the aforementioned technology will be used at each stoppage unit in pairs for detecting proximity of buses to the bus stop so that the driver can be alerted to stay within close distance of the bus stop [6] so that the flow of traffic is not obstructed as is observed in the region of observation of the National Capital Region of the state of Delhi from the analyzed study. In the scenario that the RFID reader does not detect the bus within the appropriate distance the driver would be liable for a reprimand.

LCD Screen

LCD stands for liquid crystal device. It is the device being used to get the display of the final output from the microcontroller. It has a 16-pin interface. Arduino is interfaced with the LCD and then, it displays the output accordingly. The microcontroller inside the LCD sets various pins at a time to control the output on the LCD screen (Fig. 35.2).

EPS8266 WIFI Module

The ESP8266 is a low-cost Wi-Fi microchip, with a full TCP/IP stack and microcontroller capability being used to contact the parent server in the prototype model [5].

Experimental Setup Design

For the purpose of the design of the system in question, an experimental prototype was designed with specific capabilities to emulate the important characteristics of the full-scale requirements yielding the expected results. Two bus stops were connected to a common server through WIFI modules which enables cross communication between them such that when a bus on the route docks at one bus stop, the other receives information about where the bus currently is and how much time it will take for the same to get to the respective bus stop.

Representational Diagram

The components used for the project are listed in Table 35.2 (Fig. 35.3).

Table 35.2 Component list

S. No.	Component	Feature
1	Microcontroller	Arduino
2	ESP8266	WIFI module
3	Battery	9 V
4	LCD display ($\times 2$)	1602 A
5	RFID reader	Pn532
6	Solar panels	110×40 mm

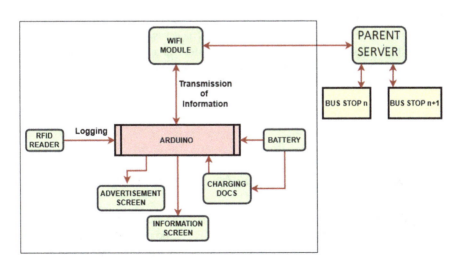

Fig. 35.3 Representational diagram

Programming Algorithm

The program used operates in two modes based on whether the bus has arrived at the bus stop or not; these modes with their algorithms are as follows.

Detection-Based Response

1. RFID tag is detected.
2. Bus number along with time of arrival is logged.
3. Information that bus has arrived is displayed on screen.
4. Information of bus's arrival is relayed to parent server.
5. Parent server is updated and relays the same information to next bus stop.
6. Estimated time of arrival is displayed on the screen of the next bus stop.

Idle Response

1. Microcontroller links with parent server.
2. Video ads are downloaded and bus route information is retrieved from the server.
3. Video ads and route information are updated on respective screens.
4. Microcontroller updates.
5. New information is displayed if received.

Experimental Results

The depicted circuit diagram was used to connect the aforementioned equipment and the different sections: RFID Reader, LCD display, battery, solar panel, and microcontroller. The result being that when an RFID tag was scanned, the LCD screen displayed the unique identification number of the bus along with the name of the bus stop while at the same time, registering a successful docking of the bus and transmitting the same information to other bus stops on the route.

Scope

Long-Distance Wireless Communication Between Bus Stops

A long-distance communication capability will be added to each bus stop so that information pertaining to the buss' arrival status can be relayed back to the parent server for displaying and processing. In doing so, the main application will be able to track bus routes and monitor real-time live location of the bus as well as with a live video feedback to the server to check up on the traffic status and to also safeguard national property in case of a civil disputes. The addition of long-distance wireless communication of buses will require an interfacing of WIFI modules to the buses and bus stops [7] as well so that when a bus comes in proximity of the bus stops, WIFI zone an information exchange can happen between the two objects which will further be relayed back to the parent server.

In the terms of modifications to the existing circuit, WIFI modules will be added to the Arduino circuit and then linked to a common hotspot on which the information will be independently uploaded and exchanged when the RFID tags are read by the sensors on the bus stop, further buses will also have small WIFI modules so that they can be live tracked and logged accordingly [6]. This will require a total of four microcontroller boards and four WIFI modules for the same along with a parent server and host linked to an executable application for monitoring the same.

Setup of a Central Parent Host Server

A central parent host server will have the most important job of interconnecting all buses and bus stops to a common data bank through which the information could later be relayed to other devices for checking the ride status and live tracking of the same. In a real practical scenario, Delhi's free WIFI which is under the talks of implementation will be the mesh interwoven between the bus stops and other consumer devices interconnecting them while also acting at the main bridge of communication and processing.

In the terms of modifications to the existing circuit, a WIFI host network will be created using a laptop to which all the WIFI modules will be connected and then monitored subsequently which will also require a change in the existing coding of the Arduino microcontroller circuits for interfacing of the individual devices along with configuration and security.

File Logging with Continuous Synchronization

With the help of the locally hosted parent server, all connected devices will be able to interlink, synchronize, and share information about the bus ETA along with traffic conditions and route progress updates [8]. All the files will be continually relayed to the server which will then make copies of the information gathered and then log it using data structures and algorithms. The same data can then be used to train AIs through machine learning and then make reliable traffic models which will give very important insights into the modern-day traffic conditions of the state.

Application for Relaying Bus Information

By employing a mobile application, the bus's location can be tracked in real time for the consumer's reference and at the same time, the application will be responsible for estimating the best of available accuracy, the estimated time of arrival.

Interface for Advertisements

For making the bus stops practically maintenance free, they will be enabled to generate revenue from installed LED screens which are cheap and are a one-time investment; these screens will be powered from the installed solar panels. These screens can then be lent to advertisers on a monthly contract and with a fixed sum as the payment for using the same facility.

CCTV Surveillance

Along with all, the other components used the bus stops will also have CCTV surveillance so that the areas near bus stops can be safer, any suspicious activities can be monitored, and the safety of the public is enhanced.

In-facility Ticketing Kiosk

Ticketing system will also be added to the bus stop to make travel convenient so that freeloaders can be prevented from entering the buses and wasting seats for other travelers. This system will be capable of issuing tickets using metro cards, debit cards, cash, and other forms of payment.

Conclusion

In this paper, we have discussed the need for the upgradation of bus stops and their effects on the overall traffic flow of the surrounding areas, along with the increasing need to improve the reliability of the same to gain back lost public confidence. This paper aims at providing a holistic approach to upgrade the public transport to make it inherently more efficient on road by minimizing its effects on traffic continuity and at the same time, increasing fuel efficiency, reducing overall traffic jams, reducing the overall commute duration, and majorly cutting down on toxic air pollution thereby also increasing the average AQI of the area by the implementation of the applications provided herein.

References

1. Sen R, Cross A, Vashistha A, Padmanabhan VN, Cutrell E, Thies W (2013). Accurate speed and density measurement for road traffic in India. In: Proceedings of the 3rd ACM symposium on computing for development. New York, USA, pp 1–10
2. Fernandez R, Tyler N (2005) Effect of passenger–bus–traffic interactions on bus stop operations. Transp Planning Technol 28(4):273–292
3. Saranraj B, Dharshini NSP, Suvetha R, Bharathi KU (2020) ATM security system using Arduino. In: 2020 6th international conference on advanced computing and communication systems (ICACCS). IEEE, pp 940–944
4. Alvanou AG, Zervopoulos A, Papamichail A, Bezas K et al (2020) CaBIUs: description of the enhanced wireless campus testbed of the Ionian University. Electronics 9(3):454
5. Patil S, Abhigna A (2018) Voice controlled robot using labview. In: 2018 international conference on design innovations for 3Cs compute communicate control (ICDI3C). IEEE, pp 80–83
6. Eken S, Sayar A (2014) A smart bus tracking system based on location-aware services and QR codes. IEEE Int Symp Inn Intell Syst Appl (INISTA) Proc Alberobello 2014:299–303
7. Rahman F, Ritun IJ, Biplob MRA, Farhin N, Uddin J (2018) Automated aeroponics system for indoor farming using Arduino. In: 2018 joint 7th international conference on informatics, electronics & vision (ICIEV) and 2018 2nd international conference on imaging, vision & pattern recognition (icIVPR). IEEE, pp 137–141
8. ARDUINO UNO REV3 The UNO is the most used and documented board of the whole Arduinofamily. https://store.arduino.cc/usa/arduino-uno-rev3,Code:A000066,Barcode:7630049200050
9. PN5180 High-performance multi-protocol full NFC frontend, supporting all NFC Forum modes. https://www.nxp.com/docs/en/data-sheet/PN5180A0XX-C1-C2.pdf

Chapter 36
Analysis of Different Aspects of Smart Buildings and Its Harmful Effects on the Ecosystem

D. K. Chaturvedi and Boudhayan Bandyopadhyay

Introduction

Buildings have developed over time from ancient to the modern times. In present days, about 50% population of world lives in city areas and projected that in 2050 it will reach to about 66% [1]. Modern buildings or smart buildings are deployed with IoT sensors which are coupled with information and communication technology for creating smarter environments. Hence, a smart building controls and integrates the various aspects like ventilation, sunlight, safety and security, easy access of things and their control, etc., and leverages the IoT data to get better efficiency in terms of energy and consumer understanding of the occupants with big data analytics [2]. This leads to generation of huge amounts of data followed by challenges with privacy and security which if compromised can be lethal for the occupants. This paper identifies the various challenges which may arise for the tenants or occupants of a smart building.

Historical Aspect of Building Automation

The gradual development of buildings from primitive buildings to smart buildings has been cited by Buckman et al. [3]. It has been categorically discussed how buildings have emerged from primitive buildings to simple buildings to automated buildings and followed by intelligent/smart buildings.

D. K. Chaturvedi (✉) · B. Bandyopadhyay
Dayalbagh Educational Institute (Deemed University), Dayalbagh, Agra, India
e-mail: dkc.foe@gmail.com

© The Editor(s) (if applicable) and The Author(s), under exclusive license to Springer Nature Singapore Pte Ltd. 2021
A. K. Singh and M. Tripathy (eds.), *Control Applications in Modern Power System*, Lecture Notes in Electrical Engineering 710,
https://doi.org/10.1007/978-981-15-8815-0_36

Building automation was in vogue right from the ancient Roman times. In ancient Rome (350 BC), there was a concept of Hypocaust, which was a scheme for building heating from a central location from where hot air is circulated through embedded pipes in the floor of a room and the walls. This heating facility was generally located in the basement and the heat was transferred to the rooms on the top.

Although heating and ventilation came down in different formats through the ages, the concept of creating a closed feedback control loop was absent until the invention of the room thermostat.

- Andrew Ure (1778–1857), a Scottish chemist, patented the bimetallic thermostat in 1830. This was a bimetallic strip that would bend with increase in room temperature and thereby cut off the energy supply.
- Warren Johnson (1847–1911) developed and patented the thermostat for room automation, which is used to ring a bell in the boiler room when the temperature of the room increased to notify the operator to adjust the furnace damper.
- Albert Butz (1845–1909) patented the "Damper-Flapper," which was a thermostat for controlling heat by furnace door opening or closing automatically.

Definition of Smart Buildings

There are various opinions on the definition of smart buildings.

- Buckman et al. [3] clearly mentioned about smart building which join together activities, artificial intelligence, planning and efficient control of energy, water, gas etc., features of adaptability, optimization, and comfort. The big data received from different sources makes system adaptable and controllable to get ready to modify at any time.
- Bajer [4] described about smart buildings whose center of attention is automation and minimize building operation and maintenance expenditure, greater satisfaction, and eco-friendly.

Smart buildings are a modified version of intelligent buildings, which not only considers the responsive behavior towards change in stimulus but also considers an adaptive behavior. This has been discussed by Buckman et al. [3] in their paper (Fig. 36.1).

Problem Space Identification

Smart buildings use ICT to connect the various devices that are intended to control the building. This puts forward the generation of huge data silos along with different channels through which data is exchanged. On the other side, as an energy system smart building also accounts for a huge consumption to meet the multi-functionality.

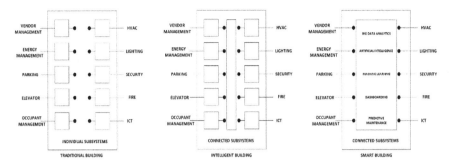

Fig. 36.1 Traditional vs Intelligent vs Smart building [15]

In big buildings, about 70% of total electrical energy is utilized and emit about 40% of greenhouse gases per year in USA [5].

The problem space can be broadly classified into the following dimensions:

- **IoT applications and IT security**—This is the area where the operation of a building can be compromised by intrusion of hackers into the network.
- **Energy system management**—This is the area where unnecessary use of energy can be identified by various building systems and problems about a grid shutdown can be identified.
- **Wireless technologies**—This is the area in which the use of different wireless devices to perform tasks comes into hindrance with human physiological functions.

IoT Applications and IT Security

IoT applications result in several security threats and it has been observed in various cases. [6–8]. The associated IoTs are vulnerable for cyberattacks. The main causes behind that are:

- IoT devices work automatically without human intervention, so for a hacker it is very easy to get access of them.
- IoT interacts with other system components wirelessly or through Internet; hence, it is very easy to hack the information.
- Strong IoT security schemes cannot be afforded due to low energy consumption and less computing resource requirements.

Buildings generally use standard communication protocols such as BACnet or KNX or LON for communicating between different devices. These protocols do have limited IT security options or are in the process of upgrading their IT Security features with new releases. This has led to identification of certain gaps which may prove to be fatal for the occupants of the buildings.

This has been discussed by Caviglione et al. in their paper [9].

The problems areas have been identified as:

- Surveillance—An intruder to the system can monitor the activity of the occupants of the buildings and that can be used to initiate active or passive attacks in the form of physical or biological assaults. This is related to unnecessary use of real-time sensor data.
- Remote control—This is related to the controlling the functions by handling the actuators remotely. For example, a hydraulic lift can be stopped by overriding the actuator of the valve of the hydraulic and causing a disturbance for the inhabitants.
- Physical exploitation—This can be done by remotely controlling an edge device by an intruder and using it for unintended use, specifically assets owned by other occupants. For example, using the parking sensors of a fixed parking lot to park someone else's car and then leave it stranded.
- Availability—Creation of false alarms by causing a denial-of-service (DOS) attack. This can raise a safety concern for the occupants during times of need.
- Smart building botnets—This is an application of both surveillance and remote control from a broader perspective, where an electricity company can manipulate the heating/cooling demands of an entire building to increase its sales pipeline.

Thus, real-time exploitation of building assets can be performed by breaching the IoT devices through the IT channels.

Energy System Management

Smart buildings house different electrical systems which include HVAC, lighting, security and access control and other miscellaneous utilities like motors for lifts and parking lots etc. The occupants of the buildings are heavily dependent on these systems, and it is almost an uncertain cloud that peeps in if there is a grid failure which will lead to failure of these systems [10].

With increase in smart buildings, the consumption of electricity is on the rise and this in turn results in increased carbon footprint on the globe.

Onat et al. [11] in their paper have divided the carbon footprint accounting into three different scopes:

- Scope 1: Emissions occurring from the activities occurring within the physical boundary of the building during the construction phase.
- Scope 2: Emissions that are created due to the generation of electricity in the power plants.
- Scope 3: Other indirect sources of emissions (Fig. 36.2).

The results that have been inferred show that Scope 2 (electricity use) accounts for 47.8% of the total carbon footprint contribution. Hence, it can be indirectly inferred that in order to make buildings smart, there is an indirect contribution to global carbon footprint.

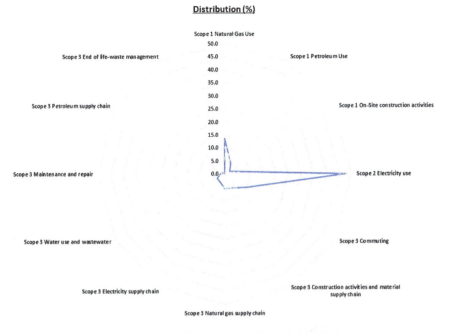

Fig. 36.2 Carbon footprint percentage distribution of US buildings [11]

Wireless Technologies

There is a huge number of devices that are used in buildings which use wireless technologies for communication. These devices include wireless routers, wireless sensors, bluetooth radios, and other handheld devices such as mobiles, laptops, and tablets. Wi-Fi exposures are one of the most common things in smart buildings and the exposures do take place without the consent of the occupants or tenants of the building.

Larik et al. [12] have discussed adverse effects of wireless devices on humans' physical and mental health. Pall [13] has reviewed and discussed the threats brought about by Wi-Fi or WLAN in detail in his paper. This paper has pointed out the different medical problems.

The issues related to IoT applications and other wireless items in smart building are listed as:

- Problems related to brain, heart, and ear
- Problems in sleep
- Human infertility
- Damaging of DNA
- Adverse effect on fetus
- Alzheimer's and Parkinson's problem (Fig. 36.3).

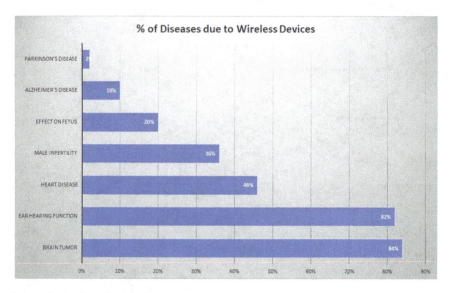

Fig. 36.3 The % of different diseases due to wireless devices [12]

Miller et al [14] discussed the health risks from RF radiation which operates in the non-ionizing frequency range. This paper has discussions about problems resulting toward carcinogenicity, effects on children, and reproduction due to RFR. Also, it has been mentioned about people developing a constellation of symptoms due to RFR exposures such as headaches, fatigue, insomnia, and appetite loss which have been termed as a syndrome "Microwave Sickness or Electro-Hyper-Sensitivity".

Conclusion

Smart buildings are dominating the building industry rapidly with inclusion of technologies such as IoT devices and other smart endpoints, but there is a lack of awareness of how these technologies are harmful not only to the occupants of the building but considering the complete ecosystem. Stress has been given to the effects of IoT devices and how this data can be used to bring about physical and cyberattacks. Also, it has been pointed out how smart buildings are contributing to the carbon footprint of the globe followed by the ill effects of the wireless technologies on the human body.

References

1. United Nations (2014) Department of economic and social affairs, population division. World Urbanization Prospects: The 2014 Revision, Highlights
2. Horban V (2016) A multifaceted approach to smart energy city concept through using big data analytics. In: 2016 IEEE first international conference on data stream mining & processing (DSMP), pp 392–396
3. Buckman AH, Mayfield M, Beck SB (2014) What is a smart building? Smart Sustain Built Environ 3(2):92–109
4. IoT for smart buildings—long awaited revolution or lean evolution. Publication is based on keynote speech lead by Marcin Bajer on 23rd International Workshop of the European Group for Intelligent Computing in Engineering June 29th–July 1st, 2016. Kraków, Poland
5. Energy E_ciency Administration. Commercial buildings energy consumption survey (CBECS). http://www.eia.doe.gov/emeu/cbecs/
6. Cha I, Shah Y, Schmidt AU, Leicher A, Meyerstein MV (2009) Trust in m2m communication. Veh Technol Mag IEEE 4(3):69–75
7. Lopez J, Roman R, Alcaraz C (2009) Analysis of security threats, requirements, technologies and standards in wireless sensor networks. In: Abomhara M, Køien GM (eds) in foundations of security analysis and design. Springer, pp 289–338
8. Roman R, Zhou J, Lopez J (2013) On the features and challenges of security and privacy in distributed internet of things. Comput Netw 57(10):2266–2279
9. Analysis of human awareness of security and privacy threats in smart environments. by Luca Caviglione, Jean-Fran,cois Lalande Wojciech Mazurczyk, and Steffen Wendzel
10. DOE (2009) Buildings energy data book. Department of Energy. http://buildingsdatabook.eren.doe.gov/
11. Scope-based carbon footprint analysis of US residential and commercial buildings: an inpute-output hybrid life cycle assessment approach by Nuri Cihat Onat, Murat Kucukvar, Omer Tatari
12. Larik RSA, Mallah GA, Talpur MMA, Suhag AK, Larik FA (2016) Effects of wireless devices on human body. J Comput Sci Syst Biol 9:119–124. https://doi.org/10.4172/jcsb.1000229
13. Pall ML (2018) Wi-Fi is an important threat to human health. Environ Res 164:405–416. https://doi.org/10.1016/j.envres.2018.01.035
14. Miller AB, Sears ME, Morgan LL, Davis DL, Hardell L, Oremus M, Soskolne CL (2019) Risks to health and well-being from radio-frequency radiation emitted by cell phones and other wireless devices. Front. Public Health 7:223. https://doi.org/10.3389/fpubh.2019.00223
15. Sinopoli J (2006) Smart buildings. Spicewood Publishing, p 2

Chapter 37
Survey and Analysis of Content-Based Image Retrieval Systems

Biswajit Jena, Gopal Krishna Nayak, and Sanjay Saxena

Introduction

The advent of the Internet facilitated the exchange and querying of information. Over the years, the methodology adopted by users to query data has witnessed considerable changes primarily because of the querying and data retrieval mechanisms on user-end getting easier, interactive, and friendly. The earlier years of the Internet era saw a greater amount of text-based data being generated, queried, and transferred but as the feature of multimedia got incorporated with the textual Web pages and applications; the shift has been toward image-based retrieval schemes. Earlier, this was brought forward by text-based image retrieval systems [1], where the visuals were annotated manually by textual phrases or words. When such a system was used to query a particular image from a large database of textually annotated images, it often would suffer from imprecision in the search results. This was chiefly because different humans might perceive an image differently. Also, the process of annotating an image was time-consuming and required a lot of human effort. To combat this issue, the CBIR was put forward in the early 1980s [2]. Since then, it has become a lively research area backed by a variety of different individual fields like pattern recognition, machine learning, computer vision, and databases to name a few.

The fundamental goal of any CBIR system is feature extraction [1, 2]. An object in such a system is described in terms of its low-level features like texture, color, and shape.

Often, human beings tend to identify an object from its color; thus, we can closely link color with the visual perception of an object in the human mind. To study this feature, a lot of techniques are applied to color perception and color spaces. Color

---

B. Jena (✉) · G. K. Nayak · S. Saxena
International Institute of Information Technology, Bhubaneswar 751003, India
e-mail: biswajit310@gmail.com

© The Editor(s) (if applicable) and The Author(s), under exclusive license to Springer Nature Singapore Pte Ltd. 2021
A. K. Singh and M. Tripathy (eds.), *Control Applications in Modern Power System*, Lecture Notes in Electrical Engineering 710,
https://doi.org/10.1007/978-981-15-8815-0_37

histograms are one of them. In this, the focus is on the distribution of color in an image regardless of the spatial location where that particular color might be found in an image. They are typically employed on three-dimensional color spaces like HSV and RGB and are extremely useful because of their flexibility, low computational complexity, and compact representation. Another important low-level feature is the texture which is responsible for defining the spatial positioning of colors or various intensities in an image. Gray-level co-occurrence matrix (GLCM), Gabor filter, wavelet transform, and curvelet transform are some ways of texture representation [6–9]. The shape-based features are extremely helpful yet a challenging problem faced by CBIR systems. They help us to index the objects. An object's shape helps it to describe it more meaningfully yielding an efficient search result. But such features have a complex implementation in CBIR systems that are desired to be 2D or 3D and must show invariance to properties like translation, rotation, and scaling. Generally, shape descriptors are of two types, i.e., contour-based (representing the boundary) and region-based (representing the entire region) [3, 14].

Keeping these low-level features into consideration, a lot of CBIR systems like QBIC, Netra, Photobook, Virage, FIRE, LIRE, etc., have been developed. The ultimate goal is to minimize the interval between these low-level numerical and statistical features and high-level perception of them by the human mind. The ability with which a CBIR system is able to reduce this gap portrays its efficiency and usability. Thus, a lot of work is ongoing so as to develop a semantically and statistically synchronized CBIR system.

The organization of our paper is such that in Sect. 37.2, we discuss the basic design of a CBIR system. In Sect. 37.3, we analyze the various CBIR systems available in the literature, and in Sect. 37.4, we conclude the paper.

A Typical CBIR System

In this section, we discuss the design of a typical CBIR system. All CBIR systems begin with users generating query images. The query image may undergo some kind of preprocessing which is essential for enhancing the quality of the query image and suppress the unwanted features like noise or distortion present in the image. If the data is voluminous and redundant, it becomes difficult for algorithms to handle it. So, a feature vector is created comprising all the important data. This phenomenon is known as feature extraction [4]. The extracted features are a set of reduced relevant features that hold all the necessary information on which all tasks are performed. Now, then there is a comparison between the extracted feature vector and feature databases. The degree of similarity between the feature vector of the query image and the corresponding images present in the feature databases defines the degree of precision with which an image is retrieved by a CBIR system.

A number of retrieval approaches are used in different CBIR systems. Query by example, iterative search, semantic retrieval, and relevance feedback [5] are some of them. In query by example, an exemplary image is fed into the CBIR system with

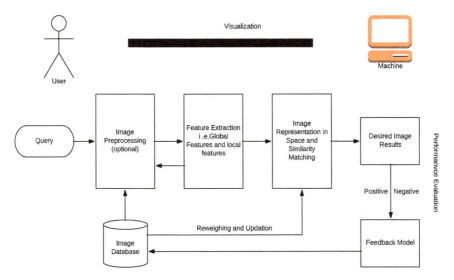

Fig. 37.1 Visualization of CBIR at the user and machine level

which it tries to match the query image. In iterative search, several machine learning methodologies are employed to retrieve the relevant data. In semantic retrieval, the user queries for images by words and phrases. This is the most difficult means of implementing a CBIR as the queries made by users are open to interpretation and it is difficult for a machine to semantically determine the correct meaning which the human user might be querying. Relevance feedback involves human intervention. The images retrieved are classified into relevant, irrelevant, and neutral by users. Then, this updated information is used to make a new search. Thus, a progressive search with relevant user feedback refines the overall search. Figures 37.1 and 37.2 show a typical CBIR system's functionality.

Analysis of Different CBIR System

1. Analysis-1 [6].

The important techniques used here are basically for feature extraction. Here, we can use both spatial and frequency domain techniques of feature extraction. The spatial domain techniques include color moments, color auto-correlogram, and HSV histogram features. Similarly, the various frequency domain feature extractor techniques are Gabor wavelet transform [15], SWT [16], and BSIF [17].

The precision was obtained using Chebyshev, Cosine, L1, and L2 distance metrics for different classes of images. The experimentation takes place considering four special feature descriptors.

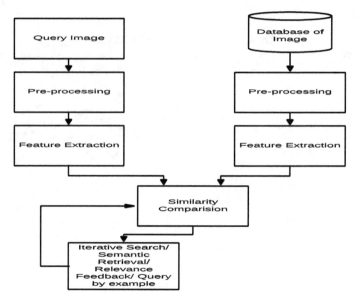

Fig. 37.2 A typical CBIR system's framework

The various distance metrics that are used in this study are Minkowski, City block, Mahalanobis, and Euclidean. These give an average accuracy near to 65 percent with respect to all the classes of images with respect to spatial, frequency domain features as well as CEDD [18], BSIF fusion features, and hybrid features.

2. Analysis-2 [7].

The theme of this article is to do the post-CBIR system, in which the large database is first partitioned into various clusters based on various image features like color, size, texture, etc., for efficient image retrieval. For clustering purposes, the ACPSO clustering algorithm [19] is chosen over the commonly used clustering algorithms like PSO [20], K-means [21], ACO [22], etc.

After that, the highly efficient ACPSO techniques are employed for image retrieval from the clustered large database of images.

The resulted accuracy for K-means is 0.91, ACO is 0.88, PSO is 0.96, and ACPSO 0.98 is noticed.

3. Analysis-3 [8].

Patterns from images were extracted using the techniques like local binary patterns, local mesh patterns, local texton XOR pattern, and local ternary co-occurrence pattern comparing their histograms and then constructing feature vectors concatenating all the histogram and comparing the query image with the database image.

To measure the performance of the system, various performance parameters used are average retrieval rate or precision, recall, and precision.

The results show that on different quantized levels, Corel-1k, Corel-k, and Corel-10k vary and gradually quantized level of 4 and 8 clearly outperforms the other level (all in multiples of 4) at each of the databases.

4. Analysis-4 [9].

Here, GLCM [23] is used as well as color histogram so that the color and texture features can be useful in the classification of cow type. The various features for GLCM are energy, contrast, homogeneity, correlation, and entropy. The angle for each is 0, 45, 90, 135° with an average of 1.

The system is trained with 100 images and tested with 20 images of five different classes. Based on these test images, various performance metrics such as recall, precision, and accuracy are measured using a confusion matrix.

The results obtained from the system show that being a robust system for image retrieval with an accuracy of 95%, precision, and recall of 100%.

5. Analysis-5 [10].

The techniques that are used for feature extraction are GLCM and discrete wavelet transform (DWT) and its combination for both texture-based and color-based features.

The database used for retrieval purposes is the WANG image database. It has 1000 color images. The average RA for GLCM texture features was 0.33, but with DWT and GLCM, the average RA increased to 0.43, wherein in both cases texture feature were taken into consideration. So, it is proved that the combined effort always produces a good result. Again, while considering both color and texture feature average RA stands with 0.77.

6. Analysis-6 [11].

The techniques that are used for feature extraction are color, texture, intersecting cortical model (ICM), and K-means clustering method.

The performance parameter considered here is precession. It is a very effective feature extraction performance measure.

The various distance metrics parameters used in this approach are Euclidean, City block, and Canberra. Out of the distance metric, the Canberra distance generates very less distance retrieved images. On the other hand, the K-means method gives the best results in terms of similarity measurements.

7. Analysis-7 [12].

The CBIR techniques are using the image transform for feature extraction such as contourlet, ridgelet, and shearlet transforms, and for also classification of the methods such as Naïve Bayes, K-nearest neighbor (k-NN), and multi-class support vector machine (multi-SVM).

The metrics are commonly used to measure the quality of the retrieval process that is false positive, true positive, false negative, and true negative. These are used in a confusion matrix which is normally a table which holds the values of true positives, false positives, true negatives, and false negatives which can be used to represent the set of test data for describing the performance of a classification model or classifier. Classifiers are compared using sensitivity, specificity, accuracy, error rate, Jacquard coefficient, F-measure.

The multi-SVM and Naïve Bayes classifiers fetch better results When the sensitivity increases, the accuracy rate also increases and resulting in the multi-SVM classifier outperforming with 90.76% accuracy for ridgelet transform. The sensitivity, specificity, and accuracy of the multi-SVM are much higher for contourlet transform when compared to the others. The multi-SVM classifier gives better results when used with shearlet transform. The sensitivity, specificity, and accuracy of the multi-class SVM are much higher than others. The error rate produced by the multi-SVM classifier is also very much low and only 2.8%. The shearlet extracts more features and thus outperforms the other two classifiers because of its ability to handle and extract more features from the images than the ridgelet and contourlet.

8. Analysis-8 [13].

In the era of computation and the Internet, creating huge multimedia databases to retrieve these multimedia data in an efficient way is always a challenge. The CBIR system is a monumental achievement in this direction. The multimedia data is retrieved based on various features such as color, size, and texture. But in this study, an efficient index-based technique is developed for retrieval purposes.

For achieving state-of-the-art results in terms of accuracy and efficiency an attempt is made by this indexing CBIR system.

For the experimentation of the index-based CBIR system, medical images are indexed on MATLAB coding. Then, the query image and corresponding results in images are compared for the retrieval process.

Conclusion

The CBIRs allow us to search and query images efficiently from a large image database. From the papers we surveyed, we can ascertain that the performance of a CBIR system depends on the accuracy and the retrieval time in which it is able to produce the results. A lot of different schemes are employed to increase precision though most of them still do not hold effective in case of large databases primarily due to the increase in the search time. Thus, the careful study of various systems will help us in developing a robust and efficient CBIR system.

References

1. Long F, Zhang H, Feng DD (2003) Fundamentals of content-based image retrieval. In: Multimedia information retrieval and management. Springer, Berlin, Heidelberg, 1–26
2. Gudivada VN, Raghavan VV (1995) Content based image retrieval systems. Computer 28(9):18–22

3. Kim W-Y, Kim Y-S (2000) A region-based shape descriptor using Zernike moments. Sig Process Image Commun 16(1-2):95–102
4. Kumar G, Pradeep KB (2014) A detailed review of feature extraction in image processing systems. In: 2014 Fourth international conference on advanced computing & communication technologies. IEEE
5. Zhang H-J, Zhong S, Zhu X (2006) Relevance maximizing, iteration minimizing, relevance-feedback, content-based image retrieval (CBIR). U.S. Patent No. 7,113,944. 26 Sep. 2006
6. Mistry Y, Ingole DT, Ingole MD (2018). Content based image retrieval using hybrid features and various distance metric. J Electri Syst Inf Technol 5(3):874–888
7. Meshram SP, Thakare AD, Gudadhe S (2016) Hybrid swarm intelligence method for post clustering content based image retrieval. Procedia Comput Sci 79:509–515
8. Bala A, Kaur T (2016) Local texton XOR patterns: a new feature descriptor for content-based image retrieval. Eng Sci Technol Int J 19(1):101–112
9. Sutojo T et al (2017) CBIR for classification of cow types using GLCM and color features extraction. In: 2017 2nd international conferences on information technology, information systems and electrical engineering (ICITISEE). IEEE
10. Atlam HF, Attiya G, El-Fishawy N (2017) Integration of color and texture features in CBIR system. Int J Comput Appl 164(3):23–29
11. Bhadoria S et al (2012) Comparison of Color, Texture and ICM Features in CBIR System. In: Advanced materials research, vol. 403. Trans Tech Publications Ltd
12. David HBF, Balasubramanian R, Pandian AA (2018) CBIR using multi-resolution transform for brain tumour detection and stages identification. Int J Biomed Biolog Eng 10(11):543–553
13. Rahmani MKI, Ansari MA, Goel AK (2015) An efficient indexing algorithm for CBIR. In: 2015 IEEE international conference on computational intelligence & communication technology. IEEE, pp 73–77
14. Jena B, Nayak GK, Saxena S (2019) Maximum payload for digital image steganography obtained by mixed edge detection mechanism. In: 2019 international conference on information technology (ICIT). IEEE, pp 206–210
15. Arivazhagan S, Ganesan L, Bama S (2006) Fault segmentation in fabric images using Gabor wavelet transform. Mach Vis Appl 16(6):356
16. Nason GP, Silverman BW (1995) The stationary wavelet transform and some statistical applications. Wavelets and statistics. Springer, New York, NY, pp 281–299
17. Kannala J, Rahtu E (2012). Bsif: Binarized statistical image features. In: Proceedings of the 21st international conference on pattern recognition (ICPR2012). IEEE, pp 1363–1366
18. Chatzichristofis SA, Yiannis SB (2008) CEDD: color and edge directivity descriptor: a compact descriptor for image indexing and retrieval. In: International conference on computer vision systems. Springer, Berlin, Heidelberg
19. Ouadfel S, Batouche M, Ahmed-Taleb A (2012) ACPSO: a novel swarm automatic clustering algorithm based image segmentation. In: Multidisciplinary computational intelligence techniques: applications in business, engineering, and medicine. IGI Global, pp 226–238
20. Cui X, Thomas EP, Paul P (2005) Document clustering using particle swarm optimization. In: Proceedings 2005 IEEE swarm intelligence symposium, 2005. SIS 2005. IEEE
21. Kanungo T et al (2002) An efficient k-means clustering algorithm: analysis and implementation. In: IEEE Trans Pattern Anal Mach Intell 24(7):881–892
22. Kao Y, Cheng K (2006) An ACO-based clustering algorithm. International workshop on ant colony optimization and swarm intelligence. Springer, Berlin, Heidelberg
23. Sebastian V, Unnikrishnan A, Balakrishnan K (2012) Gray level co-occurrence matrices: generalisation and some new features. arXiv preprint arXiv:1205.4831

Chapter 38
A Prototype Model of Multi-utility Mist Vehicle for Firefighting in Confined Areas

Sasidhar Krishna Varma, Pankaj Bhagath, and Nadakuditi Gouthamkumar

Introduction

Firefighters have been rescuing to save many lives and sometimes may not return safely due to the heavy radiant heat, improper balance, the severity of the fire, abnormal behavior of fire, inadequate of water, inaccuracy in firefighting and lagging in multitasking. Conventional firefighting techniques are inadequate for the modern fire service due to catastrophic damage, longer time to rescue, property damage, water drained wastefully, etc. In order to overcome the flaws of conventional techniques, multi-utility mist vehicle (MUMV) technology has been evolved based on cooling and smothering principles for effective maintenance of water even in oil fires also [1]. Many authors have been explored and presented the innovative technology before the commencement of MUMV such as an autonomous robot [2] to detect the indoor flames by using extinguishing media as CO_2, firefighting robot using the Arduino system [3], firefighting robot using Bluetooth model, firefighting robot with microcontroller, fire sensor, radio frequency module, GSM module, model with combination the circuit of fire sensor [4], smoke sensor, PIC microcontroller, motors followed by activation of the pump, firefighting robot to modern generation by installing two optically isolated DC geared motors, firefighting tank robot model which installed with an ultrasonic sensor, compass sensor and flame detector to sustain in such severe atmosphere [5]. Besides, few researchers have been proved

---

S. K. Varma (✉) · P. Bhagath · N. Gouthamkumar
National Fire Service College, Ministry of Home Affairs, Government of India, Nagpur, India
e-mail: sasidharkrishnavarma@gmail.com

P. Bhagath
e-mail: pankajbhagat83@gmail.com

N. Gouthamkumar
e-mail: gowthamkumar218@gmail.com

© The Editor(s) (if applicable) and The Author(s), under exclusive license to Springer Nature Singapore Pte Ltd. 2021
A. K. Singh and M. Tripathy (eds.), *Control Applications in Modern Power System*, Lecture Notes in Electrical Engineering 710,
https://doi.org/10.1007/978-981-15-8815-0_38

that the multi-utility water mist effectively attacks on cooling, smothering and attenuation of radiant heat, provides more heat absorption rate and decreases the toxic gases like CO_2, NO_x, SO_2 and CO in the atmospheric conditions by using the wireless technology for effective controls to extinguish the fire easily [6, 7]. However, all these technologies based systems are working toward to detect the fire and utilized the water, firefighting foam, retardants and co_2 as an extinguish media but failed to produce the working prototype model in confined areas such as corridors, tunnels and warehouses. Therefore, the present research work is explored toward to develop an efficient prototype model for confined areas.

This research paper presents a prototype model of multi-utility mist vehicle (MUMV) with fully mechanized to work in confined areas for firefighting. In this proposed prototype model, the extinguishing media is treated as water mist and that can be pressurized or un-pressurized. Further, the stream can be pointed to the object (fire, tank, gas cloud) with a wired remote which controls the angle of the turbine, up or down, and left to right. Thus, the results revealed that the proposed prototype model is efficient in controlling fire as well as to extinguish the fire in confined areas.

Multi-utility Mist Vehicle

Multi-utility mist vehicle (MUMV) can perform multiple tasks during fire breakout and 'search and rescue' operations and successfully provide the helping aid to firefighters in the required hour of need. The horizontal and vertical elevation of nozzle assembly can provide the flexibility for the firefighters in order to tackle the erogenous conditions of India. Further, it can be operated via remote control and fully automatic function is achieved with the provision of all the sensors in communication with the fire detection panel. The proposed vehicle consists of DC motor, AC motor, reciprocating pumps, chain drive, blower and jack. In this proposed vehicle, the input water can be given either from unpressurised water or pressurized water. Now, the pump inside the vehicle regulates the pressure required to generate the water mist. The regulated pressurized water flows through the nozzle and comes out as water mist. Now a blower is used to increase the wide range of water mist by giving positive displacement to the air. The proposed vehicle consists of a chain drive which is used to pass over the obstacles, and a jack is used to increase the nozzle head height. Now, in contact with ground, caster wheels are given to bear the load and assist the vehicle to move around 360°.

Design of MUMV Model Specifications

According to the space required and its utilization, the following dimensions have been decided for the various components used in the MUMV as (i) overall vehicle dimensions 600*450*600 and (ii) plywood dimensions 600*450*120. The skeleton

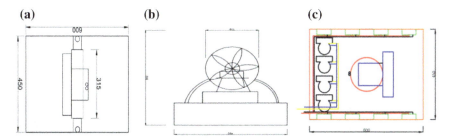

Fig. 38.1 **a** Top view, **b** Front view, **c** Electrical layout

model has drawn in auto CAD with all topographical views shown in Fig. 38.1. The chassis of the vehicle is built of mild steel angles to increase the strength of the base and avoid deflection as mild steel is hard, but also it is malleable, so it can be easily fabricated into desired shapes. Some typical values for physical properties of steel that is commonly used are density (ρ): 7.7–8.1 [kg/dm^3], elastic modulus (E): 190–210 [GPa], Poisson's ratio (ν): 0.27–0.30, thermal conductivity (κ): 11.2–48.3 [W/mK] and thermal expansion (α): 9–27 [K]. The total weight of the vehicle by considering all the components shown in Table 38.1 is 20 kg, and the fabricated prototype model of multi-utility mist vehicle is shown in Fig. 38.2.

Table 38.1 Design components of the MUMV

S. No.	Item/component	Quantity (No.)	Unit weight (grams)	Total weight (grams)
1.	Wooden ply	1	1960 + 980	2940
2.	Rectangle M.S. channel used inchassis	2 (vertical), 2 (horizontal)	820, 320	1140
3.	DC motor: -30 RPM	8 (for wheels) 1 (for blower)	90, 90	810
4.	Chain	2	100	200
5.	Sprockets	8	60	480
6.	Other fittings(including nut–bolts and foam)	–	520	520
7.	Jack	1	2000	2000
8.	Motor and blower assembly	1,1	5000	5000

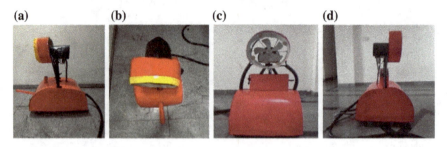

Fig. 38.2 **a** Right view, **b** Top view, **c** Front view, **d** Left view

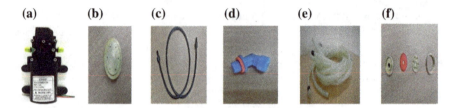

Fig. 38.3 **a** Diaphragm pump (A sub-type of the reciprocating pump) **b** Nozzle, **c** Delivery hose **d** Connector, **e** Suction (braided hose), **f** Mist (spray nozzles)

Mechanical Equipment

The diaphragm pump has been used for the model, the operating voltage of the pump is 12 V, and maximum flow is 4.0 lpm as shown in Fig. 38.3. The weight of the pump is 150 grams, and its maximum pressure generation is 0.48 Mpa. Total no. of pumps used in the model is 4. Spray nozzles find applications, where the risk of radiant heat from severe fire hazard is high, i.e., for cooling and in agricultural lands. These nozzles are special application nozzles/monitors, suitable for agricultural activities. The application rate from these nozzles is more effective in controlling such fires, resulting in lower extinguishing times. A braided hose has been used for the suction inlet to get water from the source. A total of four such hoses have been used each connecting source and pump, and the total of four mist nozzles (all spray agricultural nozzles) have been used in the model as shown in Fig. 38.3.

Electrical Components

A total of nine DC motors have been used in the vehicle. Among these nine motors, one motor has been used for the horizontal movement of the blower nozzle assembly. The 8 motors that have been used for driving the vehicle are electric DC 12 V, 30 RPM speed reduced gear motor which is providing a torque of 6 kg/cm that is equal to 0.5886 NM as shown in Fig. 38.4.

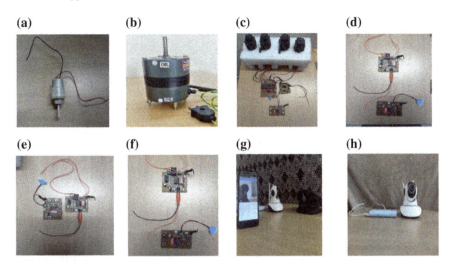

Fig. 38.4 **a** DC motor, **b** Electric AC motor, **c** RC-1 pump on/off, **d** RC-2 horizontal motion, **e** RC-3 Back n forth motion, **f** RC-4 Blower (On/Off), **g** Wireless modules, **h** Integrated Wi-Fi connected Camera Android Phone setup

An electric motor has been used for the blower assembly so that it generates high discharge wind velocity required for the water mist acceleration. The mechanism has been converted into a wireless one with all the mechanisms being controlled by wireless kit designed for the vehicle motors and pumps as (i) relay circuit—1: 4 pumps (on/off), (ii) relay circuit—2: 8 DC motors(back and forth movement), (iii) relay circuit—3: 1 DC motor(horizontal movement) and (iv) relay circuit—4: blower (on/off). In order to provide the visual access to the vehicle, a camera setup has been arranged to provide visibility to the vehicle and to act as 'eyes' of the vehicle by an integrated camera to the android where mobile phone acts as viewer for both connected via same Wi-Fi connection.

Reliability Analysis and Discussion

Since there are no standard tests, procedures are designed to analyze the affectivity of the vehicle that to bring out with some new self-designed tests for the same as shown in Fig. 38.5. A tin container was used with which we tied two steel angles and they were joined with the steel wire to hold the nozzle exactly at the center. Now we used three pumps and three nozzles, and thus, we had variations available w.r.t pressure, discharge and number of the orifices. Now what we did firstly a known volume of fuel was burnt, i.e., newspaper which was constant in all the tests. Firstly, a fire was ignited and when it reached the peak just then the water supply with nozzle was switched on with the pump and the time taken to extinguish the fire together

Top view   Before fire test   After fire test

Fig. 38.5 Apparatus setup

with the amount of water used was calculated. Similar test was repeated nine times firstly with the same pump and then with different combinations of the nozzles.

A water mist system is a fire protection system using very fine water sprays (i.e., water mist). The very small water droplets allow the water mist to control, suppress or extinguish fires by cooling of the flame and oxygen displacement by water vapor and radiant heat attenuation. There are two principles in which water mist takes role in extinguishing the fire. At first, smothering is the removal of oxygen (via evaporation steam is generated which displaces oxygen) and the other thing is cooling by absorption of heat. The water mist technology is relatively simple, and it is based on the physical principle that a combustion process cannot sustain a fire if the percentage of oxygen present in the air is less than 11%. It attacks the fire from two sides smothering and cooling parallel. Being remote controlled, there is less chance in life loss of firefighters. The motion is chain drive so that it can easily overcome the obstacles. Water mist as a firefighting media needs less content of water. With this vehicle, firefighting can be done as near as possible because the vehicle has given a sufficient protection to resist radiant heat. The model of MUMV, which we have designed, is facilitated with various firefighting equipment as well as on low discharge in irrigation purposes also. Our vehicle mainly focused on mobility of vehicle which will not be hampered, due to its chain drive technology. Our proposed vehicle will prove to be better than any other vehicle in terms of confined space fire. The working model had little problems, like chain sagging. This was, happening as we had made arrangements, to keep the chain sagging adjustable according to the terrain condition. The vehicle will successfully work in all the terrain conditions whether it be rocky terrains, plain terrain or any such unworkable regions. To ensure that the developed vehicle is working efficiently as per its development criteria, we performed several tests on the vehicle. The flow rate of a nozzle depends on the output of pump, and major and minor losses between the nozzle and pump. We carried out test of output of nozzle (N_1, N_2, N_3) by collecting the discharge in an empty container of known volume for an instant of time. The discharge rate was found to be 2.0 lpm. The reliability analysis is carried out by having different nozzles and pumps in comparison with the water consumed to extinguish the fire, and time represents the time taken to extinguish the fire and the temperatures as shown in Tables 38.2, 38.3 and 38.4.

Table 38.2 Pump-1 different nozzles

Setting	Water used (l)	Time taken (s)	Temperature (ambient) (C)	Temperature (peak) (°C)
P_1N_1	0.6673	11.20	33.48	693.5
P_1N_2	0.359	10.89	33.43	691.2
P_1N_3	0.1334	9.08	33.43	697.9

Table 38.3 Pump-2 different nozzles

Setting	Water used (l)	Time taken (s)	Temperature (ambient) (°C)	Temperature (peak) (°C)
P_2N_1	1.596	9.42	35.61	695.7
P_2N_2	1.45	8.5	35.61	698.7
P_2N_3	1.31	5.89	35.61	690.7

Table 38.4 Pump-c different nozzles

Setting	Water used (l)	Time taken (s)	Temperature (ambient) (°C)	Temperature (peak) (°C)
P_3N_1	0.283	9.87	36.43	694.7
P_3N_2	0.164	6.57	36.41	696.8
P_3N_3	0.118	5.94	36.41	691.2

Here, the velocity pressure and area of nozzle for water filled pipe are determined by the following relation:

$$P_v = 5.61 \times 10^{-7} \frac{Q^2}{D^4} \tag{38.1}$$

where P_v represents the velocity pressure (bars) and Q, D indicates the flow of water in liter/min and inner diameter in mm.

With the help of above results, we can say that as the area of nozzle decreases from N_1 to N_3 with increasing the pressure from P_1 to P_3, the quality of mist increases (the diameter of mist water droplet decreases and more number of droplets) which lead more surface area to absorb the radiant heat from fire. On absorbing the large amount of heat, the mist converts into steam and gives smothering effect. As early as smothering effect occurs, the fire extinguishment time decreases ultimately leads to less usage of water in an efficient manner according to Bernoulli's theorem. The range of a trajectory is high with the horizontal. So we calculated the maximum range which came out to be 2.5 m. The comparison between different scenarios is plotted between water consumed to extinguish the fire and time consumed to extinguish the fire to analyze the results in Fig. 38.6. It is observed that (P_1N_3, P_2N_3, P_3N_3) are more efficient than (P_1N_2, P_2N_2, P_3N_2). Also, (P_1N_2, P_2N_2, P_3N_2) are more efficient than (P_1N_1, P_2N_1, P_3N_1) and some parabolic conditions will occur at a certain point.

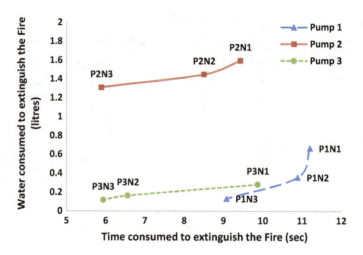

Fig. 38.6 Comparison between water consumption and time

Conclusions

This paper concludes that the proposed prototype model of multi-utility mist vehicle (MUMV) worked efficiently for firefighting operations in confined areas. The diaphragm pump is working dynamically to pressurize the water and discharge through fine orifice spray nozzles and advanced integrated technology regulated pressurized water flows through the nozzle comes out as water mist. Finally, the proposed prototype model has been tested on different pumps and nozzles combinations to identify the discharge and time taken to release the water and results revealed that it is effective model for controlling fire as well as to extinguish the fire in confine areas.

References

1. Mkanabus-kaminska J, Ping-li yen (2014) Extinguishment of cooking oil fire by water mist suppression system. Fire Technol 40(4):309–333
2. Singh HP, Akanshu MSN, Veena BSS, Amit K Anadi V (2015) Control of an autonomous industrial fire fighting mobile robot. DU J Undergraduate Res Inn 124–130
3. SakthiPriyanka S, Sangeetha R, Suvedha S (2017) Android controlled Fire Fighting Robot. Int J Sci Technol Eng 3(09):540–544
4. Mathew S, Sushant G, Vishnu KR, Vishnu Nair V, Vinod Kumar G (2016) Fabrication of fire fighting robot. Int J Inn Res Sci Technol 22(2):375–383
5. Dipali AM, Pratima SM, Shraddha KD (2015) AVR based fire fighting robot. Int J Eng Res Technol 4(03):770–773
6. Chee FT, Alkahari MR, Sivakumar DM, Said MR (2013) Firefighting mobile robot: state of the art and recent development. Aust J Basic Appl Sci 7(10):220–230

7. Gouthamkumar N, Srinivasarao B, Venkateswararao B, Narasimham PVRL (2019) Nondominated sorting-based disruption in oppositional gravitational search algorithm for stochastic multiobjective short-term hydrothermal scheduling. Soft Comput 23(16):7229–7248

Chapter 39
Security Analysis of System Network Based on Contingency Ranking of Severe Line Using TCSC

Kumari Gita and Atul Kumar

Introduction

In recent era, electricity plays an essential role in domestics and commercial field because all these are running through electricity. Therefore, the power system security plays an important role now a days. As the power system network consists of generation, transmission and distribution setups, the failure of any units of these systems effects the performance of the system. With the help of contingency, the prior line severity can be identified [1–4]. Security analysis of Power system network topology, without development the new transmission network can be studied through contingency ranking of most severe lines in the topology. Performance parameters used to detect the security of network topology can be determined based on the parameters such as voltage and active power performance index. OPI is summation of both voltage and active power performance index. According to highest OPI in the existing contingency marked as most severe line in the topology [5–7]. Contingency analysis method is widely used to find the effect of outages like equipment failures, transmission line failure, etc. To prevent this, we have to take appropriate action to keep power system secure. Practically, only ranked or choose contingencies will leads to challenging conditions in power system network. The use of AC power flow solution in contingency analysis gives active, reactive power flows in lines and magnitudes at bus voltage. Therefore, making these lines to be secure in unusual operation or in incremental load situation is main concern to prevent these lines in unwanted situations. Proper switching is one of the main problems in the network

---

K. Gita (✉) · A. Kumar
IEEE Member, Uttar Pradesh, India
e-mail: Kumarigita01july@gmail.com

A. Kumar
e-mail: kmat1594@gmail.com

© The Editor(s) (if applicable) and The Author(s), under exclusive license to Springer Nature Singapore Pte Ltd. 2021
A. K. Singh and M. Tripathy (eds.), *Control Applications in Modern Power System*, Lecture Notes in Electrical Engineering 710,
https://doi.org/10.1007/978-981-15-8815-0_39

topology, so finding the correct line, which improves the system network security need to be measure properly [8]. The power system network comes into emergency state when some of the components violate the limits or the system frequency starts decreasing. By incorporation of many FACTS devices like STATCOM, TCSC in shunt and series connection, we can achieve power flow control, voltage control and stability improvement [9, 10]. The main factors by which the voltage instability occurs are:

(i) Increased demand leads heavily weight on networks.
(ii) Shortage of reactive power.
(iii) Unexpected switching operations which cause decrease in voltage drop.
(iv) Transmission line outages or generator outages.

So attaining secured system is a big question, and therefore, to find the solution TCSC used in the network to solve these problems and load disturbance should be less for the improved secured system. Load flow analysis is done for assessing overall performance indices of the power system in less repeated steps. With the help of iterative steps, we can find the Jacobin matrix for the parameters like voltage angle, reactive power and voltage magnitude on buses. The change in these parameters with respect to corresponding variable leads to find the load flow result, and on that basis, we can evaluate the ranking of contingency [11]. TCSC is also one of the most frequently used flexible devices to control the line flow which can results in controlling the OPI to preserve the most severe line. The study of contingency of lines in power system in terms of security is one of the significant features in network topology planning and operation [12]. This paper comprises the four section; Sect. 1 describes introduction, Sect. 2 explains methodology used to achieve the objective, Sect. 3 enlightens the results analysis, and Sect. 4 clarifies conclusion of the paper.

Methodology

The method to evaluate the performance index of the system is deliberated in this section. Voltage and active power performance parameter is calculated for analysis of security of network topology. Weighting factor and penalty factor are used in this paper to scale the values of performance index [13, 14]. Weighting factor is three, and penalty factor is taken one for the evaluation of OPI. Equations (39.1–39.3) shows the mathematical formula for evaluation of performance index. In power system, contingency ranking method is used to rank the line based on the severity measured using the performance index. These indices are calculated with the help of Newton–Raphson method for each line outage. OPI is the summation of active performance and voltage performance index and increasing any one parameter or both, we can improve the OPI of the system. By incorporation of STATCOM at bus reactive voltage, violation can be improved, and by using TCSC, the violation of active power flow in the line can be improved.

$$PI_v = \sum_{i=1}^{nb} \left(\frac{W}{M}\right) \{(|V_i|-|V_i^{sp}|)/\Delta V_i^{\lim}\}^{2M} \quad (39.1)$$

$$PI_p = \sum_{n=1}^{nl} (W/M)(P_l/P_l^{\max})^{2M} \quad (39.2)$$

$$OPI = PI_v + PI_p \quad (39.3)$$

PI_v	Performance index of bus voltage
W	Weighting factor
M	Penalty factor
V_i	Bus voltage at bus i
V_i^{sp}	Specified voltage at bus i
ΔV_i^{lim}	Voltage deviation at bus i
PI_p	Performance index of active power
P_l	Active power flow in line l
P_l^{max}	Max power flow in line l
OPI	Overall performance index of the system

About TCSC: TCSC is the FACTs device, connected in series with the transmission line conductors, which improve system stability and reduces the losses. As we know that TCSC is a device used to dynamically control the reactance of line which provide adequate load compensation. There are many other static electronics devices like STATCOM, static VAR compensator (SVC), thyristor controlled in AC power system to increase power transfer capacity also, these are used for overcrowding management and loss optimization.

Results

Contingency ranking-based security analysis becomes an important aspect to find out the most severe line in the existing network topology. TCSC integration in the system has better advantages to secure the power system with observation of performance index. The load flow performance studies on the standard 24-bus system using MATLAB software. The system contains thirty-two (32) generating units, thirty-eight (38) transmission lines, twenty-four (24) buses and load of 2850 MW at the different location of buses in the system. TCSC is located to those lines which cause more severity by opening the particular line in base case. As from the results, line 6 has more severity by its switching, so TCSC is connected to L6 to improve the power flow through this line and which results the improvement in severity by opening the line 23. When line 6 chance will come for open, then second most severe line in base case is used to connect the TCSC to perform the load flow and evaluate the performance index. By this way, each line switching can be tested by adding the

TCSC to the most severe line, and results show that the severity of line gets improved with incorporation of TCSC.

Security Analysis Without TCSC

Security of power system network topology with identification of most severe line has an important objective towards deregulated power operation. As the line will be loaded in load incremental situation or any irregular situation arises in the system, security gives a prior solution to identify those lines and make them secure with appropriate remedial action. Table 39.1 shows the highest OPI lines that can be calculated through ranking of selected contingencies. Line 6, 1 and 33 are top three severe line in the system without integration of TCSC in the system. Line six is highly loaded as from the results aspect so to make them secure from its peak loading time, TCSC integration plays an important role. Figure 39.1 shows the values of OPI at every probable contingency in the system. The difference between the top two severe

Table 39.1 Top five most severe line contingency without and with integration of TCSC

Without TCSC		With TCSC	
Line contingency	OPI	Line contingency	OPI
L6	5.1867	L23	4.1541
L1	3.9081	L10	3.8928
L33	3.3375	L15	3.1623
L27	3.1026	L28	2.9652
L29	2.8131	L1	2.9304

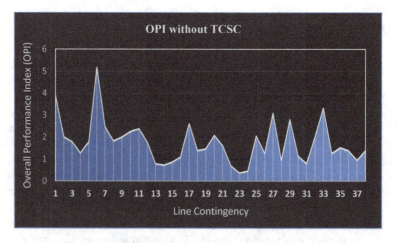

Fig. 39.1 Overall performance index during each line contingency without TCSC

lines in the network topology without using TCSC is 1.2786 that is 24.65% deviation from top one severe line, i.e. from L6.

Security can be restrained in several objectives for carefully operation of system. Every contingency is accomplished to calculate the OPI in the system and rank them according to highest to lowest order ways. The higher value of OPI gives the most severe line in network topology and need to control these lines from any unusual situation or heavily loading situation.

Security Analysis with TCSC

TCSC integration in the system network can control the active power flow in the line as per the requirements. Active power performance index will be changing with proportional change in the line flow using TCSC so OPI is also controlled. This controlled OPI can be used to secure the most severe line in the network topology.

From Table 39.1, top five most severe line using TCSC is 23, 10, 15, 28 and 1. With comparison with base case, it can be seen the most severe line gets changed, but severity gets increased because OPI reduces by 1.0326, i.e. 19.9% with use of TCSC. From results, it can be analysed that contingency ranking may change with integration of TCSC in the network topology, but severity gets increased. Figure 39.2 shows the variation in performance index with integration of TCSC in the topology.

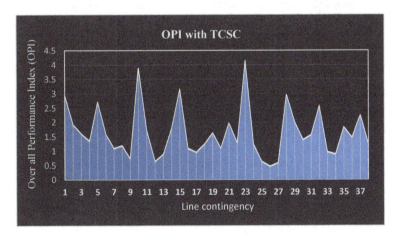

Fig. 39.2 Overall performance index during each line contingency with TCSC

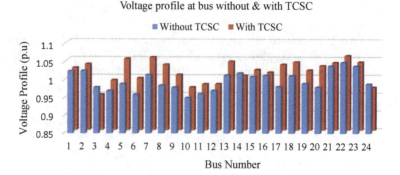

Fig. 39.3 Voltage profile at different buses with & without TCSC

Voltage Profile at Buses with and Without TCSC

Voltage profile at buses is compared with and without incorporation of TCSC in the system line. Standard 24-bus system is used for the study of performance index of the system. As from Fig. 39.3, we can conclude that the voltage profile at maximum buses gets improved when TCSC is connected in the system lines. TCSC can improve the line flow so accordingly the voltage profile at the buses gets change proportionally. This improved voltage profile causes the voltage security of the system. Active power flow gets enhanced through TCSC incorporation in system lines so under the situation of improving the transfer capability of lines TCSC is more thinkable FACTS device used in the system fit operation.

Conclusion

Security analysis becomes an important aspect in power system operation as world approaches deregulated power system. Contingency ranking for analysis of most severe line has an aspect to investigate the security of network topology. TCSC has capability to control the power flow in the line so accordingly OPI can be controlled. This paper deliberates the advantage of TCSC in network topology to make the network secure by securing the most severe line in the network. TCSC is connected to the line to control the active power flow through that line which can balanced the OPI according to make them secure without customer load interruption. The conclusion of the paper is using TCSC the most severe line can be changed and system becomes more secure from previous one analysis.

References

1. Albuquerque MA, Castro CA (2003) A contingency ranking method for voltage stability in real time operation of power systems. In: IEEE Bologna power tech conference proceedings, vol 1, p 5
2. Cruz EFD, Mabalot AN, Marzo RC, Pacis MC, Tolentino JHS (2016) Algorithm development for power system contingency screening and ranking using voltage-reactive power performance index. In: IEEE region 10 conference (TENCON), pp 2232–2235
3. Ferreira CM, Pinto JD, Barbosa FM (2002) On-line security of an electric power system using a transient stability contingency screening and ranking technique. In: 11th IEEE Mediterranean electrotechnical conference (IEEE Cat. No. 02CH37379), pp 331–335
4. Gongada SR, Rao TS, Rao PM, Salima S (2016) Power system contingency ranking using fast decoupled load flow method. In: 2016 international conference on electrical, electronics, and optimization techniques (ICEEOT), 2016, pp 4373–4376
5. Hasan KN, Preece R, Milanović JV (2016) Priority ranking of critical uncertainties affecting small-disturbance stability using sensitivity analysis techniques. IEEE Trans Power Syst 32:2629–2639
6. Jian Z, Yiwei Z, Feng C (2002) A comprehensive and practical approach for power system security assessment. In: Proceedings. international conference on power system technology, 2002, pp 2336–2339
7. Jmii H, Meddeb A, Chebbi S (2018) Newton-Raphson load flow method for voltage contingency ranking. In: 15th international multi-conference on systems, signals & devices (SSD), 2018, pp 521–524
8. Musirin I, Rahman TA (2003) Fast automatic contingency analysis and ranking technique for power system security assessment. In: Proceedings. student conference on research and development, SCOReD, 2003, pp 231–236
9. Naik P (2014) Power system contingency ranking using Newton Raphson load flow method and its prediction using soft computing techniques
10. Sekhar P, Mohanty S (2013) Power system contingency ranking using Newton Raphson load flow method. In: Annual IEEE India Conference (INDICON), 2013, pp 1–4
11. Srinivas T, Reddy KR, Devi V (2009) Application of fuzzy logic approach for obtaining composite criteria based network contingency ranking for a practical electrical power systems. In: IEEE student conference on research and development (SCOReD), 2009, pp 389–391
12. Subcommittee PM (1979) IEEE reliability test system. In: IEEE Transactions on power apparatus and systems, pp 2047–2054
13. Varshney S, Srivastava L, Pandit M (2011) Optimal location and sizing of STATCOM for voltage security enhancement using PSO-TVAC. In: International Conference on Power and Energy Systems, 2011, pp 1–6
14. Wu L, Venayagamoorthy GK, Harley RG, Gao J (2018) Cellular computational networks based voltage contingency ranking regarding power system security. In: Clemson University Power Systems Conference (PSC), 2018, pp 1–8

Chapter 40
A Renewable Energy-Based Task Consolidation Algorithm for Cloud Computing

Sanjib Kumar Nayak, Sanjaya Kumar Panda, Satyabrata Das, and Sohan Kumar Pande

Introduction

Cloud computing has established itself a remarkable technology trend, and it is reforming the global information technology (IT) industry and its marketplace [1]. It is indeed attributed to the fact that cloud computing provides numerous benefits, such as accessibility, availability, flexibility, friendliness, cost-effectiveness and many more [2]. Accordingly, several firms are migrating to cloud for their business requirements by hosting cloud-oriented applications and reducing IT costs. According to the United States-based global research and advisory firm, called Gartner Inc., the revenue of the public cloud is expected to increase 17% in 2020 (i.e., $266.4 billion from $227.8 billion in 2019) and its adoption is mainstream [3]. On the other hand, several CSPs are engaging different firms to use their services and provisioning the services without any user intervention. Consequently, CSPs are addressing various challenges, such as ensuing infrastructure, monitoring, EC of datacenter, transparency, security and compliance, and seeking cost-effective solutions [4–8]. One resembling challenge is EC by geographically distributed datacenters, failing which it substantially affects our environment by generating carbon footprints, particulate matters and atmospheric heat [9]. Here, datacenter runs using NRE sources like

---

S. K. Nayak (✉) · S. Das · S. K. Pande
Veer Surendra Sai University of Technology Burla, Burla, Odisha 768018, India
e-mail: fortunatesanjib@gmail.com

S. Das
e-mail: teacher.satya@gmail.com

S. K. Pande
e-mail: ersohanpande@gmail.com

S. K. Panda
National Institute of Technology Warangal, Warangal, Telangana 506004, India
e-mail: sanjayauce@gmail.com

fossil fuels (i.e., oil, gas, coal and orimulsion) to generate electricity [4]. These fuels are finite and would eventually run out with their ever-increasing demand. Therefore, CSPs are planning to use RE sources, such as solar, wind and hydropower to run their datacenters in order to abate the above challenge. According to Wired, an American magazine [10], many CSPs like Google and Microsoft claim that their datacenters are fully powered (100%) by RE sources. However, these providers still use NRE sources to generate electricity as RE sources may not available around the clock. Recently, the meaning of 100% RE is redefined for Google by the GreenBiz group, which does not consider all the time of a day or a season [11]. Many researchers [4–9, 12–14] have provided various solutions to assign the user applications/requests to the datacenters by using both NRE and RE sources. These solutions are based on different factors, such as the highest available RE sources, cost, static ordering and many others. However, these solutions have not emphasized on utilization as a factor. This phenomenon inspired us to consider utilization as a potential factor for reducing EC.

In this paper, we consider the utilization of user requests and resources, and propose a RE-based TC algorithm, called MinUtil to assign the user requests to the datacenters. Here, TC is used to minimize the usage of the number of resources and improve the utilization of resources. The utilization of NRE resources is restricted to a pre-determined threshold in order to reduce the over generation of carbon footprints. However, the utilization of RE resources is not limited to maximize the profit and environmental benefit of these resources. We simulate the proposed algorithm using MATLAB and carry out the simulation runs on four generated datasets. To the best of our knowledge, cloud computing-based TC algorithms are not comparable to MinUtil. Therefore, we carry out the simulation runs on two existing algorithms, namely RR [4, 5] and random [6, 12]. We illustrate the outcomes of MinUtil, RR and random algorithms in terms of EC, OC and |URE| resources, which clearly shows the efficacy of MinUtil.

The rest of this chapter is organized as follows. Section 'Related Work' summarizes the related work. Section 'Problem Formation' formulates the TC problem. Section 'Proposed Algorithm' presents the MinUtil with its complexity analysis and illustration. Section 'Performance Metrics, Datasets and Simulation Results' focuses on presenting the performance metrics, datasets and simulation results. Finally, section 'Conclusion and Future Work' summarizes the work and presents remarks for future research directions.

Related Work

Many NRE and/or RE-based task scheduling algorithms [4–9, 12–14] have been developed for cloud computing. They are briefly discussed as follows. Le et al. [5] have studied the temperature of geographically distributed datacenters, and listed three baseline algorithms, RR, worst fit and static cost-aware ordering for comparison with their two developed algorithms, namely cost-aware distribution and cost-aware distribution with migration. However, they have mentioned that it is crucial for their policies to consider all types of costs. Lee et al. [6] have introduced TC in cloud computing to increase resource utilization and reduce the EC. It also reduces the usage of resources by minimizing idle resources. They have developed two TC algorithms, namely energy-conscious TC and MaxUtil to achieve the above objectives. However, these algorithms do not reduce the electricity bills directly. Hsu et al. [7] have noticed the EC with respect to utilization and seen that the EC rate increases between 70% to 100% greatly. Therefore, they have set 70% as CPU utilization threshold (CUT) and developed an energy-aware TC algorithm. However, the CUT is set manually in their implementation. Panda and Jana [12] have considered both utilization value of resources and processing time of tasks and presented a multi-criteria based TC algorithm. Here, resource utilization of resources is not shown explicitly in the simulation results. Again, Panda and Jana [13] have addressed the demerits of both TC and task scheduling, and developed an energy-efficient task scheduling algorithm. However, they have not considered the energy and execution cost in their algorithm. Most of the above-discussed algorithms have not considered any RE sources and they have failed the requirements of RE-based algorithms.

Chen et al. [8] have considered RE sources, such as solar and wind, and presented a scheduling algorithm for minimizing the NRE (brown) consumption of datacenters. However, they have not considered price and cost analysis of their algorithm. Nayak et al. [9] have presented a review of RE-based resource management. Moreover, they have formulated a load balancing problem and discussed several solutions for the same. They have suggested to consider both overall cost and number of used renewable energy resources for developing a more efficient algorithm. Toosi and Buyya [4] have used fuzzy logic to develop a load balancing algorithm in which future knowledge about the resources and workload is not required. They have suggested to tune the window size for the favorable outcome of their algorithm. Pierson et al. [14] have developed a DATAZERO project in which the objective is to run the datacenter smoothly and develop a negotiation process (between IT and power control) to handle the unexpected events. In future work, they have suggested to tune the developed negotiation process using approaches like game theory. The above-discussed algorithms have not considered the utilization of the user requests and resources.

The proposed algorithm is a novel algorithm with respect to the following things. (1) To the best of our knowledge, this algorithm is the first and foremost TC algorithm, which considers RE sources into account. It considers the utilization of the user requests and resources, which is not used in the recently developed RE-based algorithms [4, 8, 14]. (2) The proposed algorithm uses the concept of TC as reported in [6, 7, 13]. However, they have not considered RE resources in their work. (3) We introduce a pre-determined threshold as reported in [7] to restrict the usage of NRE resources, which in turn reduce the amount of EC. However, the existing algorithms [4, 8, 14] have not considered such threshold, hence they may increase the amount of EC beyond this threshold as seen in [7].

Problem Formation

Consider a set of U user requests and a set of D datacenters. Each user request U_i, $1 \leq i \leq n$ is a 4-tuple and they are start time (ST), duration (D), nodes (N) and utilization (U). The user requests are placed in a global queue Q as per the non-decreasing order of STs. Each datacenter D_j, $1 \leq j \leq m$ consists of a set of R_j resources/nodes. Similarly, each resource R_{jk}, $1 \leq k \leq p$ consists of a set of NRE/brown or RE/green energy resource slots. The future knowledge about NRE or RE resource slots is represented using a time window o, i.e., $\max(ST[i] + D[i] - 1)$, $1 \leq i \leq n$. Here, max is a function to determine the maximum of a set of time windows. We assume that the cost is predetermined and set to a dynamic value in case of NRE resource slots and the cost is predetermined and set to a static value in case of RE slots. However, the determination of cost is restricted to the time window as stated in [4, 9]. We also assume that the data transfer time between the scheduler and the NRE and RE resources is negligible for the simplicity of the problem. Further, the utilization of the user requests is predetermined and identifiable, and the interference among the tasks on the NRE and RE resources is negligible as adopted in [6, 7, 13]. The TC problem is to assign the user requests to the datacenters, such that EC and OC are minimized and |URE| resources is maximized.

Proposed Algorithm

In this section, we present a RE-based TC algorithm for cloud computing, MinUtil. The objectives are to minimize the EC and OC, and maximize the |URE| resources. For this, it goes through a two-phase process, namely mapping and assigning. In the mapping phase, a user request is mapped to all the datacenters. It determines the total utilization in each datacenter and finds the minimum of all the total utilization. The rationality behind this minimum total utilization is that it reduces the amount of carbon dioxide generation and the EC. In the assigning phase, the user request is assigned to that datacenter, which results in the minimum total utilization. This process is repeated until all the user requests are assigned to one of the possible datacenters.

Algorithm Description

The pseudo-code for the proposed algorithm, along with input and output, is presented in Algorithm 1. The algorithm contains a global queue Q to keep the user requests as per their arrival/start time (Line 1 of Algorithm 1). It sets the ST, D, N and U in line 3 as per the user request and initializes the total utilization of datacenter to zero in line 5. Then it checks whether the resource slot is NRE or not in line 8. We represent NRE resource slots by 1. If the resource slot is NRE, it determines the sum of the utilization of the resource slot and utilization of the user requests is less than a pre-determined threshold ($\tau\%$) or not in line 9. Note that finding the optimal threshold is a trade-off, as stated in [1, 2] and beyond the scope of this chapter. Here, $UTIL$ is a function to determine the utilization of resource slots or user requests. If it is less than the threshold (Line 9), then it updates the total utilization in line 10. On the contrary, if the resource slot is RE (Line 12), it determines the sum of the utilization of the resource slot and utilization of the user requests is less than 100% or not. Note that 100% is the maximum allowable utilization of any resource slot. If it is less than 100% (Line 13), then it updates the total utilization in line 14. This process is repeated from ST of user request to the sum of ST and $D-1$ (Line 7 to Line 17), and for p and m number of resources and datacenters, respectively (Line 4 to Line 19). Next, the algorithm determines the minimum of all the total utilization in line 20 using a min function and selects the datacenter that results in the minimum total utilization in line 21. Then it assigns the user requests to the selected datacenter by following a similar process as follows. It checks whether the resource slots is NRE or not in line 24. If the resource slot is NRE (or RE), it determines the sum of the utilization of the resource slot and utilization of the user requests is less than a pre-determined threshold (or 100%) or not in line 25 (line 29). If it is less than the threshold (or 100%) (Line 25 or Line 29), then it updates the utilization of resource slots in line 26 (or 30). This process is repeated from ST of user request to sum of ST and $D-1$ (Line 23 to Line 33), and for p number of resources (Line 22 to Line 34). The overall process is repeated for n number of datacenters (Line 2 to Line 35) and until the Q is empty (Line 1 to Line 36). At last, the algorithm calculates the EC, OC and |URE| resources and produces these outputs (Line 37).

Time Complexity

Line 3 and Line 5 take constant time, i.e., $O(1)$. The conditional statement in line 8 to line 16 (or line 9 to line 11 or line 13 to line 15) takes $O(1)$ time. Line 7 to Line 17 and Line 23 to Line 33 loop require $O(o)$ time. Line 6 to Line 18 and Line 22 to Line 34 loop require $O(po)$ time. Line 4 to Line 19 loop requires $O(mpo)$ time and Line 2 to Line 35 loop requires $O(nmpo)$ time. Line 20, Line 21 and Line 37 take $O(1)$ time. However, Line 1 iterates K times (Line 1 to Line 36). Therefore, time complexity of the proposed algorithm is $O(Knmpo)$.

Illustration

We illustrate the proposed algorithm using nine tasks (i.e., U_{1-9}) and two datacenters (i.e., D_{1-2}) as shown in Table 40.1. Each datacenter consists of five nodes in which NRE and RE resource slots are represented in gray color and white color, respectively.

The cost of datacenters with respect to time is represented above the datacenters in the form of numeric values. Note that these costs are only applicable for NRE resource slots and a constant cost of 0.1 is applicable for RE resource slots irrespective of time. The time window is set to 9 and the threshold is fixed at 70%.

Algorithm 1 Pseudo-code for MinUtil

Input: 1-D matrices: $n, m, p, ST, D, N, UTIL, R$ and τ **Output:** EC, OC and |URE| resources

---

1: while $Q \neq$ NULL do
2:   for $i = 1, 2, 3, \ldots, n$ do
3:     Set $ST[i], D[i], N[i]$ and $UTIL[i]$
4:     for $j = 1, 2, 3, \ldots, m$ do
5:       Set $TUTIL[j] = 0$
6:       for $k = 1, 2, 3, \ldots, p$ do
7:         for $l = ST[i], ST[i] + 1, ST[i] + 2, \ldots, ST[i] + D[i] - 1$ do
8:           if $R_{jkl} == 1$ then
9:             if $(UTIL[R_{jkl}] + UTIL[i]) \leq \tau\%$ then
10:               $TUTIL[j] += UTIL[R_{jkl}]$
11:             end if
12:           else
13:             if $(UTIL[R_{jkl}] + UTIL[i]) \leq 100\%$ then
14:               $TUTIL[j] += UTIL[R_{jkl}]$
15:             end if
16:           end if
17:         end for
18:       end for
19:     end for
20:     $MTUTIL = min(TUTIL)$
21:     Determine the datacenter j' that results $MUTIL$
22:     for $k = 1, 2, 3, \ldots, p$ do
23:       for $l = ST[i], ST[i] + 1, ST[i] + 2, \ldots, ST[i] + D[i] - 1$ do
24:         if $R_{j'kl} == 1$ then
25:           if $(UTIL[R_{j'kl}] + UTIL[i]) \leq \tau\%$ then
26:             $UTIL[R_{j'kl}] += UTIL[i]$
27:           end if
28:         else
29:           if $(UTIL[R_{j'kl}] + UTIL[i]) \leq 100\%$ then
30:             $UTIL[R_{j'kl}] += UTIL[i]$
31:           end if
32:         end if
33:       end for
34:     end for
35:   end for
36: end while
37: Calculate the EC, OC and |URE| resources

At time $t = 1$, user requests U_{1-3} are available in Q and these are mapped to two datacenters D_{1-2}. User request U_1 requires 4 units of time of three different nodes. Let us assume that the initial cost of both datacenter is zero. As a result, user request U_1 can be assigned to both datacenter. However, it is assigned to datacenter D_1 by assuming chronological order. The cost of this datacenter is updated to 2.6 (i.e., (0.1 × 5) for RE slots + (0.4 × 2 + 0.1 × 2 + 0.3 + 0.4 × 2) for NRE slots). Next, the user request U_2 requires 1 time unit of two nodes. Here, the algorithm finds the total utilization of both datacenter and these are 105% and 0%, respectively. Note that

Table 40.1 A set of nine tasks with their properties

User request	U_1	U_2	U_3	U_4	U_5	U_6	U_7	U_8	U_9
Start time	1	1	1	5	5	7	8	8	9
Duration	4	1	3	3	2	1	2	1	1
Nodes	3	2	3	1	3	2	2	1	3
Utilization (%)	35	30	56	63	61	51	55	36	42
D_1		0	105	315	0	126	63	0	110	110
D_2		0	000	060	0	000	00	0	000	000

total utilization is calculated at $t = 1$, irrespective of the resources and resource slots. As minimum total utilization results in datacenter D_2, user request U_2 is assigned to datacenter D_2. The cost of this datacenter is updated to 0.2. Next, the user request U_3 requires 3 units of time of three nodes. Here, the total utilization of both datacenter is calculated as 315% and 60%, respectively. As minimum total utilization is resulted in datacenter D_2, user request U_3 is assigned to datacenter D_2. The cost is updated to $0.2 + 0.7 = 0.9$. Here, user requests U_2 and U_3 are using the same resource slots at $t = 1$ and it is permitted as this resource slot is RE.

At time $t = 5$, user request U_4 requires 3 units of time of a node. Here, the total utilization of both datacenter is calculated as 0%. Therefore, user request U_4 is assigned to datacenter D_1. The cost of this datacenter is updated to 2.9. The partial Gantt chart after assigning four tasks is shown in Table 40.2. In the similar way, user requests U_5 to U_9 are assigned to datacenters D_2, D_2, D_1, D_2 and D_2, respectively. The total utilization before scheduling decision is shown in the last two rows of Table 40.1. The overall cost of the two datacenters is $3.3 + 2.1 = 5.4$ cost units. Note that the CSP earns these costs. The final Gantt chart is shown in Table 40.3. Here, |URE| resources is 29 and EC is 13715 units of energy. We also present the Gantt chart for RR and random algorithms in Tables 40.4 and 40.5, respectively. The comparison of various performance metrics for the proposed and existing algorithms is shown in Table 40.6. The summary shows the superior performance of the MinUtil over the RR and random algorithms.

Performance Metrics, Datasets and Simulation Results

In this section, we discuss three performance metrics, generation of datasets and simulation results of the proposed and existing algorithms.

Table 40.2 Gantt chart after assigning four tasks for MinUtil algorithm

	0.4	0.1	0.3	0.4	0.4	0.2	0.2	0.1	0.4
Data center D_1	35% (1)	35% (1)	35% (1)	35% (1)					
	35% (1)	35% (1)	35% (1)	35% (1)					
	35% (1)	35% (1)	35% (1)	35% (1)	63% (4)	63% (4)	63% (4)		

	0.1	0.1	0.5	0.3	0.2	0.1	0.4	0.5	0.1
Data center D_2	56% (3)	56% (3)	56% (3)						
	30% (2)+ 56% (3)	56% (3)	56% (3)						
	30% (2)+ 56% (3)	56% (3)	56% (3)						
	$t=1$	$t=2$	$t=3$	$t=4$	$t=5$	$t=6$	$t=7$	$t=8$	$t=9$

Table 40.3 Final Gantt chart for MinUtil algorithm

	0.4	0.1	0.3	0.4	0.4	0.2	0.2	0.1	0.4
Data center D_1	35% (1)	35% (1)	35% (1)	35% (1)					
	35% (1)	35% (1)	35% (1)	35% (1)				55% (7)	55% (7)
	35% (1)	35% (1)	35% (1)	35% (1)	63% (4)	63% (4)	63% (4)	55% (7)	55% (7)

	0.1	0.1	0.5	0.3	0.2	0.1	0.4	0.5	0.1
Data center D_2	56% (3)	56% (3)	56% (3)		61% (5)	61% (5)			42% (9)
	30% (2)+ 56% (3)	56% (3)	56% (3)		61% (5)	61% (5)	51% (6)		42% (9)
	30% (2)+ 56% (3)	56% (3)	56% (3)		61% (5)	61% (5)	51% (6)	36% (8)	42% (9)
	$t=1$	$t=2$	$t=3$	$t=4$	$t=5$	$t=6$	$t=7$	$t=8$	$t=9$

Table 40.4 Final Gantt chart for RR algorithm

	0.4	0.1	0.3	0.4	0.4	0.2	0.2	0.1	0.4
Data center D_1	35% (1)+ 56% (3)	35% (1)+ 56% (3)	35% (1)+ 56% (3)	35% (1)	61% (5)	61% (5)			42% (9)
	35% (1)+ 56% (3)	35% (1)+ 56% (3)	35% (1)+ 56% (3)	35% (1)	61% (5)	61% (5)		55% (7)	55% (7)+ 42% (9)
	35% (1)+ 56% (3)	35% (1)+ 56% (3)	35% (1)+ 56% (3)	35% (1)	61% (5)	61% (5)		55% (7)	55% (7)+ 42% (9)

	0.1	0.1	0.5	0.3	0.2	0.1	0.4	0.5	0.1
Data center D_2							51% (6)		
	30% (2)						51% (6)		
	30% (2)				63% (4)	63% (4)	63% (4)	36% (8)	
	$t=1$	$t=2$	$t=3$	$t=4$	$t=5$	$t=6$	$t=7$	$t=8$	$t=9$

Table 40.5 Final Gantt chart for random algorithm

	0.4	0.1	0.3	0.4	0.4	0.2	0.2	0.1	0.4
Data center D_1	56% (3) 56% (3) 35% (1)+ 56% (3) 35% (1)+ 30% (2) 35% (1)+ 30% (2)	35% (1)+ 56% (3) 35% (1)+ 56% (3) 35% (1)+ 56% (3)	35% (1)+ 56% (3) 35% (1)+ 56% (3) 35% (1)+ 56% (3)	35% (1) 35% (1) 35% (1)				36% (8)	42% (9) 42% (9) 42% (9)

	0.1	0.1	0.5	0.3	0.2	0.1	0.4	0.5	0.1
Data center D_2					61% (5) 61% (5) 61% (5) 63% (4)	61% (5) 61% (5) 61% (5) 63% (4)	51% (6) 51% (6) 63% (4)	55% (7) 55% (7)	55% (7) 55% (7)
	$t=1$	$t=2$	$t=3$	$t=4$	$t=5$	$t=6$	$t=7$	$t=8$	$t=9$

Table 40.6 Comparison of various performance metrics for MinUtil, RR and random algorithms

Performance metrics	EC	OC	|URE|
MinUtil	13715	5.4	29
RR	15990	6.9	18
Random	15535	6.8	19

Performance Metrics

We use three performance metrics, namely EC, OC and |URE| resources. These are defined as follows. The EC of a datacenter D_j (i.e., E_j) is as follows.

$$E_j = (p_{\max} - p_{\min}) \times \sum_{k=1}^{p} \sum_{l=1}^{o} UTIL[R_{jkl}] + p_{\min} \quad (1)$$

where $p_{\max}$ and $p_{\min}$ values are power consumption at 100% utilization (max) and 1% utilization (min), respectively, and these values are taken as 30 (300 W) and 20 (200 W), respectively as adopted in [6, 7, 13]. However, we assume that the power consumption of NRE resources is up to $\tau\%$ utilization, i.e., $p_{\tau\%} = 25$ and the power consumption value at 100% utilization of RE resources is taken as 25. The overall EC is as follows.

$$EC = \sum_{j=1}^{m} E_j \quad (2)$$

The OC of a datacenter is the sum of the cost of NRE and RE resource slots. Here, the cost of NRE resource slots is doubled if it exceeds 70%. The |URE| resources is the number of resources that are assigned with the user requests.

Datasets and Simulation Results

We generated four datasets, namely 200 × 10 (i.e., 200 denotes the number of user requests and 10 denotes the number of datacenters), 400 × 20, 600 × 30 and 800 × 40 using MATLAB and setting the range of *ST*, *D*, *N* and *U* as [1∼300], [10∼100], [10∼50] and [10∼50], respectively. The slot of NRE and RE resources is unlimited and [10∼50], respectively and the cost is [5∼10] and 1, respectively.

The simulation results are carried out in terms of three performance metrics for MinUtil, RR and random algorithms using four datasets, as shown in Figs. 40.1, 40.2 and 40.3, respectively. We considered $\tau = 70\%$ as adopted in [6, 7]. The MinUtil performs better than the RR and random algorithms. The rationality behind this performance is that the user requests are assigned to the datacenters by calculating the total utilization and further finding the minimum of it.

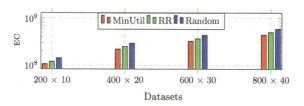

Fig. 40.1 Comparison of EC

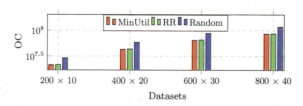

Fig. 40.2 Comparison of OC

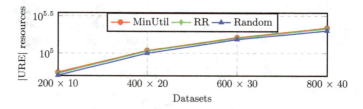

Fig. 40.3 Comparison of |URE| resources

Conclusion and Future Work

In this chapter, we have presented a RE-based TC algorithm for cloud computing. This algorithm goes through a two-phase process and is shown to require $O(Knmpo)$ time. The simulation results have been shown using four datasets and compared using three performance metrics. The results have shown that the proposed algorithm outperforms than the RR and random algorithms. However, it has taken the threshold as 70%, which will be further investigated.

References

1. Panda SK, Jana PK (2015) Efficient task scheduling algorithms for heterogeneous multi-cloud environment. J Supercomput 71(4):1505–1533
2. Panda SK, Jana PK (2018) Normalization-based task scheduling algorithms for heterogeneous multi-cloud environment. Inform Syst Front 20(2):373–399
3. Gartner. Gartner forecasts worldwide public cloud revenue to grow 17% in 2020. https://www.gartner.com/en/newsroom/press-releases/2019-11-13-gartner-forecasts-worldwide-public-cloud-revenue-to-grow-17-percent-in-2020. Accessed on 10, 2020
4. Toosi AN, Buyya R (2015) A fuzzy logic-based controller for cost and energy efficient load balancing in geo-distributed data centers. In: Proceedings of the 8th international conference on utility and cloud computing. IEEE Press, pp 186–194
5. Le K, Bianchini R, Zhang J, Jaluria Y, Meng J, Nguyen TD (2011) Reducing electricity cost through virtual machine placement in high performance computing clouds. In: Proceedings of 2011 international conference for high performance computing, networking, storage and analysis. ACM, p 22
6. Choon Lee Y, Zomaya AY (2012) Energy efficient utilization of resources in cloud computing systems. J Supercomput 60(2):268–280
7. Hsu C-H, Slagter KD, Chen S-C, Chung Y-C (2014) Optimizing energy consumption with task consolidation in clouds. Inf Sci 258:452–462
8. Chen C, He B, Tang X (2012) Green-aware workload scheduling in geographically distributed data centers. In: 4th IEEE international conference on cloud computing technology and science proceedings. IEEE, pp 82–89
9. Nayak SK, Panda SK, Das S (2020) Renewable energy-based resource management in cloud computing: a review. In: International conference on advances in distributed computing and machine learning. Springer, pp 1–12
10. Oberhaus D Amazon, Google, Microsoft: here's who has the greenest cloud. https://www.wired.com/story/amazon-google-microsoft-green-clouds-and-hyperscale-data-centers/. Accessed on 11, 2020
11. Golden S Google redefines what it means to be 100. https://www.greenbiz.com/article/google-redefines-what-it-means-be-100-renewable. Accessed on 15, 2020
12. Panda SK, Jana PK (2016) An efficient task consolidation algorithm for cloud computing systems. In: International conference on distributed computing and internet technology. Springer, pp 61–74
13. Panda SK, Jana PK (2019) An energy-efficient task scheduling algorithm for heterogeneous cloud computing systems. Cluster Comput 22(2):509–527
14. Pierson J-M, Baudic G, Caux S, Celik B, Da Costa G, Grange L, Haddad M, Lecuivre J, Nicod J-M, Philippe L et al (2019) Datazero: datacenter with zero emission and robust management using renewable energy. IEEE Access 7:103209–103230

Chapter 41
Multi-objective Optimization for Hybrid Microgrid Utility with Energy Storage

Kapil Gandhi and S. K. Gupta

Abbreviations

PV	Photovoltaic modules
wt	Wind turbine modules
bat	Battery bank
bio	Biogas power plant
mop	Modules of solar PV
mow	Modules of wind turbines
rat	Rated
ci	Cut-in
co	Cut-out
wp	Wind power available
ES	Maximum energy stored
CP	Charging power of the energy storage units
DCP	Discharging power of the energy storage units
η_{inv}	Efficiency of the inverter
η_{bat}	Efficiency of the battery bank
η_{pv}	Efficiency of the solar PV modules
η_{wt}	Efficiency of the wind turbine modules
η_{bio}	Efficiency of the biogas power plant
δ	Self discharging rate of the battery bank
G_i	Global Irradiance (W/m^2)

K. Gandhi (✉)
Department of EN, KIET Group of Institutions, Delhi NCR, India
e-mail: kapilkiet@gmail.com

S. K. Gupta
Department of EE, DCRUST Murthal, Murthal, India

© The Editor(s) (if applicable) and The Author(s), under exclusive license to Springer Nature Singapore Pte Ltd. 2021
A. K. Singh and M. Tripathy (eds.), *Control Applications in Modern Power System*, Lecture Notes in Electrical Engineering 710,
https://doi.org/10.1007/978-981-15-8815-0_41

TRS   Total running cost of the system ($/day)
TES   Total Greenhouse gas emissions of the system (tonne/day)
TGD   Total generation and demand matching (MW/day)
IC_i Initial cost per power plant ($/day)
OM_i Operation and maintenance cost ($/day)
w_1 Weight of cost minimization objective function (%)
w_2 Weight of minimum generation demand mismatch objective function (%)
GE_i Total gas emissions through the plant
GE_{bat} Total gas emissions through the battery bank

Introduction

According to the action plan of the Indian government, the electrical energy demand of the country will rapidly grow up to 2025, which will rapidly accelerate the CO_2 emission in the environment because the most power generation depends on fossil fuels till now. Due to adverse and dangerous effects on the environment, increasing cost, dependency on gulf countries, and limited fossil fuel reserves, renewable energy generation is the only possible solution to fulfil the electrical energy demand of the country [1, 2]. However, renewable energy sources are not reliable due to their dependency on atmospheric conditions. The stand-alone wind turbine or solar PV cannot provide a reliable solution due to the intermittent nature of wind speed or solar radiance. Due to this scenario, the authors provide an acceptable hybrid renewable energy system to deliver reliable power to consider green energy constraints.

The new possibilities for effective management of a smart grid are exposed to the new field of research. The minimum price of electricity to consumers with the maximum reliability of electricity is one of them. The price of electricity is high due to the fixed tariff for the whole time. Due to fixed tariffs, the demand is uncertain at any time of the day, and generation is also uncertain due to renewable penetration. It will affect grid reliability and stability. The price of per unit electricity can decrease when the time-based tariff introduced. It helps to control the load demand during peak hours. It leads to overconsumption during off-peak hours. The United States has only 80–90 h of the peak load during the whole year [3, 4].

The residential buildings generate 42% of carbon prints, and the commercial buildings generate 36% of carbon footprint according to the report of World Bank, 2018 [5]. The energy consumption is exponentially increased from the last decades. It is expected to increase further to 50% by 2035 [6].

When the energy demand increases more than the generation, the power shortage and blackouts are more frequent during peak hours of the day. It will also increase the price of electricity. Demand-side management (DSM) can help to reduce the negative environmental effect, i.e., greenhouse gas emission, and also decrease the price of electricity by managing the consumer load in the manner that optional or unnecessary load will operate during off-peak hours. It also reduces the requirement of new systems, i.e., generation, transmission, and distribution. DSM improves the system energy efficiency by modifying the load profile through load shifting, energy conservation, load growth, valley filling, flexible load shift, and peak clipping. It also includes changing the low efficient equipment or appliances with highly efficient equipment or appliances, i.e. LED bulb can replace the CFL [7, 8].

The demand response is the technique to shape the demand to manage the problem of dynamic pricing. Demand-side management is the tool to reduce the peak electric load and flexible demand to manage the random generation. In this research paper, the author considers the various types of load, i.e. controllable load, uncontrollable loads, and dump load to balance the excess generation, if any. The generation of electric power is done through Wind turbines, solar PV, and biogas plants. The battery and the grid are also connected to the system. DSM is also introduced to maximize the available utilities. Each load will assign the load profile depending on the load rating and working hours. Every domestic and commercial consumer wishes to maximize its utilities to optimal schedule the electrical power consumption. The overall system will be beneficial if the utility company uses dynamic pricing of the electricity to manage the demand response [9]. However, Real-time dynamic prices (time-based) of electricity are not extensively used in the retail electricity sector [10].

The organization of the remaining paper is as follows: Sect. 41.2 discusses the detailed pricing of retail electricity in the open market. Section 41.3 mathematical modelling of the system and renewable energy systems. The load profile of the district is also discussed in this section. Section 41.4 discusses the methodology and control system of a microgrid. The optimized results of the microgrid system are discussed in Sect. 41.5. Finally, the conclusions of this research paper are drawn in Sect. 41.6.

Pricing in Retail Electricity

In the open market of electricity, there is a lack of policies and guidelines until now. However, the reliable electricity at minimum price is the main challenge and objective for energy market regulators. The aware consumers are responsible for the investment in the energy sector for the investors [11].

The private power producer and seller are providing more reliable electricity to the consumers than a regulated utility by a government agency. Due to this, the number of consumers is attracting private suppliers. Many states in the US and India have opened retail markets for private participants to sell the electricity to different customers. The consumers are not interested in sellers; they are interested in reliable power at a low price. The two-part tariff is more popular among the sellers and consumers. The adoption of dynamic pricing is suitable for social welfare. The power requirement of residential consumers is low as compared to commercial and industrial consumers. The retail price competition will decrease the tariff price of electricity among competitive energy sellers [12] (Table 41.1).

Table 41.1 Comparison of various tariff schemes [10]

Tariff	Economic benefit	Equity	Bill generation	Revenue scope
Flat rate	1	1	5	2
PTR	4	2	5	1
CPP	4	3	3	4
TOU	4	4	3	3
Only variable cost	5	5	1	4
Fixed and variable cost	5	5	2	5

[a]5 represent the 'very good' and 1 represent the 'very bad.'

Mathematical Modelling of System

Gajraula, the district in UP, faces the shortage of electricity to the various industries in the district. In Gajraula, approximately 6700 homes and 72 industries are established. The average load of the residential consumers is 27.3 MW, commercial consumers are 18.1 MW, and industrial consumers are 68 MW. The author proposed the grid-connected microgrid for the district for a reliable power supply. The hybrid microgrid structure is shown in Fig. 41.1. The load profile of the district for a year is available and plotted in Fig. 41.2. The author supposed the load is constant for all days of any month. The load data is available for a day of each month on an hourly basis.

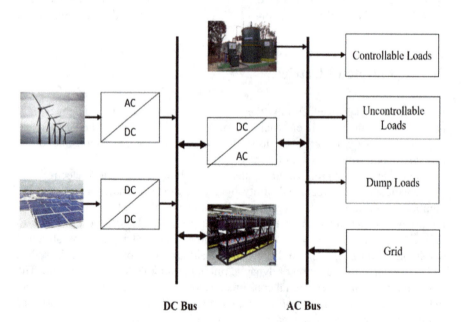

Fig. 41.1 Microgrid system with Solar PV, wind turbine and biogas plant [13]

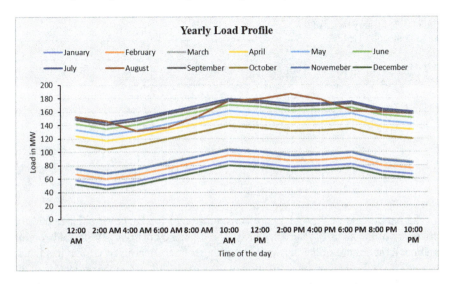

Fig. 41.2 The load profile of the district for a day for every month

The authors formulate the multi-objective optimization problem to optimize the proposed microgrid with three objectives and corresponding weights [14]

$$\text{Objective} = w_1 \cdot \text{TRS} + w_2 \cdot \text{TGD} + (1 - w_1 - w_2) \cdot \text{TES} \qquad (41.1)$$

here

$$\text{TRS} = \sum_{i=\{pv,wt,bg\}} \left[N_i \left(\frac{IC_i}{T_i} + OM_i \right) \right] + \frac{IC_{\text{bat}}}{T_{\text{bat}}} \cdot \frac{\sum_{i=1}^{24} \sum_{m=1}^{12} CP_{\text{bat}}(t,m)}{12 \cdot ES_{\text{bat}}}$$

$$\text{TES} = \sum_{i=\{pv,wt,bg\}} \left[\frac{N_i \cdot GE_i}{T_i} \right] + \frac{GE_{\text{bat}}}{T_{\text{bat}}} \cdot \frac{\sum_{i=1}^{24} \sum_{m=1}^{12} CP_{\text{bat}}(t,m)}{12 \cdot ES_{\text{bat}}}$$

$$\text{TGD} = \sum_{i=\{pv,wt,bg\}} [P_i(t,m)] + DCh_{\text{bat}} - \sum_{\text{All type}} DL(t,m) - Ch_{\text{bat}(t,m)}$$

Modelling of Solar PV System

The output power of the solar PV module can be calculated through Eq. (41.2)

$$P_{pv} = \eta_{pv} \cdot N_{pv} \cdot A_{\text{mop}} \cdot G_i \qquad (41.2)$$

Table 41.2 Ideality factor of the solar modules [15]

Material	Ideality factor
Si-mono	1.2
Si-poly	1.3
a-Si–H	1.8
a-Si–H Tandem	3.3
a-Si–H triple	5.0
cdTe	1.5
CTs	1.5
AsGa	1.3

The parameters of Eq. (41.1) are constant except for Global irradiance and the efficiency of the solar PV modules [13]. The efficiency of solar PV depends on the ambient temperature of the particular hour to generate electricity. Global irradiance depends upon the geographical location of the solar PV plant and the projection of the panels. The Ideality factor of the material of solar module also affects the efficiency of the system (Table 41.2).

Modelling of the Wind Power System

The wind is also the form of solar energy. The output power of the wind turbine modules depends upon the swept area and rotor angle of the rotor, velocity, and direction of the wind and air density. It is given by the following equation

$$P_{wt} = \begin{cases} \alpha \cdot V^3 - \beta \cdot P_{rat}, & V_{ci} < V < V_{rat} \\ P_{rat}, & V_{ci} < V < V_{rat} \\ 0, & \text{else} \end{cases} \quad (41.3)$$

Here

$$\alpha = \frac{P_{rat}}{V_{rat}^3 - V_{ci}^3}$$

$$\beta = \frac{V_{ci}^3}{V_{rat}^3 - V_{ci}^3}$$

The net power available for transmission and distribution from wind turbine generators is given by,

$$P_{wp}(t) = \eta_{wg} \cdot N_{wg} \cdot A_{mow} \cdot P_{wt} \quad (41.4)$$

Here η_{wg} is the combined efficiency of wind turbine generators and required converters [16].

Modelling of Biogas Power Plant

Biogas is another type of renewable energy source with many advantages as compared to solar PV and wind generators. The biogas power plant can start or generate power as per requirement. The output power does not depend on the weather conditions. The energy from the gas converts in mechanical energy through the gas turbine. The output power of the biogas power plant mainly depends on the Total solid waste available from the waste, and it is given by [17],

$$P_{\text{bio}} = \eta_{\text{bio}} \cdot T S_{\text{bio}} \cdot A_{\text{bio}} \tag{41.5}$$

Modelling of Battery

The current development of the renewable energy system requires the integration of energy storage units because the output of the solar PV and wind turbine generators depend on the atmospheric conditions. In micro-grid, multiple functions can perform by energy storage units, e.g. control of power quality, as well as voltage regulation and frequency, smoothing renewable energy output, providing system emergency power, and cost Optimization. The storage unit may be batteries, flywheels, and Ultra-capacitors. Generally, the electrochemical batteries are used in the power system. The battery is charged when the excess power is available or off-peak time and used during power shortage or peak hours. It is quite challenging to identify the exact state of the charge condition of the battery bank. The capacity of a battery bank at any instant is given by [18].

During charging

$$C_{\text{bat}}(t) = C_{\text{bat}}(t-1) \cdot (1 - \delta) - P E_{\text{DC}} \cdot \eta_{\text{bat}} \tag{41.6}$$

During Discharging

$$C_{\text{bat}}(t) = C_{\text{bat}}(t-1) \cdot (1 - \delta) + P D_{\text{DC}} \cdot \eta_{\text{bat}} \tag{41.7}$$

Availability of AC Power

The price of the electricity will decrease when the electric power does not waste on dump loads. It means the power generation by any means is equal to the power consumed by all loads except the reservoir. The power output through solar and wind cannot control. However, the power available through a biogas power plant and the battery will be optimized in the way that the loss of power will be minimum.

$$P_T = \left(P_{pv}(t) + P_{wt}(t) + C_{bat}(t)\right) \cdot \eta_{inv} + P_{bio} \qquad (41.8)$$

Power Management Methodology

The power management methodology of the microgrid is shown in Fig. 41.3. The main aim of the microgrid central controller is to maintain the balance between power availability and demand. All controllers are automatic, but automation works on predefined algorithms or decisions. Many different situations are discussed and explained through the algorithm in this paper [19]. The algorithm decides the operation biogas power plant or storage units. It also decides whether the excess power is required from the grid to purchase or not.

Results

After applied the multi-objective problem approach for the proposed microgrid, the author configures the system and optimizes the proposed objectives. The cost of the system will be minimum, with the least greenhouse gas emissions in the environment. Figure 41.4 shows that each node on the Pareto front is the unique and optimal solution for the proposed problem. The Pareto front technique is robust as compared to other techniques. The calculations performed by authors are based on various assumptions that are influenced by the literature survey (Table 41.3).

Conclusion

The Pareto front optimization technique is used to optimize the problem of an AC–DC hybrid microgrid. The mathematical modelling of solar PV, wind turbine, biogas power plant, and battery storage as done by the author and the power management algorithm was also developed to minimize the human interface in the decision making. These situations can be the standard mode of operation, overloaded

41 Multi-objective Optimization for Hybrid Microgrid Utility …

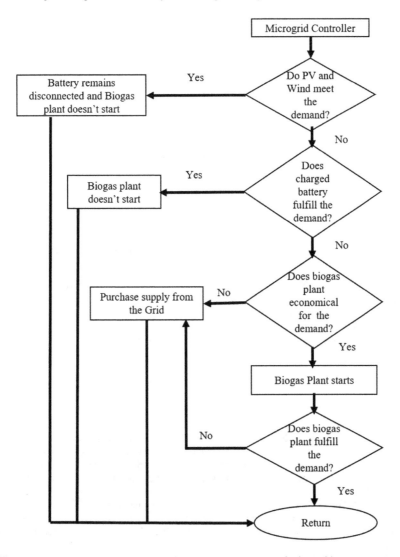

Fig. 41.3 Algorithm for decision making for power management of microgrid

microgrid, or contingency management. The other techniques, Archive-based Micro Genetic Algorithm (AMGA) and Neighborhood Cultivation Genetic Algorithm (NCGA) can also be used instead of the Pareto front optimization technique.

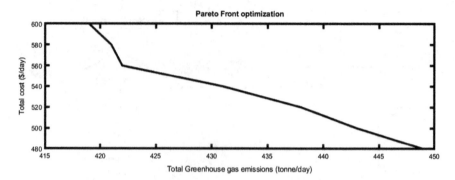

Fig. 41.4 Optimization result for Pareto front optimization techniques

Table 41.3 Weight management of optimization function

S. No.	Weight w_1	Weight w_2
1	0.25	0.50
2	0.30	0.45
3	0.35	0.40
4	0.40	0.35
5	0.45	0.30
6	0.50	0.25

References

1. European Commission (2007) Communication from the commission to the European Council and the European Parliament—an energy policy for Europe. Comm Eur Commun
2. Lau LC, Lee KT (2012) Global warming mitigation and renewable energy policy development from the Kyoto protocol to the Copenhagen accordant comment. Renew Sustain Energy Rev 16(7):5280–5284
3. Faruqui A, Hledik R, Tsoukalis J (2009) The power of dynamic pricing. Electri J 22:42–56
4. Wang Q, Liu M, Jain R (2012) Dynamic pricing of power in smart-grid networks. In: IEEE 51st IEEE conference on decision and control (CDC), pp 1099–1104. https://doi.org/10.1109/CDC.2012.6426839
5. UNEP SBCI (2009) Buildings & climate change—a summary for decision-makers.UNEP DTIE, Paris
6. Birol F (2012) World Energy Outlook 2012. International Energy Agency (IEA), Paris
7. AboGaleela M, El-Marsafawy M, El-Sobki M (2013) Optimal scheme with load forecasting for demand-side management (DSM) in residential areas. Energy PowerEng 5:889
8. Abushnaf J, Rassau A, Górnisiewicz W (2015) Impact of dynamic energy pricing schemes on a novel multi-user home energy management system. Electric Power Syst Res 125:124–132
9. Li N, Chen L, Low SH (2011) Optimal demand response based on utility maximization in power networks. In: IEEE power and energy society general meeting
10. Dutta G, Mitra K (2017) A literature review on dynamic pricing of electricity. J Oper Res Soc 68(10):1131–1145.https://doi.org/10.1057/s41274-016-0149-4
11. Bublitz A, Keles D et al A survey on electricity market design: Insights from theory and real-world implementations of capacity remuneration mechanisms. Chair of Energy Economics,

Institute for Industrial Production (IIP), Karlsruhe Institute of Technology (KIT), Hertzstraße 16, 76187 Karlsruhe, Germany.
12. Puller SL, West J (2013) Efficient retail pricing in electricity and natural gas markets: a familiar problem with new challenges
13. Tafreshi S, Zamani H, Baghdadi M, Vahedi H (2010) Optimal unit sizing of distributed energy resources in microgrid using genetic algorithm. In: IEEE Confenece, 2010, pp836–841
14. Saif A, Gad Elrab K, Zeineldin HH, Kennedy S (2010) Multi-objective capacity planning of a PV-wind-diesel-battery hybrid power system. In: IEEE international energy conference, pp 217–222
15. Tsai H-L, Tu C-S, Su Y-J (2008) Development of generalized photovoltaic model using MATLAB/SIMULINK. In: Proceedings of the world congress on engineering and computer science WCECS. San Francisco, USA
16. Manish K, Harish K, Narveer (2019) Impact of wind units in congestion management for hybrid electricity market. In: International conference on advancements in computing & management (ICACM-2019), pp 605–612
17. Methuku SR, Srivastava AK, Schulz NN (2009) Comprehensive modeling and stability analysis of biomass generation. In: 41st North American power symposium. Starkville, MS, pp 1–6. https://doi.org/10.1109/NAPS.2009.5484005
18. Sidorov DN, Muftahov IR, Tomin N, Karamov DN, Panasetsky DA, Dreglea A, Liu F, Foley A (2019) A Dynamic analysis of energy storage with renewable and diesel generation using volterra equations. IEEE Trans Ind Inf 16(5):3451–3459
19. Milczarek A, Główczyk M, Styński S ()2019 Advanced power management algorithm in DC microgrid subsystem controlled by smart transformer. In: 2019 IEEE 28th international symposium on industrial electronics (ISIE). Vancouver, BC, Canada, pp 2335–2342

Chapter 42
Damaged Cell Location on Lithium-Ion Batteries Using Artificial Neural Networks

Mateus Moro Lumertz, Felipe Gozzi da Cruz, Rubisson Duarte Lamperti, Leandro Antonio Pasa, and Diogo Marujo

Introduction

In recent years, battery technology is evolving fast, and many studies are being done on the subject. The latter is due to the high demand for emerging technologies such as electric vehicles and intelligent solar energy systems. Many of the recent applications of power batteries are using lithium-ion ones because they have high energy density and no memory effects [1, 2].

The cost of batteries used in electric vehicles is a large part of the product's total cost, being today a factor that hinders the popularization of the technology. However, the batteries have a finite life cycle, and their degradation can be accelerated by several factors, such as charge speed, temperature, and depth of discharge (DOD) [2, 3].

The state of the degradation of batteries can be evaluated with the state of health (SOH) [3]. When the SOH decreases, the cells lose the capacity to store energy. At deficient levels of SOH, the latter can be called 'dead' [4, 5]. A dead cell practically does not contribute to the battery's total current, so it is necessary to identify the problem to limit the battery's charge/discharge speed, thus avoiding damaging other cells.

Battery management systems (BMSs) are used to monitor the battery and avoid these adverse conditions. These devices have several sensors that monitor each cell or stack of cells, which compose the battery [6–8]. Since each cell is not produced

---

M. M. Lumertz (✉)
School of Engineering of São Carlos, University of São Paulo, São Paulo, Brazil
e-mail: mateuslumertz@usp.br

F. G. da Cruz
Western Parana State University, Foz Do Iguaçu, Brazil

R. D. Lamperti · L. A. Pasa · D. Marujo
Federal University of Technology–Paraná, Medianeira, Paraná, Brazil

© The Editor(s) (if applicable) and The Author(s), under exclusive license to Springer Nature Singapore Pte Ltd. 2021
A. K. Singh and M. Tripathy (eds.), *Control Applications in Modern Power System*, Lecture Notes in Electrical Engineering 710, https://doi.org/10.1007/978-981-15-8815-0_42

identically, they can charge, discharge, die, or degrade differently [7]. The BMS acts when it identifies an abnormal condition in the battery, switching relays, and removing damaged cells from the electrical circuit, also balancing the charge and discharge of cells [3].

Electric-vehicle batteries have hundreds of cells, and then monitoring each of them is a complicated and expensive process, because each one commonly uses a single analog-to-digital converter for voltage measurement or voltage-level converters with multiplexed outputs [4, 8].

In the literature, some works have implemented artificial intelligence techniques, such as artificial neural networks (ANNs), in the battery management systems without relying on knowledge of the internal battery parameters, which may vary with temperature and SOH. However, most of the developed proposals are only observer algorithms to estimate the state of charge (SOC) or the SOH of the battery, without acting directly in protecting the system [9, 10].

ANN is used in several other engineering areas, such as in protecting electrical power systems, classification, and location of faults in transmission lines. In such applications, it is more efficient to use data processing tools, such as the Fourier transform (FT) or the wavelet transform (WT), to create the ANN entries rather than to use electrical signals in the time domain [11, 12]. This work proposes the use of artificial neural networks (ANNs) for the identification and location of damaged cells within $LiCoO_2$ lithium-ion batteries, enabling a BMS to switch off a cell faster, and the proposed method does not require voltage measurements in each cell, only the output signal from the battery.

Methodology

The proposed method is illustrated in the flowchart of Fig. 42.1. Initially, a high-power pulse is applied to the battery terminals, lasting for half a second. Subsequently, data processing tools are used to extract information from the measured voltage signal.

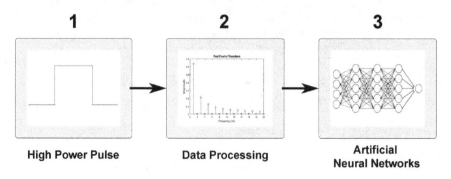

Fig. 42.1 Process flowchart

In this step, the Fourier transform and the wavelet transform are applied. Finally, the data captured is used as inputs for an ANN, which will identify if there are dead cells on the battery. If so, a second ANN will locate the defective cell's position in the battery's electrical circuit.

Impedances between each cell and the battery terminal are different because of their arrangement within the battery, as exemplified in Fig. 42.2, where the cell-to-cell connections of an electric car battery are exposed.

Figure 42.3 shows the necessary database for the training of the ANNs, considering a set of five cells. Among these cells, there may be a defective one, and if it exists, it can be in any one of the five positions shown.

Fig. 42.2 Battery cells

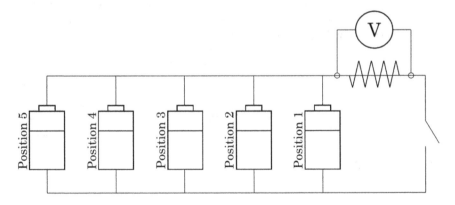

Fig. 42.3 Cells position

Wavelet Transform

Wavelet transform (WT), in its discrete form, the wavelet multi-resolution analysis (MRA), is a tool that analyzes the signal in both the time domain and frequency domain. The latter is done by decomposing the input signal in the function of a mother wavelet's orthonormal bases. MRA applies convolutions with a filter bank, containing a high-pass and low-pass filters to obtain this decomposition's coefficients. Results obtained by the low-pass filter are the scale coefficients (An), which capture the signal trends. High-pass filtering results in the expansion coefficients (Dn), which extract fluctuations and transients.

The convolution process with the filter bank can be repeated using the previous low-pass filter output as the new input signal, increasing the transform's decomposition level. This procedure is limited only by the initial signal sampling rate, which decreases by half at each convolution with the filters [13].

Artificial Neural Networks

ANNs of feedforward type are tools that can be used to model systems with unknown mathematical relationships. In most applications, they have three layers of neurons, the input layer and the output layer with a settled number of neurons, depending on the number of inputs and outputs of the system, and a hidden layer.

Neurons in the hidden layer define the linearity of the model and directly influence the ANN performance. Many free parameters increase the degree of freedom of the system and can even extrapolate the complexity of the problem being modeled, causing vague answers in situations not inserted in the training process.

A valid number of neurons are the smallest that guarantees good accuracy in the training set because this will make the ANN have a higher capacity of generalization. Meaning it will maintain its accuracy for samples different from those used in training [14].

Also, there are more sophisticated methods such as the regularizations, which seek to minimize the size of the ANN using an algorithm that penalizes its complexity. Bayesian regularization is an algorithm that uses this principle to perform the network training, based on the backpropagation method and the Levenberg-Marquardt Optimization, being an algorithm with a large generalization capacity [15].

Simulation Results

To simulate battery cells considering the effects of temperature and load, SimPowerSystems Toolbox was used in the MATLAB®, considering a set of five $LiCoO_2$

cells with 40Wh each connected in parallel. They represent the battery cells or stacks of cells, and there may or may not have damaged cells. For this work, damaged cells have a maximum rated capacity of 10% of the healthy cells.

The ANN input data simulations were performed with the set of cells, during which the cells were not in the process of charge or discharge. The analysis consists of a load pulse during 0.5 s. The battery pack is connected to a high-power resistor (100 W) through a MOSFET, according to Fig. 42.3.

Using only voltage measurements on this resistor, a multi-layer feedforward ANN was used to analyze the electric signal's transient behaviour and distinguish the defective sets from the healthy ones. Thus, the expected output of the network is 1 if dead cells are present in the set and 0 otherwise.

Subsequently, a second ANN was used to locate the dead cell's position in the electric circuit. Thus, this network's expected output varies between 1 and 5, with 1 being the closest to the terminals of the set of cells and 5 the farthest, as shown in Fig. 42.3. These two networks, the ANN of identification and the ANN of location, have 50 neurons in the hidden layer and are trained by the Bayesian regularization method.

Two data preprocessing methods were tested to convert the voltage samples in the continuous time to inputs for the ANN: Fourier and wavelet transforms.

In this work, 2736 samples were collected to identification ANN training and 2280 for location ANN training. 10% of these samples were randomly chosen for training validation and another 10% for the tests. The varied parameters in the simulations were as follows:

- Charge of healthy cells (between 50 and 100%);
- Dead cell charge (between 0.01 and 10%);
- Dead cell position;
- Battery temperature (between 20 and 30° C).

Figures 42.4, 42.5, 42.6, and 42.7 show the influence on voltage pulses when a parameter is changed during the data acquisition. Figure 42.4 shows the simulations' results varying batteries' temperature, considering a range of 20–30 °C. All cells are considered with full SOC and healthy. This temperature range was chosen for the simulation because lithium power batteries require thermal control to operate correctly since they lose their SOH quickly at high temperatures; in extreme cases, they may come to combustion.

Figure 42.5 shows the influence of the SOC of cells, with 20 °C and only healthy cells. Since SOC is the parameter that most influences the signal in the time domain, because the load interferes directly in the cells' voltage in an exponential relation, it affects the rate of reduction of the battery voltage and, therefore, in the slope of the pulse curve.

Figure 42.6 shows the influence of the maximum load capacity of the dead cell, in percentage relative to healthy cells, being a parameter that can be used to measure the state of health (SOH). Healthy cells with full SOC and dead one in position 1 were considered. This parameter presented a high non-linearity, because, besides the pulses having different inclinations, the decrease of SOH from 10% to 5% caused the

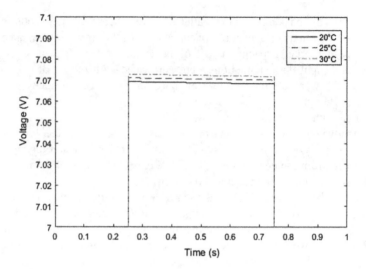

Fig. 42.4 Influence of temperature

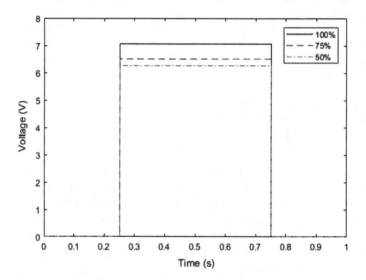

Fig. 42.5 Influence of SOC

level of voltage to rise, while the decline from 5% to 1% caused the level of voltage to decrease.

Considering the presence of a dead cell with 10% SOH in the circuit and varying its position, the simulated results are shown in Fig. 42.7 were obtained for a constant temperature of 20 °C. Damaged cell position is the parameter that needs to be diagnosed by location ANN. In the time domain, its influence on the signal is very similar to the other parameters' impact, but it is much less significant than SOC.

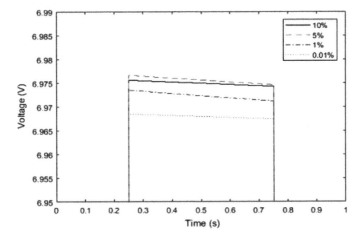

Fig. 42.6 Influence of damaged cell SOH

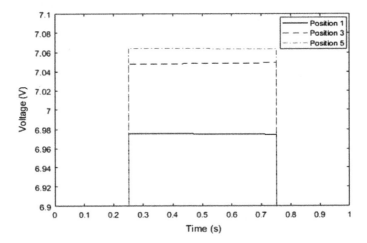

Fig. 42.7 Influence of damaged cell position

Applying the WT with the Haar function as the mother wavelet, all ten first decomposition levels had few variations due to the change in SOC, SOH, and position of the dead cell (Table 42.1). It is impossible to use knowledge of the system's thermal variations as the only data-preprocessing tool to identify and locate dead cells.

On the other hand, the FT was sensitive to all parameters. However, the dead cell position's influence on this signal is minimal, especially in the locations farthest from the battery terminals. Figure 42.8 shows the magnitude of the test's pulse, varying the position of the dead cell, and the enlarged graph shows the fundamental frequency values, being the region with the most significant amplitude and variation in the system.

Table 42.1 Wavelet transform outputs

Temperature (°C)	SOC (%)	Wavelet transform
20.6	80	$4.4512e^{-12}$
20.6	90	$4.4484e^{-12}$
20.6	100	$4.2816e^{-12}$
22.8	80	$1.1116e^{-10}$
22.8	90	$1.1591e^{-10}$
22.8	100	$0.9033e^{-10}$
24.3	80	$2.8205e^{-10}$
24.3	90	$2.7318e^{-10}$
24.3	100	$2.0866e^{-10}$

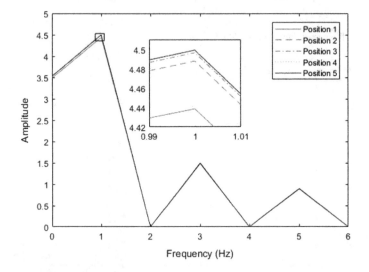

Fig. 42.8 FFT magnitude for different dead cell positions

ANN Training Results

Initially, both ANNs were trained using only the FT as a data-preprocessing tool. The best structure found for both ANNs was with three inputs: the magnitude of the fundamental wave, the peak voltage of the load pulse, and the average temperature of the cells. However, this structure presented a low performance, with a mean squared error (MSE) in the training of 0.387224 for the location ANN, as described in Table 42.2. The latter represents very high value, especially in positions 4 and 5, where the dead cell location became indistinguishable. Besides, the correlation between the location ANN training data was low, indicating that there is no specific pattern between input and output data. In this sense, new ANNs were tested to solve this problem. Now four inputs were used, including the first level decomposition response of the wavelet

Table 42.2 Results with three inputs ANN

	Identification ANN	Localization ANN
Samples	2736	2280
Training MSE	$1.222194e^{-1}$	$3.87224e^{-1}$
Correlation	0.348356	0.898645
Testing MSE	$1.29242e^{-1}$	$4.25188e^{-1}$

Table 42.3 Results with four inputs ANN

	Identification ANN	Location ANN
Sample	2736	2280
Training MSE	$7.54264e^{-3}$	$7.84597e^{-3}$
Correlation	0.97240	0.98044
Testing MSE	$1.18273e^{-2}$	$2.43561e^{-2}$

MRA, with the Haar function as the mother wavelet. Although this input varies only with changes in temperature since it is already one of the network's inputs and, therefore, could be a redundant modification, the relation between these two parameters is not linear, and the knowledge of both can simplify the pattern recognition.

Results are presented in Table 42.3, and the identification ANN obtained a training error of $7.5426e^{-3}$ and a correlation coefficient of 0.9724, with an accuracy of 95%. Performance with the wavelet MRA proved to be much higher than the network with only three inputs. All responses with errors greater than 0.5 occurred when the dead cell was in position 5, so the ANN was able to identify the presence of dead cells in 100% of cases in the other positions.

Mean training error of $7.84597e^{-3}$ and a correlation coefficient of 0.98044 were obtained with location ANN (Table 42.3). Note that all errors with a module greater than 0.5 were located at position 1.

Conclusion

The artificial neural networks, for identification of defective battery packs, and location the dead cell's position, obtained excellent training and testing results using the wavelet MRA and the Fourier transform simultaneously for data processing. Identification ANN presented lower accuracy when the dead cells are distant from the terminals of the battery, while the localization ANN presented higher error when they are close to the terminals.

Wavelet MRA was sensitive mostly to thermal variations. However, although the temperature was already one of the ANN inputs, this tool significantly increased the results' accuracy. For future works, intelligent temperature estimators can be created using only voltage measurements with the wavelet MRA. These estimators would eliminate the need to use sensors to estimate the batteries' average temperature, being useful in several functions performed by BMS.

References

1. Boicea V (2014) Energy storage technologies: the past and the present. Proc IEEE 102:1777–1794. https://doi.org/10.1109/JPROC.2014.2359545
2. Nicolaica MO, Tarniceriu D (2016) Analysis platform for energy efficiency enhancement in hybrid and full electric vehicles. Adv Electri Comput Eng 16:47–52. https://doi.org/10.4316/AECE.2016.01007
3. Haq IN et al (2014), Development of battery management system for cell monitoring and protection. In: International conference on electrical engineering and computer science, Kuta, pp 203–208. https://doi.org/10.1109/iceecs.2014.7045246
4. Collet A, Crébier JC, Chureau A (2011) Multi-cell battery emulator for advanced battery management system benchmarking. In: IEEE international symposium on industrial electronics, Gdansk, pp 1093–1099. https://doi.org/10.1109/isie.2011.5984312
5. Oriti G, Julian AL, Norgaard P (2014) Battery management system with cell equalizer for multi-cell battery packs. In: IEEE energy conversion congress and exposition, Pittsburgh, pp 900–905. https://doi.org/10.1109/ecce.2014.6953493
6. Man XC, Wu LJ, Zhang XM, Ma TK, Jia W (2016) A high precision multi-cell battery voltage detecting circuit for battery management systems. In: IEEE 83rd vehicular technology conference, Nanjing, pp 1–5. https://doi.org/10.1109/vtcspring.2016.7504072
7. Hommalai C, Khomfoi S (2015) Battery monitoring system by detecting dead battery cells. In: 12th International conference on electrical engineering/electronics, computer, telecommunications and information technology, Hua Hin, pp 1–5. https://doi.org/10.1109/ecticon.2015.7206976
8. Buchmann I (2017) Batteries in a portable world: a handbook on rechargeable batteries for non-engineers, 4 edn. Cadex Electronics Inc., pp 280–286
9. Chen X et al (2016) Robust adaptive sliding-mode observer using RBF neural network for Lithium-Ion battery state of charge estimation in electric vehicles. IEEE Trans Veh Technol 65:1936–1947. https://doi.org/10.1109/TVT.2015.2427659
10. Chaoui H, Ibe-Ekeocha CC (2017) State of charge and state of health estimation for lithium batteries using recurrent neural networks. In: IEEE transactions on vehicular technology, vol 66. pp 8773–8783. https://doi.org/10.1109/tvt.2017.2715333
11. Abdollahi A, Seyedtabail S (2010) Transmission line fault location estimation by Fourier and wavelet transforms using ANN. In: 4th International power engineering and optimization conference, Shah Alam, pp 573–578. https://doi.org/10.1109/peoco.2010.5559253
12. Costa FB, Silva KM, Souza BA, Dantas KMC, Brito NSD (2006) A method for fault classification in transmission lines based on ann and wavelet coefficients energy. In: IEEE international joint conference on neural network proceedings, Vancouver, pp 3700–3705. https://doi.org/10.1109/ijcnn.2006.247385
13. Murguía JS, Rosu HC (2011) Discrete wavelet analyses for time series. In: Olkkonen (ed) Discrete wavelet transforms—theory and applications. InTech, pp 3–20, 2011
14. Hagan MT, Demuth HB, Beale MH, Jesús O (2014) In: Neural network design, 2nd edn. pp 468–483
15. Nguyen NV, Jeon KS, Lee JW, Byun YH (2008) Design optimization process using artificial neural networks, bayesian learning and hybrid algorithm. In: 49th AIAA/ASME/ASCE/AHS/ASC structures, structural dynamics, and materials conference, Shaumburg. https://doi.org/10.2514/6.2008-1976

Chapter 43
Fuzzy Logic-Based Solar Generation Tracking

Anish Agrawal and Anadi Shankar Jha

Introduction

With the already humungous and rapidly growing demand for electrical power, there has been an increase in exploitation of the natural fossil fuels for power generation leading to increased pollution. It has become extremely vital for the integration of renewable and non-polluting energy sources with the electric grid for sustainable development keeping in mind the needs of the future. Solar energy is one of the vast inexhaustible energy sources available for use. However, the uncertainty in solar power generation leads to a critical problem in grid integration of such systems. Solar generation varies with the solar irradiance humidity temperature as well as other weather conditions.

Chakraborty et al. [1] discuss techniques for forecasting their applications, advantages, and their requirements. Neves et al. [2] show a study on an isolated microgrid system based on its demand response, and its performance has been studied keeping in mind the uncertainties in solar power. Use of hybrid solar irradiance forecasting methodology for microgrid with electric vehicles as load has been presented in [3]. Corsetti et al. [4] discuss the short-term forecasting of solar generation for the erection of microgrid system which has been analyzed using neural networks.

Vermaak et al. [5] provide a hardware module for the logging of load power to an external storage device with the help of a smart power meter. The data logged is then transferred to an android application which is capable of visually displaying in the form of graphs. Hussain et al. [6], in his paper, reveal the process of bidirectional energy flow using smart meters and emphasize the role of rooftop system of solar panels for grid-connected power generation.

A. Agrawal (✉) · A. S. Jha
Delhi Technological University, New Delhi 110042, India
e-mail: anish.agrawal.2009@gmail.com

© The Editor(s) (if applicable) and The Author(s), under exclusive license to Springer Nature Singapore Pte Ltd. 2021
A. K. Singh and M. Tripathy (eds.), *Control Applications in Modern Power System*, Lecture Notes in Electrical Engineering 710,
https://doi.org/10.1007/978-981-15-8815-0_43

Kenner et al. [7] present a software for data collection and analyses for a reference smart grid. The Modbus TCP/IP protocol a RESTful web service is studied for the basis of collecting data. Geetha and Jamuna [8] offer a digital meter model that shows real-time power usage to evaluate energy use and expenditure using different graph, tabulated and manipulated data forms.

Gaga et al. [9] address the design and implementation of a photovoltaic system based on an enhanced P&O algorithm and verify its efficacy by simulating PowerSim simulator and by using MPPT algorithms, classical and enhanced perturb and observe algorithms under the developed framework. Bonganay et al. [10] use an automatic meter reading system interface. The platform, via python, uses the integration of ZigBee protocol into Raspberry Pi single-board machine.

Chugh et al. [11] offer a simplified fuzzy logic model developed for short-term solar energy forecasts using solar irradiance data. The model's efficiency is measured on the basis of a mean absolute percentage error (MAPE). Mbarek and Feki [12] provide a novel approach to solar irradiance forecasting using flexible and accurate fuzzy logic and robust multi-linear regression.

Saez et al. [13] propose fuzzy interval prediction models which integrate uncertainty representation of future predictions. The suggested models of forecast cycles would help build a reliable microgrid energy management framework. Hippert [14] advises the use of the artificial neural networks (ANNs) for load prediction. The paper looks objectively at the ways the proposed NNs are built and checked.

Tee et al. [15] study the development of exogenous multivariable input (NARX) and nonlinear autoregressive artificial neural networks (ANNs) to efficiently predict short-term loads. In this article, with the introduction of the discussed method, the mean actual percent errors in the prediction were reached in the range of 1%. Liu et al. [16] discuss the practical techniques, namely fuzzy logic, neural networks, and autoregressive model for very short-term load forecasting his paper.

This paper describes a Mamdani, fuzzy logic-based prediction system trained on a vast dataset of values to obtain accurate and reliable results. The power generation data from the data logger system is used to validate the result. Section 43.1 comprises of the introduction to the research. Section 43.2 describes solar setup and measuring devices. The design of the fuzzy logic system has been discussed in Sect. 43.3. The results obtained have been discussed in Sect. 43.4 with the conclusion in Sect. 43.5.

Data Collection System

A data logger system has been implemented for the generation and gathering of data. The variables being logged are temperature (°C), solar irradiance (W/m2) and humidity (%) as the inputs and power(W) as output for the fuzzy system. The block diagram of the data logger system has been shown in Fig. 43.1.

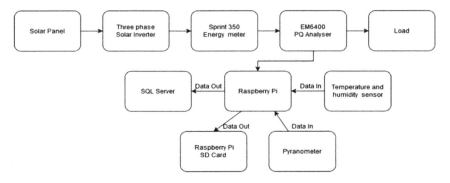

Fig. 43.1 Data logger block diagram

Hardware Devices

Solar Panels: A total of 5-kilowatt power generation capacity has been provided by 20 solar panels each of 250 watts. The system comprises of two strings of equal number of solar panels connected in parallel with the max voltage of 307.2 volts and 16.3 A DC.

Solar Inverter: The generated solar power is converted into three-phase AC power. Considering the present system and keeping in mind potential upgrades to the system, 10-kilowatt three-phase grid connected luminous solar inverter has been used which converts DC power to three-phase AC power at 50 Hertz. As the laboratory has various equipment running at three-phase voltage, this particular inverter has been used.

Energy Meter: An energy meter has been connected after the solar inverter to keep track of solar power generation. A Sprint 350 energy meter has been used for this application. It is suitable for residential, business, and industrial applications. It is connected in a system of three-phase wiring. This meter has four effective current ranges which are of 5–30 A, 10–40 A, 10–60 A, and 20–100 A. This meter has a voltage range of 230 V phase or 415 V line. This meter operates at 50 Hz $\pm$ 5% key frequency.

Raspberry Pi: The data logger system requires a processing unit for data handling capabilities. A Raspberry pi 3 is a small debit card-sized microprocessor-based development board. The Pi also has wireless communication capabilities such as Wi-Fi and Bluetooth support. The Raspberry Pi provides 40 GPIO pins for data input and output, along with 4 USB ports. A user-defined python script can be installed on the Raspberry Pi in this setup which can be modified to meet future expansion needs.

Temperature Sensor: The DHT22 is a digital temperature and humidity sensor which is simple and low cost. It uses a capacitive humidity sensor for humidity measurement and a thermistor to measure the ambient temperature of the air around the panels and sends out a digital signal on the output pin.

Pyranometer: A pyranometer sensor SR03 is used for the measurement of sun irradiance. It has an angle of view of 180°. The pyranometer monitors the hemispheric solar radiation using concentric disks and has a flat response throughout the entire solar spectrum of irradiation.

Data Collection

The devices mentioned in Sect. 43.2.1 constitute the data logger system and solar setup. Using this setup, the required data necessary for the prediction system has been acquired. The data is stored in external SD card, and the data can be transferred to the laptop in the form of an excel sheet. Radiance has been measured by the pyranometer. Temperature and humidity are measured by the DHT22 sensor.

The data for solar generation has been measured at electrical engineering department at Delhi Technological University. The data logger logs data at a sample rate of once per 2 min. A random sample set of logged data for a single day has been used for validity of fuzzy output.

Fuzzy Logic System

The fuzzy system has been designed with four uncertain inputs which are time of day, irradiance, temperature, and humidity. These inputs are fuzzified with the membership functions having a range between 0 and 1. The distribution of membership functions for the four inputs has been done on the basis of the observed maximum and minimum value for each parameter. Figure 43.2 shows the various inputs and outputs for the system.

The linguistic variable for temperature and humidity has five membership functions, whereas for irradiance there are nine membership functions. The division for the membership functions is based on producing a minimal error in the prediction system.

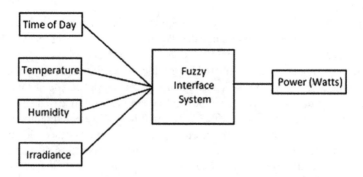

Fig. 43.2 Layout of the proposed fuzzy system

43 Fuzzy Logic-Based Solar Generation Tracking

The membership function of temperature has five triangular fuzzy functions which are V low, Low, Optimum, High, and V high. The other membership functions are described in a similar manner. In this proposed system, solar power generation has thus been predicted upon different parameters.

Figure 43.3 shows the input and output variable memberships for the proposed fuzzy interface system. The inputs are temperature, humidity, and irradiance whereas power is the output for the system.

Rule Base: A total of 256 viable rules have been formed using the logged data. Out of all the possible rules, many rules were deemed to be unfeasible and were not included in the rule base. The below given Fig. 43.4 depicts the surface view of input variables, irradiance and temperature, and output variable, power associated with each other.

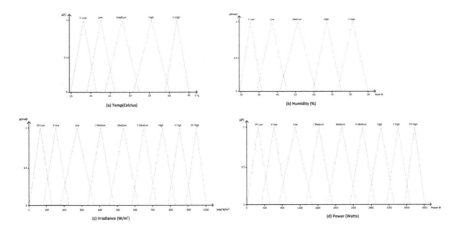

Fig. 43.3 Input-output membership functions

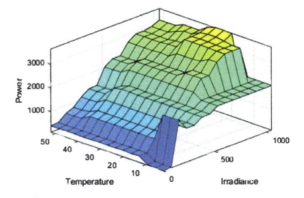

Fig. 43.4 Surface view of FIS rule base

Result

Dataset from National Renewable Energy Laboratory (NREL) was used to obtain the output of the system of the fuzzy system. In this dataset, data is logged at an interval of 1 h, so the output for the same is obtained. The data chosen was for a typical sunny day. Figure 43.5 shows the input irradiance and temperature conditions.

The input temperature conditions varied between a minimum of 23.9 °C and a maximum of 38.3 °C. The irradiance conditions for the power-producing hours varied between 220 and 1052 W/m2. The power output for the 5 kW system varied from 920 watts to 3.34 kW as shown in Fig. 43.6. Figure 43.6 also shows the predicted solar output versus the actual solar generation output.

Figure 43.7 shows the magnitude of the error percentage of the fuzzy system. The error varies between a maximum of 3.25% and a minimum of 0.45%. The maximum percentage error point is during sunset hours, where less solar power is being generated by the system.

The minimum percentage error point is present during the maximum power generation hours with average error for the same being less than 2%. However, the magnitude of error is more or less similar with an error of 40 50 watts in magnitude of the entire day.

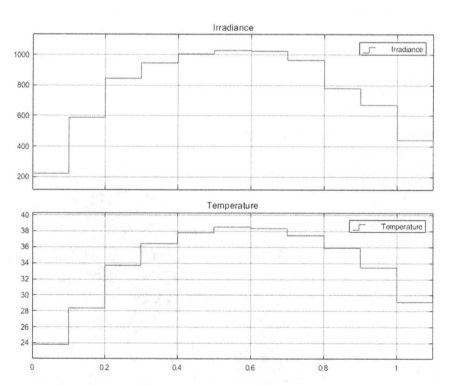

Fig. 43.5 Input temperature and irradiance curve

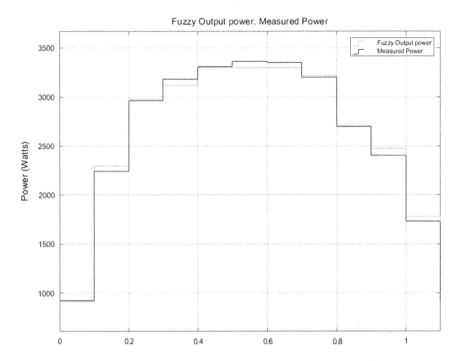

Fig. 43.6 Output power curve

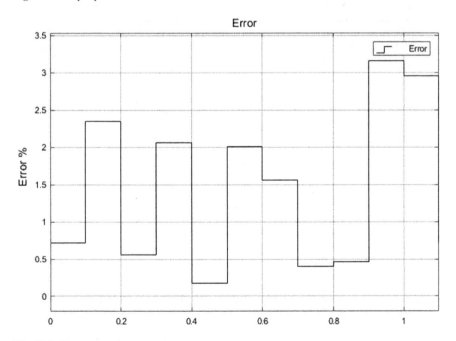

Fig. 43.7 Error curve

Conclusion

A fuzzy-based solar power generation forecasting model has been successfully implemented. A further addition to this model for a solar irradiance prediction system using fuzzy analysis is being performed. Cost versus performance analysis for the solar system is underway using the data logged by the data logger system.

Acknowledgements We thank our project supervisor and mentor Dr. M. Rizwan for his constant support and guidance for making this research possible. We would also like to thank Prof. Uma Nangia, Head of the Electrical Engineering Department, Delhi Technological University for the same.

References

1. Chakraborty S, Weiss MD, Simoes MG (2007) Distributed intelligent energy management system for a singlephase high-frequency AC microgrid. IEEE Trans Ind Electron 54(1):97–109
2. Neves D, Brito MC, Silva CA (2016) Impact of solar and wind forecast uncertainties on demand response of isolated microgrids. Renew Energ 87:1003–1015
3. Duverger E, Penin C, Alexandre P, Thiery F, Gachon D, Talbert T (2017) Irradiance forecasting for microgrid energy management. In: 2017 IEEE PES innovative smart grid technologies conference Europe (ISGT-Europe), Torino, pp 1–6
4. Corsetti E, Guagliardi A, Sandroni C (2016) Recurrent neural networks for very short term energy resource planning in a microgrid. In: Mediterranean conference on power generation, transmission, distribution and energy conversion (MedPower 2016), Belgrade, pp 1–9
5. Vermaak N, Gurusinghe N, Ariyarathna T, Gouws R (2018) Data logger and companion application for time-of-use electricity. In: 2018 IEEE international conference on industrial electronics for sustainable energy systems (IESES), Hamilton, pp 465–470. https://doi.org/10.1109/ieses.2018.8349921
6. Hussain SMS, Tak A, Ustun TS, Ali I (2018) Communication modelling of solar home system and smart meter in smart grids. IEEE Access 6:16985–16996. https://doi.org/10.1109/ACCESS.2018.2800279
7. Kenner S, Thaler R, Kucera M et al (2017) EURASIP J Embedded Syst Springer Int Publishing, 12. https://doi.org/10.1186/s13639-016-0045-7
8. Geetha A, Jamuna K (2013) Smart metering system. In: 2013 International conference on information communication and embedded systems (ICICES), Chennai, pp 10471051. https://doi.org/10.1109/icices.2013.6508368
9. Gaga A, Errahimi F, Es-Sbai N (2014) Design and implementation of MPPT solar system based on the enhanced P and O algorithm using Labview. In: 2014 International renewable and sustainable energy conference (IRSEC), Ouarzazate, pp 203–208. https://doi.org/10.1109/IRSEC.2014.7059786
10. Bonganay ACD, Magno JC, Marcellana AG, Morante JME, Perez NG (2014) Automated electric meter reading and monitoring system using ZigBee-integrated raspberry Pi single board computer via Modbus. In: 2014 IEEE students' conference on electrical, electronics and computer science, Bhopal, pp 1–6. https://doi.org/10.1109/sceecs.2014.6804531
11. Chugh A, Chaudhary P, Rizwan M (2015) Fuzzy logic approach for short term solar energy forecasting. In: Annual IEEE India conference (INDICON), New Delhi, pp 1–6
12. Mbarek MB, Feki R (2016) Using fuzzy logic to renewable energy forecasting: a case study of France. Int J Energ Technol Policy 12(4):357–376

13. Saez D, Avila F, Olivares D, Canizares C, Marin L (2015) Fuzzy prediction interval models for forecasting renewable resources and loads in microgrids. IEEE Trans Smart Grid 6(2):548–555
14. Hippert HS, Pedreira CE, Souza RC (2001) Neural networks for short-term load forecasting: a review and evaluation. IEEE Trans Power Syst 16(1):44–55
15. Tee CY, Cardell JB, Ellis GW (2009) Short-term load forecasting using artificial neural networks. In: North American power symposium, Starkville, MS, USA, pp 1–6
16. Liu K et al (1996) Comparison of very short-term load forecasting techniques. IEEE Trans Power Syst 11(2):877–882

Chapter 44
Airborne Manoeuvre Tracking Device for Kite-based Wind Power Generation

Roystan Vijay Castelino and Yashwant Kashyap

Introduction

Electric power generation through wind resources goes into the favors of reduction in pollution levels and climate control. Although horizontal axis wind turbines being developed on a large scale, it cannot trap the high altitude wind power, which is more powerful and steadier. Therefore, the tethered kite-based power generation is the solution that traps the enormous power at higher altitude. These systems generated electricity from persistent winds at an altitude of 200 m–10 km above ground level. Parafoil kites can quickly achieve high altitude easily and theoretically generate more power than turbine-based windmills [1].

Research in airborne wind technology is rapidly growing in recent decade. The kite-based wind power system consists of a parafoil wing, ground station, and the lines which tether the kite to the base station. Two line kite is controlled by varying the line length of any of the lines. The change in the line length varies the angle of attack of the kite [1, 2]. To effectively actuate the parafoil kite with respect to the wind, the control system requires information about the orientation of the kite. Along with the orientation of the kite, the altitude of the kite from ground level is also essential [3]. In a wind farm with many such kite generators, it will become essential to monitor the kites individually. This can be achieved by tagging each kite with its unique latitude and longitude location. In kite-based wind conversion system, the kite operates at an altitude which is more than 200 m. Therefore, it is necessary to use a wireless communication link with the lowest latency and faster sensor data

---

R. V. Castelino (✉) · Y. Kashyap
EEE Department, NIT Karnataka, Surathkal, Mangaluru, India
e-mail: roycas3@gmail.com

Y. Kashyap
e-mail: yashwant.kashyap@nitk.edu.in

© The Editor(s) (if applicable) and The Author(s), under exclusive license to Springer Nature Singapore Pte Ltd. 2021
A. K. Singh and M. Tripathy (eds.), *Control Applications in Modern Power System*, Lecture Notes in Electrical Engineering 710,
https://doi.org/10.1007/978-981-15-8815-0_44

acquisition. Research on a small prototype suggests that the wireless communication link with a latency of 100 ms is adequate for controlling the kite [4].

The optimum reel-out speed can be predicted based on the apparent wind velocity which the relative velocity between the velocity of wind and kite velocity. The orientation of the kite, wind velocity, and kite's apparent wind velocity can be used to calculate the aerodynamic force and speed at which the tether is getting pulled [5, 6]. Kite extracts the high-elevation wind power and converts it into a linear pull. This linear pull exerted by the kite lines in the form of oscillations can be extracted by a mechanism which converts linear to rotary motion [7]. The control of a kite can be done in two methods viz. ground-based control and on-board control. Ground-based control needs multi-tethers and actuators on the ground station whereas the on-board controller also called as control pod is situated near the kite which maneuvers the kite, and only a single tether is required between the control pod and ground station to produce traction force. Having multi-tether, increases the line sag due to tether weight and also increases the tether drag. The sensors can be mounted on the control pod for concurrent control of the kite [8].

The lift-to-drag ratio changes continuously as the kite is maneuvered. Using line angle sensor for calculating, the angle of attack is not reliable as the tether sag affects the calculations; hence, the on-board sensors provide accurate position and orientation with respect to earth's surface to estimate the changing lift coefficient and drag coefficient [9]. The mathematical modeling of the kite is very complex, and without modeling of the system, a feedback controller can be implemented with the sensor data obtained from the kite which mimics the human control of the kite. The controller needs to control the control lines of the kite which needs to adapt to the changes in wind direction and countering the different wind conditions. So there is a need to design an algorithm which mimics the learning behavior of human being [10].

The tethered kite flying at an altitude of 300 m is controlled by a winch control system. The control system requires real-time information about the state of the kite to effectively control it. The parameters required for autonomous control of the kite are—kite position and speed vector which must be measured or must be estimated by using triaxial accelerometer and gyroscope [11]. Sensor fusion algorithm can estimate the position of the kite irrespective of the dimensions of the kite as it depends entirely on the kinematic laws [12]. Turbulence in the wind causes the kite to change its orientation drastically. The closed loop setup with a sensor data-driven algorithm can predict these changes and stabilize the kite in the turbulent conditions. The predictive functional control scheme can predict the changes in the kite due to turbulence in the wind and make the kite stable in the sky [13]. The maximum efficiency of a kite-based system depends on the sensor data propagation delay. There is a need for estimation of kite position as the delay persists in the data transmission system. To estimate the position of the kite, the position data from the kite is necessary. By using the kinematic models of the kite, the experimentally gathered data can be used to estimate the kite's position. The control strategy can be implemented using a kite's position data obtained from the sensor unit, and tracking of figure eight in both the cycles is possible [14].

In this paper, the designed device logs the different patterns of the kite's orientation by which the controller can be designed. As a wired method of communication cannot be used, we have chosen Sub-1 GHz radio frequency communication to transmit the sensor values to the ground station. All these system specifications need to be met while ensuring the embedded system is low powered. Texas instruments MCU we have chosen is an ultra-low-power device which has low-powered RF communication link built-in [15].

Methodology

The setup for the parafoil kite-powered wind energy system can be divided into two broad hardware parts viz. the on-board sensor array and the ground unit.

The on-board sensor array contains the microcontroller unit (MCU) hub, GPS sensor, and IMU unit. The ground unit contains an identical MCU connected to a PC via serial communication for the data logging. The entire setup can also be divided based on the RF communication parameters. The CC1310 MCU on the parafoil kite acts as the RF transmitter, and the CC1310 MCU at the ground station acts as the RF receiver.

Transmitter Side Operation

The block diagram shown in Fig. 44.1 describes the full setup for the sensor array for parafoil kite-powered wind generator. It consists of an MCU with a built-in transmitter, GPS, IMU module.

A GPS or global positioning system unit is required to accurately determine the latitude, longitude, and altitude above the ground unit (GU) of the para-foil kite. Once connected to the MCU and powered on, the GPS sensor initially takes about a minute to get a position fix. The accuracy of the position fix is also improved over time. As the data is captured by the GPS and converted to NMEA string, it sends it over the UART channel to the MCU at the data rate of 9600 k baud. On the MCU unit, end special string parsing functions convert the NMEA string to float values of latitude, longitude, and altitude and pass these as 8-bit unsigned integer data values to the RF packet.

Inertial measurement unit is used here to get kite's orientation with respect to the earth's surface. IMU consists of accelerometer to measure the sudden changes in the orientation and a gyroscope which can give the rotation of the kite in degrees and a compass sensor to calculate the movement direction of the kite. IMU is required to determine the tilt of the parafoil kite to estimate the angle of impact of the wind and provide appropriate control commands. It should also provide the force experienced by the kite in all its axis. The IMU, MPU6050 communicates with the MCU via I2C communication which will be detailed in the following section.

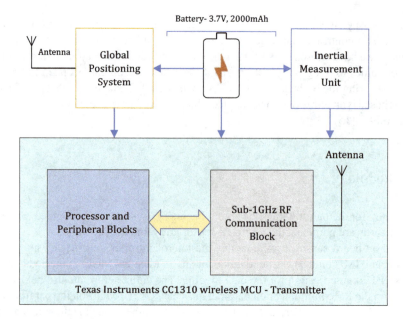

Fig. 44.1 Block diagram of the transmitter

The microcontroller unit (MCU) is connected to the IMU, the GPS, and the communication module. It is the confluence of the data acquired from all sensors. The MCU is responsible for processing the incoming data, compiling them into a given packet structure, and then transmitting them via the communication link to the ground module. The technical requirements of the microcontroller unit are high speed with low power consumption and UART/SPI communication capabilities.

Figure 44.2 shows the hardware realization of the transmitter system which acquires the sensor data and transmits it to the base station. Texas instruments CC1310 SimpleLink™ microcontroller unit has a built-in sub-1 GHz wireless module. Besides, it is a low power device in both RX and TX modes and has a wide operating voltage between 1.8 and 3.8 V which is ideal for this application.

Receiver Side Operation

Figure 44.3 shows the block diagram of receiver side MCU with sensor data logging. The receiver MCU is identical to the transmitter MCU which consists of Sub-1 GHz wireless module built-in. The received data at the ground station is then processed and sent to the PC through serial communication. The RF receiver consists of a CC1310 MCU placed at the ground station and connected to a PC via serial communication. The receiver implements an interrupt-based system. As soon as an RF packet is received an interrupt is generated which calls a callback function. This function

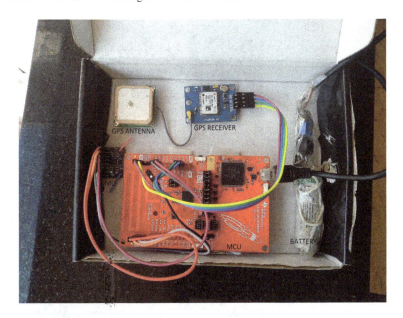

Fig. 44.2 Hardware setup of transmitter unit

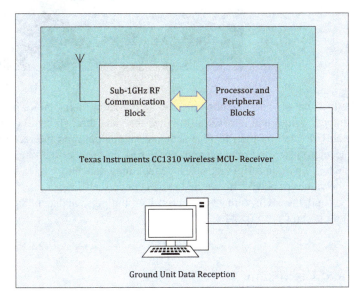

Fig. 44.3 Block diagram of the receiver unit

Fig. 44.4 Hardware setup of the receiver unit

has subroutines that process the incoming data into floating-point values from 8-bit unsigned integer values and also serially writes this data through serial communication which is logged by the PC. One such subroutine converts the received raw values of accelerometer and gyroscope to pitch, roll, and yaw of the kite. The data logger system is a python script that saves the incoming data into a spreadsheet and also plots the values of accelerometer and gyroscope in real-time.

On the ground, the MCU with a built-in receiver unit logs the data in the PC. The hardware unit is shown in Fig. 44.4.

Results and Discussion

Figure 44.5 shows the experimental yaw, roll, and pitch values when the sensor kept at a stationary position. The raw sensor values obtained are varying even when the sensor is stationary due to sensitivity. The sensor values are calibrated. The MCU, sensors, and the battery are kept in a small box securely and mounted on the center of the kite. The kite is flown in three different axes, and the obtained sensor values are plotted as follows.

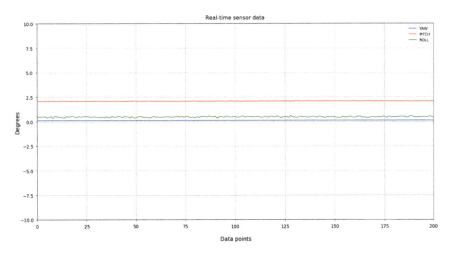

Fig. 44.5 Yaw, pitch, roll stationery (Scale: 250°/sec—Gyro, 2G—Accel)

Figure 44.6 shows the experimental values of yaw, roll, and pitch when the sensor is mounted on the kite, and the motion is in $Zs =$ axis.

Figure 44.7 shows the plotted sensor values when X-axis shows the motion values.

Figure 44.8 shows the Y-axis movements from the obtained sensor values

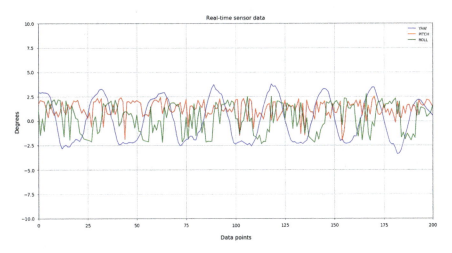

Fig. 44.6 Plot consisting of yaw, pitch, and roll with Z-axis motion

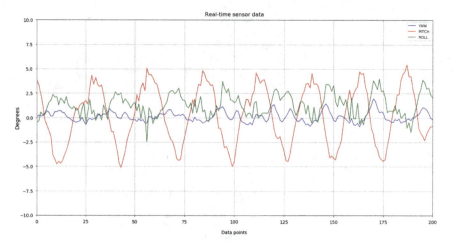

Fig. 44.7 A Plot consisting of yaw, pitch, and roll with X-axis motion

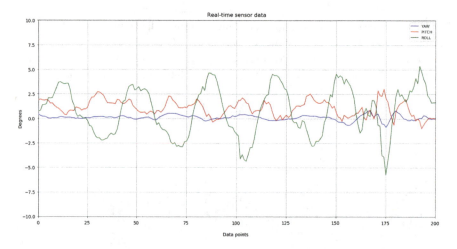

Fig. 44.8 Plot consisting of yaw, pitch, and roll with Y-axis motion

Sensor Data Plots

Figures 44.5, 44.6, 44.7, 44.8 contain the sensor data plotted in real-time. The X, Y, Z axes mentioned in the graphs are about the fixed axes marked on the MPU6050 device. The pitch of the kite is the angle at which air hits the kite. Roll gives the rotations of the kite, and the yaw gives the movement of the kite in the left and right direction about the earth's surface. These three parameters are necessary to track a kite's orientation in the sky. The three-axis raw values received from each sensor are plotted using a python script on the computer.

Table 44.1 Time delays in sensors

Component	Delay(ms)	Remarks
GPS/IMU	20	Time to refresh
Wired Connections	5	Kite to controller
Wireless latency	30	Latency
Kite state estimation	5	Calculation
Kite controller	20	Control time
Motor controller	20	Control time
Total	**100**	

The time delays involved in the control system is as shown in Table 44.1.

Field Testing

The setup consisting of two CC1310 launchpads was tested. The board acts as the transmitter was interfaced with the sensors and made to transmit the packets to the other board, acting as the receiver at the base station. The time delays involved in the control system is as shown in Table 44.1.

The packet transmission versus various distances, frequencies, and distances were tested in the field. Error corresponds to the packets missed in one minute. The results are as shown in Table 44.2.

Table 44.2 Experimental results

Sl. No	Frequency (MHz)	Data rate (Kbps)	Distance (m)	Packets per min	Error (%)
1	433	50	N/A	N/A	N/A
2	433	625	N/A	N/A	N/A
3	490	50	N/A	N/A	N/A
4	490	625	N/A	N/A	N/A
5	868	5	90	61	0
6	868	5	130	59	0
7	868	5	270	60	3.2
8	868	50	70	60	0
9	868	50	90	60	0
10	868	50	130	59	1.3
11	868	50	270	58	38
12	915	240	90	61	0
13	915	240	130	57	30
14	915	240	270	57	78

Conclusions

Position data is the sole requirement for the autonomous flight of airborne wind technology. This experimental setup provides the data which helps develop on-board kite controllers. The designed setup with Texas Instruments MCU has low power consumption and lower weight as the wireless modules are embedded in the MCU itself. The kite was flown at a height of a maximum of 270 m, the three readings at 90 m, 130 m, and 270 m of specified thread length. The experimental analysis concludes that a kite-based wind power generation system that can reach up to 270 m height can transmit the data efficiently using an 868 MHz band at a data rate of 50 Kbps. The inertial data provided by IMU and the longitude and latitude values provided by the GPS are received at the base station and logged using python script on the computer. The data obtained can be used for controlling the position of the kite by designing a controller for the obtained data. The data transmission rate has to be faster for the faster response time of the controller.

Acknowledgements We would like to thank the National Institute of Technology Karnataka, Surathkal for providing the facilities to test our experiments and also to the Electrical and Electronics Engineering department for their support for carrying out the experiments.

References

1. Loyd ML (1980) Crosswind kite power (for large-scale wind power production). J. Energy 4(3):106–111
2. Castelino RV, Jana S, Kumar R (2018) Analytical and experimental study of Parafoil kite force with different angle of attack in ANSYS environment for wind energy generation. In: 4th international conference on electrical energy systems (ICEES), pp 381–385
3. Canale M, Fagiano L, Milanese M (2010) High altitude wind energy generation using controlled power kites. IEEE Trans Control Syst Technol 8(2):279–293
4. Zgraggen AU, Fagiano L, Morari M (2015) Real-time optimization and adaptation of the crosswind flight of tethered wings for airborne wind energy. IEEE Trans Control Syst Technol 23(2):434–448
5. Fechner U, Schmehl R (2014) Feed-forward control of kite power systems. J Phys Conf Ser 524(1):012081
6. Fechner U (2016) A methodology for the design of kite-power control systems
7. Mahanta H, Rajput D, Bollam S (2016) Gandhian young technological innovation awards (GYTI), New Delhi, pp 64–65
8. Fagiano L, Marks T (2014) Design of a small-scale prototype for research in airborne wind energy. IEEE/ASME Trans Mechatron 20(1):166–177
9. Oehler J, Schmehl R et al. (2018) Experimental investigation of soft kite performance during turning maneuvers. J Phys Conf Ser 1037(5):052004
10. Fagiano L, Novara C (2014) Automatic crosswind flight of tethered wings for airborne wind energy: a direct data-driven approach. IFAC Proc 47(3):4927–4932
11. Fechner U, Schmehl R (2012) Design of a distributed kite power control system. In: IEEE international conference on control applications, IEEE, pp 800–805
12. Fagiano L, Huynh K, Bamieh B, Khammash M (2013) On sensor fusion for airborne wind energy systems. IEEE Trans Control Syst Technol 22(3):930–943

13. Sun Q, Wang Y, Sun, Q, Yong-yug-yu W (2012) Data-driven predictive functional control of power kites for high altitude wind energy generation. In: IEEE Electrical power and energy conference, IEEE, pp 274–279
14. Polzin M, Wood TA, Hesse H, Smith RS (2017) State estimation for kite power systems with delayed sensor measurements. IFAC-PapersOnLine 50(1):11959–11964
15. Instruments Texas (2016) CC1310 Simplelink^{TM} Ultra-Low Power Sub-1 GHz Wireless MCU. CC1310 Datasheet (SWRS184)

Chapter 45
Integration of Electronic Engine and Comparative Analysis Between Electronic and Mechanical Engine

Sapna Chaudhary and Rishi Pal Chauhan

Introduction

Nowadays, four-stroke diesel engine is widely used in the various applications. Four-stroke cycle includes intake, compression, combustion, and exhaust. These strokes are repeated over and over to convert chemical energy into mechanical power [1]. There are various players which manufactures engine for different application of off-highway like in construction, mining, oil and gas, rail, compressor, pumps, and more. Development process in the field of off-highway applications like excavator, wheel loader, soil compactor is going on. This arises as a high demand of customer and various regulations put on off-highway applications, so the existing engine system with advance electronic used to improve the engine performance, reliability, durability, energy efficiency, and to reduce engine emission. It can be further integrated with the state of art monitoring system using telematics devices. It means for the development of new architecture of electronic engine a huge requirement of planning and testing is needed. This paper describes the testing method for feature testing, comparative analysis between mechanical and electronic engine used for off-highway applications, results, and conclusion.

S. Chaudhary (✉) · R. P. Chauhan
Department of Physics, National Institute of Technology, Kurukshetra, India
e-mail: sapna.chaudhary946@gmail.com

R. P. Chauhan
e-mail: chauhanrpc@nitkkr.ac.in

© The Editor(s) (if applicable) and The Author(s), under exclusive license to Springer Nature Singapore Pte Ltd. 2021
A. K. Singh and M. Tripathy (eds.), *Control Applications in Modern Power System*, Lecture Notes in Electrical Engineering 710,
https://doi.org/10.1007/978-981-15-8815-0_45

Existing System Versus Modified System

As shifting from mechanical engine to electronic engine in the existing machines, vast modifications are required in the existing architecture of the engine. As existing system is controlled mechanically, therefore, only a few sensors and actuators are present but in the case of the electronic engines, a controller called as Engine Control Module (ECM) along with huge number of sensors and actuators are present which enables real-time data monitoring of attributes related to engine. ECM is an embedded electronic system which controls the subsystems of a vehicle. In existing mechanical system, after treatment of exhaust emission is not efficient in comparison to the modified system; as it is having doser, Diesel Exhaust Fluid (DEF), i.e., urea solution is injected over the exhaust gases to cut down the engine out NOx level and break into atmospheric nitrogen (N2) and water (H2O) [2]. With the advancement in the electronics, the engine can be converted easily into different power rating and can be made suitable for different applications. For example, if engine has ratings of 2000 RPM at 173 HP is needed to convert into 2200 RPM at 173 HP, it can be easily achieved by electronic engine up to the limited range of ratings. This can be considered as one kind of feature and advantage of the electronic engine over mechanical engine. There are more than 73 features available which are utilized for better integration of electronic engine as compared to mechanical engine. As to validate these modified changes, various tests need to be performed [3].

Problem Statement

ECM contains the software logic to control the functionality of the engine. To find out the software bugs, it is required to verify the software logic and test the electronic engine feature repeatedly. As there are 73 features which needs to be tested. The traditional way of verifying software logic and testing electronic features by using the complete single prototype software platform that outcomes in huge cost and extended development time affecting the efficiency. If any changes occurred in the program at a later stage, it also leads to an increase in the development cost and time.

Therefore, Hardware-In-the-Loop (HIL) simulation technique can be implemented for validation of software logic and electronic features. This technique ensures testing is in house testing prior to the actual test to be performed on the machine. This helps in cost reduction and improves human risk factor while testing. This method of testing also optimizes the development time and speed up the design [4, 5]. Hence, this technique is used to help in meeting the stringent emission norms and enhances the engine performance and protection by testing the various feature of electronic engine.

Hardware-in-the-Loop Bench Setup

Overview of Test Bench Setup

To test the new software and hardware logic developed for electronic engine off-highway applications, the real-time diesel engine along with its sensors and actuators is replaced by a mathematical simulation model running on a real-time processor. The input/output (I/O) interface is done between virtual engine simulator and Electronic Control Module (ECM). By the development of virtual engine simulator, it is easier to develop, test, and validate the functionalities in comparison to a conventional development. In industrial applications, it is observed that validation of control logics via HIL is simpler, effective, and efficient than testing on real vehicles [6]. Use of HIL also reduces human risk factor of accidents in actual environment, i.e., "On-Vehicle" faults testing in development and validation of control logics. Hardware used to validate the electronic engine features and developed software are universal load box (virtual engine), breakout box, load, box wiring harness, peak adapter, Aftertreatment Control Module (ACM)/Engine Control Module (ECM) with software that is tuned to meet the emissions and performance requirements and an engineering tool that will communicate to the ECM/ACM [7].

Test bench setup consists of following hardware and software tools as shown in Fig. 45.1:

Electronic Control Module (ECM): The center of the control system is the electronic control module (ECM). The ECM is a computer which analyzes information and determines how the engine will operate. The ECM also stores information that can be used to diagnose engine problems or to inform the customer about the maintenance and operation of equipment. Vehicle and engine information are conveyed to the ECM by sensors, switches, and actuator. These devices are installed throughout the vehicle and provide the ECM with speeds, temperatures, pressures, operator requests, and a wide variety of other information. This data is used to understand how the engine is performing and what is expected from the operator [8].

Peak Adapter: Peak adapter is also known as PCAN-USB. It is used to deliver communication between ECM, load box, and other controlling software. For automotive application, it works according to CAN protocol of J1939 SAE standard. It is a two-wire bus CAN high and CAN low and must be terminated with 120 ohms on both sides. PCAN-View is a software through which received messages can be viewed, and new messages can be transmitted [9, 10].

Wiring Harness: The engine wiring harness that supports the engine control module (ECM) and connects the ECM to the engine sensors, actuators, data communications, power connections, etc.

LUIS Gen2: LUIS stands for load box user interface system. It is a digital bench. LUIS is an engine simulator used for testing engine control system hardware, software, and electronic engine feature. The LUIS Gen2 has a main module that is connected to the PC via USB. Additional modules can be used like wave maker,

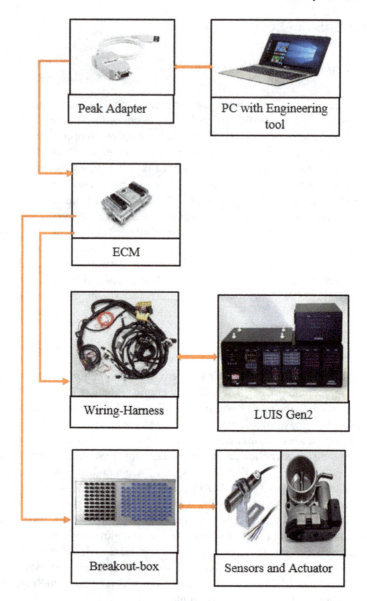

Fig. 45.1 Block diagram of test bench setup using HIL

analog, switch, and resistive loads to modify the system to fulfill the user's specific requirements [11].

Breakout Box: There are several pins on the breakout box which behaves as an input or output to the ECM. Breakout box helps in checking the functionality and continuity of different sensor and actuators by connecting to the pins of this box.

Sensors and Actuators: A sensor is a device which measures a physical quantity of the engine like coolant temperature, oil pressure, fuel pressure and converts it into a signal which can be read by the ECM. ECM controls the actuator action accordingly.

Bench Configuration

There are various types of ECM depending upon the requirement of the user for which application that ECM going to be used. There are numerous pins in the ECM, and the number of pin is different for different ECM. Corresponding to every pin various signal of sensors and actuator are received. So, to know which signal going to be received on which pin, for that develop the bench configuration tables.

- Input/output matrix
- Resource location Table.

Input/output matrix is defined by the application engineer that describes which signal is being received on which pin of the ECM. In resource location table, a resource location number is assigned corresponding to each pin of the ECM. According to input/output matrix, the signal must assign the same resource location number corresponding to the respective pin of the ECM.

Developed Methodology

Electronically controlled engine consists of an ECM. ECM is a controller having a memory in which a calibration file is required to be flashed. Calibration file is developed with the help of software in which approximately twenty thousand parameters are present corresponding to every component of engine like sensors, actuators, and after-treatment system.

For testing these software logics, the actual combustion engine with its sensors, actuators, and switches are swapped by a mathematical simulation model running on a real-time processor, i.e., is called LUIS Gen2. With the help of LUIS develop a graphic user interface on which all the sensor and switches are created as shown in Fig. 45.2. A specific input/output (I/O) interface is developed to connect the virtual engine simulator to the ECM with the help of wiring harness. It helps in creating communication between engine simulator and ECM. The software logic which is dumped in ECM interacts with the engine simulator by means of graphic user interface (GUI) developed using LUIS software. The engineering tool allows to monitor all the parameter of the input/output, and it is an engineering development tool used to monitor the ECM data and used to alters the calibration data according to the requirement. It is used to modify the calibration parameter and feature settings in an engineering development and test environment. Varying tunings for different conditions can be provided with the help of engineering tool.

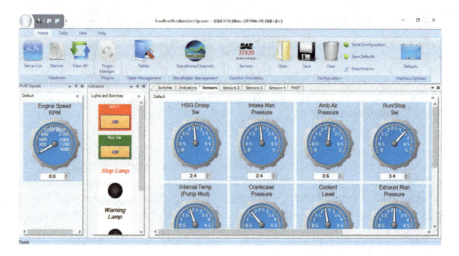

Fig. 45.2 LUIS GUI

This HIL test bench setup system able to yield the correct and consistent outputs corresponding to various sensor and actuators signals of a combustion engine. It can produce the same output behavior as of engine for the same inputs in the real-time application [12].

Tests

In the proposed system what modification can be done in comparison to the existing system, so for that feature testing need to be done which helps in optimizing the high energy efficiency, reliability, and safety of the engine.

Various test is performed on the HIL test bench setup.

(1) Circuit continuity fault condition check: This test shows that the I/O signals connected to the ECM are, if

- Open-circuit
- Short-circuit to supply voltage(5 V/24 V)
- Short-circuit to supply ground.

(2) Functionality fault conditions: As a part of functionality fault conditions simulation, out of range fault conditions for sensors are simulated, and set or reset time of the faults are calculated in this test [13, 14].
(3) Feature testing: Under this type of test, various diagnostics related features, engine performance tests, integrated system check, durability, and reliability are tested to check the synchronized functionality of the software developed for the new system.

Results and Implementation

In the electronic system, at first the functionality of the software is checked and validates the logics which are dumped into the ECM in the form of configuration and calibration files. It is done on Hardware-In-The-Loop test bench set up which allows the tester to ensure feasibility of the engines before doing it on actual engines. So, it enhances safety and reduces human risk factor for, and optimizes the cost of testing.

Circuit Continuity Fault Condition Check for Sensors

There are around thirty sensors mounted on the electronic engine. In software, a corresponding parameter of count for lower threshold and higher threshold is set for every sensor like oil pressure, coolant temperature, exhaust pressure, ambient pressure, etc. Whenever a sensor crosses their set threshold a fault condition is mapped, which indicates that sensor is out of its range.

As a case study, the exhaust pressure sensor is considered; this sensor is used for the measurement of exhaust engine pressure.

Figure 45.3 describes the sequences for circuit continuity fault condition check for the engine exhaust pressure sensor.

i. When exhaust pressure signal is connected to ground pin using breakout box then real-time exhaust pressure count decreases to zero which is lower than lower set threshold count value, and fault condition gets mapped, which indicates sensor is shorted to low source.
ii. When exhaust pressure signal is connected to supply pin using breakout box then the real-time exhaust pressure count increases to its highest value of 16,457 count which is greater than higher set threshold count; a fault condition is mapped, which indicates sensor shorted to high source, out of range.
iii. When exhaust pressure sensor real-time value is in between these set thresholds, it indicates sensor is in range.
iv. If fault code does not map even after all the pre-conditions are satisfied for setting the fault code, then it is a fail condition which arises due to the presence bugs in the software, and further troubleshooting is required.

As shown in Fig. 45.4, whenever exhaust pressure sensor real-time count becomes greater than the high threshold the fault code gets activated after some delay, i.e., set time of fault condition; and when fault condition gets deactivated, it is deactivate after some delay, i.e., reset time of fault condition. Similarly, when exhaust pressure signal is connected to the ground, we can observe the activation of fault condition. Set and reset time of fault code for exhaust pressure sensor can be determined by this test; hence, the observed value of set and reset time is 10 and 10 s.

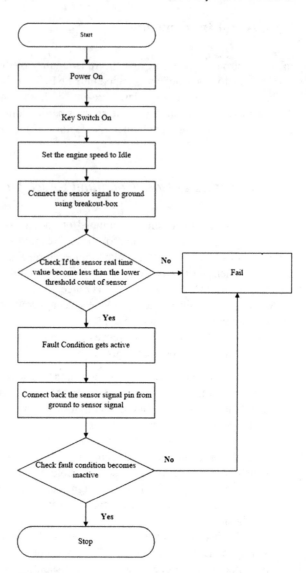

Fig. 45.3 Flow chart for checking the circuit continuity

Conclusion

As discussed in the Sect. 45.4, the electronic engine uses more sensors and actuators, whenever these sensor values get out of range a fault condition mapped, which indicates faulty state of engine this helps in optimizing the engine performance and protection. So, all the tests related to sensor out of range and electronic engine features can be easily performed on the HIL setup. Hence, integration of electronic engine can be easily done using Hardware-In-the-Loop (HIL) simulation testing technique. It is an essential and effective way in the development of new software logic and

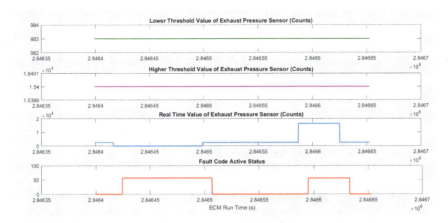

Fig. 45.4 Circuit continuity check for exhaust pressure sensor

electronic engine features for off-highway applications. It optimizes the cost and development time, hence increases the efficiency of engine.

Therefore, this research explained how efficiency increases by improving the techniques to test and validate the software platform. Some features are tested, and results are shown in Sect. 45.4 that describes its behavior on physical machine which shows the pre-validation of electronic engine features. Accordingly, all features tested and validated by using HIL and which forms the final software platform. This final platform may lead to have defined feature bunched in the single component called as software calibration.

References

1. Narwade NV, Rajani PK, Pathak M (2018) Rail engine safety through digitization and interfacing. In: 2018 Fourth international conference on computing communication control and automation (ICCUBEA), pp 1–6. https://doi.org/10.1109/iccubea.2018.869766
2. Wetzel P, McCarthy J, Kulkarni M, Mohanta L, Griffin G (1944) Diesel exhaust aftertreatment system packaging and flow optimization on a heavy-duty diesel engine powered vehicle. SAE Int J Commercial Vehicles 3:143–155. https://doi.org/10.4271/2010-01-1944
3. Tonetti M, Rustici G, Buscema M, Ferraris L (2017) Diesel engine technologies evolution for future challenges. SAE Technical Paper 2017–24-0179. https://doi.org/10.4271/2017-24-0179
4. Gehring J, Schütte H (2002) A Hardware-in-the-Loop test bench for the validation of complex ECU networks. SAE Technical Paper 2002–01-0801
5. Schulze T, Wiedemeier M, Schuette H (2007) Crank angle—based diesel engine modeling for Hardware-in-the-Loop applications with in-cylinder pressure sensors. SAE Technical Paper 2007–01-1303
6. Zhou J, Ouyang G, Wang M (2010) Hardware-in-the-Loop testing of electronically-controlled common-rail systems for marine diesel engine. In: 2010 International conference on intelligent computation technology and automation, Changsha, pp 421–424. https://doi.org/10.1109/icicta.2010.40

7. Pandey AK, Jessy S, Diwanji V (2012) Cost effective reliability centric validation model for automotive ECUs. In: 2012 IEEE 23rd international symposium on software reliability engineering workshops, Dallas, TX, pp 38–44. https://doi.org/10.1109/issrew.2012.29
8. Hui D, Bo H, Dafang W, Guifan Z (2011) The ECU control of diesel engine based on CAN. In: 2011 Fourth international conference on intelligent computation technology and automation, Shenzhen, Guangdong, pp 734–736. https://doi.org/10.1109/icicta.2011.191
9. https://www.peak-system.com/produktcd/Pdf/English/PCAN-USB_UserMan_eng.pdf
10. Sutar D, Shinde SB (2017) ECU diagnostics validator using CANUSB. In: 2017 International conference on inventive computing and informatics (ICICI), Coimbatore, pp 856–860. https://doi.org/10.1109/icici.2017.8365257
11. http://gartechenterprises.com/luisnextgen_v2.0.pdf
12. Fajri P, Lotfi N, Ferdowsi M (2018) Development of a series hybrid electric vehicle laboratory test bench with hardware-in-the-loop capabilities. In: 2018 International power electronics conference (IPEC-Niigata 2018-ECCE Asia), Niigata, pp 3223–3228. https://doi.org/10.23919/ipec.2018.8507697
13. Raul NM, Wagdarikar NM, Bhukya CR (2018) Development of hardware-in-loop automated test bench for liquid-assisted after-treatment controls system's regression tests. In: International conference on computing communication control and automation, Pune, India, August 16–18
14. Mane M, Kulkarni SJ, Gupta V (2018) Automation of FMET testing. In: International conference on computing communication control and automation. Pune, India, August 16–18

Chapter 46
Monitoring Cyber-Physical Layer of Smart Grid Using Graph Theory Approach

Neeraj Kumar Singh, Praveen Kumar Gupta, Vasundhara Mahajan, Atul Kumar Yadav, and Soumya Mudgal

Introduction

A cyber-physical infrastructure is a platform that unites the cyber layer consisting of information infrastructure and physical infrastructure like remote terminal units, sensors, etc. [1]. The cyber-physical layer of the smart grid is vulnerable to cyber intrusions, which is a combination of traditional cyber attack with a physical dimension [2, 3]. Still security tools are not as advanced as those for cyber layer protection. Researchers have been using graph theory for designing cyber-physical infrastructure [4, 5]. Graph theory uses both continuous and discrete parameters making the graph approach suitable for the cyber-physical domain [6].

The attack graph simplifies the modeled and uses it to identify how a target can be attacked [7, 8]. The traditional attack graph approach reduces the conversation of continuous data to discrete step data thus reducing the accuracy. The smart grid infrastructure is a cyber-physical system that contains many components which either belong to the cyber layer or physical layer [9]. The main advantage of the graph theory

---

N. K. Singh (✉) · P. K. Gupta · V. Mahajan · A. K. Yadav · S. Mudgal
Electrical Engineering Department, Sardar Vallabhbhai National Institute of Technology, Surat, India
e-mail: neerajksssingh@gmail.com

P. K. Gupta
e-mail: praveenpragati12@gmail.com

V. Mahajan
e-mail: vmahajan@eed.svnit.ac.in

A. K. Yadav
e-mail: yatul76@gmail.com

S. Mudgal
e-mail: mudgalsoumya@gmail.com

© The Editor(s) (if applicable) and The Author(s), under exclusive license to Springer Nature Singapore Pte Ltd. 2021
A. K. Singh and M. Tripathy (eds.), *Control Applications in Modern Power System*, Lecture Notes in Electrical Engineering 710, https://doi.org/10.1007/978-981-15-8815-0_46

Table 46.1 Research work based on graph theory

Technique	Objective	Advantage	Disadvantage
Attack graph [12]	Model attack against substation transformer	• Easier to model • Handle continuous variable	Architecture is not well defined
Attack tree [13]	To evaluate the risk of a connected vehicles	Provide link between cyber and physical layer	Required large amount of data to process
Bayesian attack graph [7]	Construct attack path for power system evaluation	It mitigates the damping effect of cyber attack	A more probabilistic and realistic model needed
Distributed attack graph [14]	Construct distribute multi-agent platform for vulnerability evaluation	Real-time monitoring and detection	Security analysis is needed for multi-agent platform

approach is, it simplifies the complex network and hence aiding researchers for risk analysis studies. In [10], researcher has introduced a novel approach by introducing graph theory and Markov chain. It helps to reduce the risk of false-positive alert during normal operation. In the research work [11], a threat to cyber-physical system is designed and its nature is studied. Some research work is highlighted in Table 46.1.

In this paper, power system properties are modeled using graph theory. It is utilized infrastructure (Cyber layer) and remote terminal units (physical layer).

System Modeling and Problem Formation

System Model

In this section system modeling is done using a graph-based model, to include a wide range of cyber-physical interactions between a pair of nodes. Each bus of the smart grid is modeled as a node, monitoring both the physical operation and the communication operations. The physical layer of the smart grid mainly consists of a transmission line used for power transfer and helping communication networks through power line communication. In the physical layer, any kind of fault will affect the synchronous machine swing equation. During a cyber intrusion, the attacker will try to delay the stability control, making the system unstable or loses synchronism.

In order to link two different parameters, i.e., power flow and communication flow, a simple signal flow strategy is used. Signal will flow according to the flow of actual parameters making the system more realistic. Figure 46.1 shows the cyber and physical layers of the smart grid. A simple graph is modeled, using the smart grid diagram that reflects the design of the smart grid. Figure 46.2 reflects the graph of

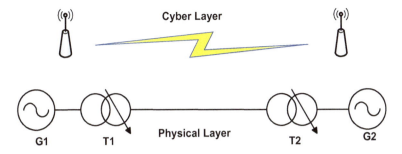

Fig. 46.1 Cyber-physical layer of smart grid

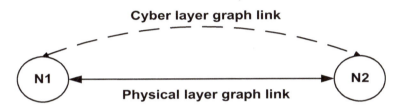

Fig. 46.2 Graphical model of Fig. 46.1

smart grid, where two nodes indicate the two generators and the link between them highlights the physical link and communication link.

Effect on synchronization can be easily related to the second-order swing equation of synchronous generator as written below:

$$\frac{2H}{w_0}\frac{\partial^2 \delta}{\partial t^2} + wD\frac{\partial \delta}{\partial t} = P_a = P_m - P_e \qquad (46.1)$$

The above equation can be represented as a two dimensional uniformly distributed using the approach developed in many researches. Using the continuum approach, the above equation becomes distributed parameters over two dimensions as:

$$P_m \rightarrow \Delta p_m \rightarrow f(x, y) \qquad (46.2)$$

$$H \rightarrow \Delta h \rightarrow f(x, y) \qquad (46.3)$$

$$D \rightarrow \Delta d \rightarrow f(x, y) \qquad (46.4)$$

Now using equations from 46.2–46.4, then the swing equation is rewritten as a nonlinear signal by considering $\Delta \rightarrow 0$:

Table 46.2 Overall system evaluation matrix

Cyber layer	Physical layer	System score	System status
0.5	0.5	0	Hacked
0.5	1	0	Cyber layer hacked
1	0.5	0	Physical layer fault/hacked
1	1	1	Healthy

$$\frac{\partial^2 \delta}{\partial t^2} + m \frac{\partial \delta}{\partial t} - k^2 \nabla^2 \delta + n^2 (\nabla \delta)^2 = P \tag{46.5}$$

where

$$m = \frac{\omega^2 d}{2h}, \ k^2 = \frac{\omega V^2 \sin\theta}{2h|z|}, \ n^2 = \frac{\omega V^2 \cos\theta}{2h|z|}, \ P = \frac{\omega(p_m - GV^2)}{2h}$$

where θ represent angle associate with transmission line impedance, h is the inertia constant of generator, ω is bus frequency, V is bus or generator voltage, ∇^2 is laplacian operator. The communication link already follows Eq. 46.5 for its propagation in terms of signal transfer.

Problem Formation

Considering a system, consisting of N generators and B buses which comprises sensors that monitor the condition of generator and buses. The graph G_{phy} will act as a replica G_{cyb} in which all elements of the physical graph are mapped with a cyber graph. During proper operation, the weight assigned to nodes which represent buses and generator is 1. During unhealthy operation the weight will change according to the following logic:

$$W = \begin{cases} 0.5 : \text{when cyber layer unhealthy} \\ 0 : \text{when physical layer unhealthy} \end{cases} \tag{46.6}$$

When the cyber and physical layer both are combined in relation to overall system evaluation it follows AND gate logic as stated in Table 2.

Simulation Result and Discussion

To evaluate the proposed method, IEEE-9 bus system is modeled using a cyber-physical graph as shown in Fig. 46.3. Every bus and generator are modeled as graph

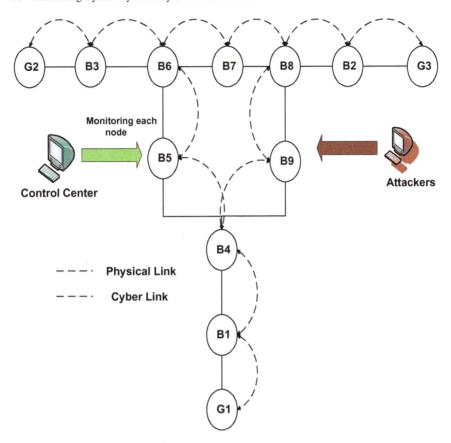

Fig. 46.3 Cyber-physical graph model of IEEE-9 bus system

nodes. The Figure helps to visualize the test system, where a specked line represents a communication link and the solid line indicates a physical link. Using the proposed method, the test system is evaluated for three different scenarios. All three cases are discussed below:

- **Case 1:** When only the physical layer gets faulty or hacked.
- **Case 2:** Only the cyber layer gets hacked.
- **Case 3:** Both cyber and physical layers get hacked.

Table 46.3 shows the simulation results for different cases stated above. Attackers may select any arbitrary node for cyber intrusion. Detecting intrusion in the physical layer is easy as compared with the cyber layer of cyber-physical infrastructure. Results may vary according to the first node which gets affected by the intruders. Figure 46.4 compiles the simulation result.

Table 46.3 Simulation results

	Nodes affected due to cyber intrusion	Nodes affected due to physical intrusion	Detection time (s)
Case 1	NA	B5, B6, B4, G2	2.47
Case 2	B8, B9, B7, B2, G3	NA	4.13
Case 3	G1, B1, B4	G3, B2, B2, B8, B9	6.21

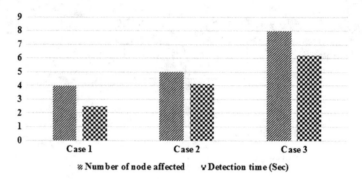

Fig. 46.4 Simulation result

Conclusion

In this paper, a smart cyber-physical graph strategy is highlighted for monitoring the smart grid. It quantifies the impact of the cyber layer on power system dynamics for effective monitoring. The swing equation plays a crucial part in assessing the performance of the smart grid. It is converted to a nonlinear equation which indicates the propagation of a signal from one point to another. During an intrusion, the attacker changes the basic parameters of the system which leads to instability. The proposed method monitors any changes in the system and changes the weight of the nodes accordingly. From the simulation, it can be seen that the method is very fast and detect both cyber and physical intrusion. Detection time totally depends on the number of generators and buses present in the system. With a higher number of nodes, detection time will increase accordingly.

References

1. Singh NK, Mahajan V, Aniket A, Pandya S, Panchal R, Mudgal U et al (2019) Identification and prevention of cyber attack in smart grid communication network. In: 2019 International conference on information and communications technology (ICOIACT), pp 5–10
2. Singh C, Sprintson A (2010) Reliability assurance of cyber-physical power systems. In: IEEE PES general meeting, pp 1–6

3. Sridhar S, Hahn A, Govindarasu M (2011) Cyber–physical system security for the electric power grid. Proc IEEE 100:210–224
4. Pasqualetti F, Dörfler F, Bullo F (2013) Attack detection and identification in cyber-physical systems. IEEE Trans Autom Control 58:2715–2729
5. Srivastava A, Morris T, Ernster T, Vellaithurai C, Pan S, Adhikari U (2013) Modeling cyber-physical vulnerability of the smart grid with incomplete information. IEEE Trans Smart Grid 4:235–244
6. Gross JL, Yellen J (2003) In: Handbook of graph theory, CRC press
7. Zhang Y, Wang L, Xiang Y, Ten C-W (2015) Power system reliability evaluation with SCADA cybersecurity considerations. IEEE Trans Smart Grid 6:1707–1721
8. Noel S, Jajodia S (2004) Managing attack graph complexity through visual hierarchical aggregation. In: Proceedings of the 2004 ACM workshop on visualization and data mining for computer security, pp 109–118
9. Singh NK, Mahajan V (2019) Cyber attack detection in smart grid substation using virtual range increment and trust weight. In: 2019 8th international conference on power systems (ICPS), pp 1–6
10. Moonesinghe H, Tan P-N (2008) Outrank: a graph-based outlier detection framework using random walk. Int J Artif Intell Tools 17:19–36
11. Semenov S, Davydov V, Engalichev S (2012) Mathematical modelling of the spreading of software threats in computer network. In: Proceedings of international conference on modern problem of radio engineering, telecommunications and computer science, pp 329–329
12. Hawrylak PJ, Haney M, Papa M, Hale J (2012) Using hybrid attack graphs to model cyber-physical attacks in the smart grid. In: 2012 5th International symposium on resilient control systems, pp 161–164
13. Karray K, Danger J-L, Guilley S, Elaabid MA (2018) Attack tree construction and its application to the connected vehicle. In: Cyber-physical systems security, Springer, pp 175–190
14. Kaynar K, Sivrikaya F (2015) Distributed attack graph generation. IEEE Trans Dependable Secure Comput 13:519–532

Chapter 47
New Active-Only Impedance Multiplier Using VDBAs

A. Hari Prakash Reddy, R. N. P. S. S. Charan, and Mayank Srivastava

Introduction

Active implementation of passive elements/impedance networks has been a very popular research domain for the last three decades. In the open literature, a huge amount of research work available in this area. An impedance multiplier is a very useful circuit idea, which can convert a passive element or impedance network into a tunable one. Such multipliers can be very useful in various classes of analog circuits like active filters, oscillators, etc. It is well understood that the on-chip implementation of high-value capacitance or resistance is very difficult. Therefore, by employing an impedance multiplier circuit one can achieve the higher value of capacitance or resistance even on using low-value resistors or capacitors. In the literature, several impedance multipliers for capacitance multiplication using various active building blocks have been reported [1–16]. An op-amp based capacitance multiplier has been presented in [1] but this circuit does not provide a wide range of capacitances. The operational trans-conductance amplifier (OTA) based electronically tunable capacitance multipliers have been reported in [2–4]. The impedance multiplier circuits in [6–9] and [12–16] exhibit some disadvantages, features like the use of three or more active components, employment of additional passive elements, and lack of electronic tunability feature, used active elements are bulky, highly sensitive multiplication factor, poor non-ideal behavior. There are so many papers that realize the

---

A. Hari Prakash Reddy (✉) · R. N. P. S. S. Charan · M. Srivastava
National Institute of Technology, Jamshedpur, Jharkhand, India
e-mail: hariprakash.allipur@gmail.com

R. N. P. S. S. Charan
e-mail: saicharan.randhi@gmail.com

M. Srivastava
e-mail: mayank2780@gmail.com

© The Editor(s) (if applicable) and The Author(s), under exclusive license to Springer Nature Singapore Pte Ltd. 2021
A. K. Singh and M. Tripathy (eds.), *Control Applications in Modern Power System*, Lecture Notes in Electrical Engineering 710, https://doi.org/10.1007/978-981-15-8815-0_47

behavior of passive elements and their multiplication values have been discussed in [17–21] also.

The main purpose of this article, therefore, is to report a new grounded impedance multiplier block VDBAs. The constructed circuit structure employs only two VDBAs and able to multiply any arbitrary grounded impedance or passive element value. Moreover, the presented circuit offers various advantages attributed to electronically tunable multiplication factor, suitability for on-chip realization, no requirement for trans-conductance matching, the low sensitivity of realized multiplication factor, and up to the mark non-ideal behavior.

The paper is structured in five different sections. The very first section is the introduction, the second section describes the concept of VDBA, and the third section contains the proposed structure, in the fourth section non-ideal analysis has been carried out while in the fifth section a filter application example is proposed. The sixth section describes the simulation results while the last section conclusion followed by the references.

VDBA Concept

The VDBA is a very popular active building block (ABB) firstly introduced in [22] which can be realized by an OTA followed by voltage buffer. The single block symbol of ideal VDBA is depicted in Fig. 47.1. The mathematical relationships describing the voltage/current associations between various ports of VDBA can be denoted in Eq. (47.1–47.3).

$$I_Z = g_m(V_P - V_N) \tag{47.1}$$

$$I_N = I_P = 0 \tag{47.2}$$

$$V_Z = V_W \tag{47.3}$$

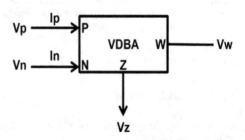

Fig. 47.1 Single block representation of VDBA

47 New Active-Only Impedance Multiplier Using VDBAs

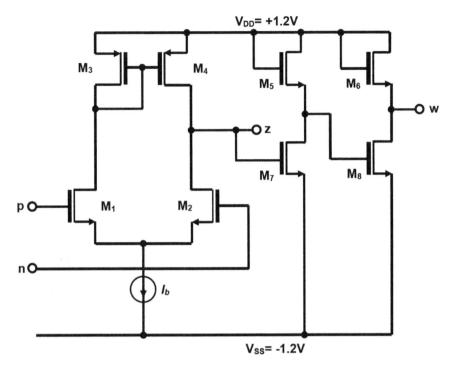

Fig. 47.2 Ideal VDBA implementation using MOS technology

The implementation of the ideal VDBA block using MOS technology has been depicted in Fig. 47.2. It can be noticed here that a VDBA has a very simple design with eight MOS transistors only.

Proposed VDBA-Based Impedance Multiplier

Proposed VDBA-based grounded impedance multiplier configuration has been depicted by Fig. 47.3, here "Z" represents the impedance to be multiplied.

By simple circuit investigation, the impedance "Zin" of the circuit is shown in Fig. 47.1 can be obtained as

$$Z_{in} = \frac{V_{in}}{I_{in}} = Z\frac{g_{m2}}{g_{m1}} \qquad (47.4)$$

Or

$$Z_{in} = \frac{V_{in}}{I_{in}} = KZ \qquad (47.5)$$

Fig. 47.3 Proposed VDBA-based active-only impedance multiplier

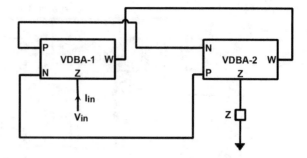

Fig. 47.4 Equivalent configuration of proposed VDBA-based active-only impedance multiplier shown in Fig. 3

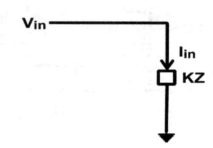

where

$$K = \frac{g_{m2}}{g_{m1}} \quad (47.6)$$

From Eqs. (47.4)–(47.6), it can be noticed, that the configuration given in Fig. 47.3, simulates the working of a grounded impedance multiplier circuit with multiplication index K. The "K" can be altered electronically through trans-conductance gm1 of VDBA1 and/or trans-conductance gm2 of VDBA2. Therefore, the proposed multiplier enjoys electronic tunability. The proposed multiplier employs only two VDBAs. Hence, it can be termed as active-only impedance multiplier. Depending on the nature of impedance Z, the presented circuit can act as a grounded resistance multiplier (If $Z = R$) or grounded capacitance multiplier (If $Z = 1/Sc$) or arbitrary impedance multiplier (if Z is the impedance of a network of R and C). Any grounded impedance whether the equivalent circuit of developed multiplier circuit depicted in Fig. 47.3 is illustrated in Fig. 47.4.

Non-ideal Analysis

The presented impedance multiplier configuration is also revisited using the Non-ideal VDBAs. The mathematical model of non-ideal VDBA can be described by Eqs. (47.7–47.9). On mathematical analysis of presented VDBA-based impedance multiplier using Eqs. (47.7–47.9), the input impedance is

$$I_Z = g_m \beta (V_P - V_N) \tag{47.7}$$

$$I_P = I_N = 0 \tag{47.8}$$

$$V_W = \alpha V_Z \tag{47.9}$$

where β is the trans-conductance gain error and α is voltage-transfer error, which can be called as the non-ideal parameter of VDBA.

On mathematical analysis of presented VDBA-based impedance multiplier using Eqs. 47.7–47.9, we can get

$$Z_{\text{non-ideal}} = \frac{V_{\text{in}}}{I_{\text{in}}} = \frac{\alpha_2 \beta_2 g_{m2}}{\alpha_1 \beta_1 g_{m1}} Z \tag{47.10}$$

Here α_1, β_1 are the non-ideal parameter of VDBA-1 while the α_2, β_2 are the non-ideal gain terms of VDBA-2.

From Eq. (47.10) it can be said that in case of using non-ideal VDBAs also, the offered configuration realizes the working of a grounded impedance multiplier. The multiplication factor in this case is slightly different from the non-ideal situation.

Application Example

To check the performance of the developed impedance multiplier, an active high-pass filtering circuit (HPFC) has been constructed. In this circuit, the presented circuit is used as resistance. This developed HPFC is illustrated in Fig. 47.5.

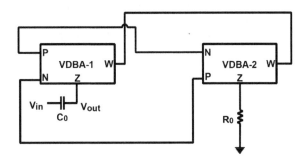

Fig. 47.5 High pass filtering design example employing proposed configuration as grounded resistance

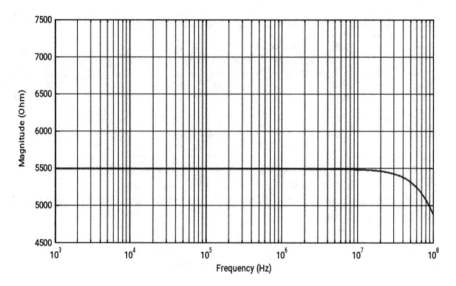

Fig. 47.6 Demonstration of proposed circuit as a grounded resistance multiplier

Simulation Results

The working of the reported impedance multiplier configuration can be studied by running simulations in PSPICE simulation tool. During the simulation, the power supply voltage has been selected as ± 12 V DC. To express the working of the designed grounded impedance multiplier as a grounded resistance amplifier the Z has been taken as a resistance of value of 5 kΩ with biasing current of VDBA-1 is 50μA and VDBA-2 is 75μA. The simulated impedance response has been shown in Fig. 47.6 which clearly indicates the conversion of 5 kΩ resistance into around 5.5 Ω.

Similarly, to express the working of presented arbitrary impedance multiplier as a capacitance multiplier the Z has been taken as a capacitance of value 0.1nF. The simulation generated magnitude plot has been depicted in Fig. 47.7

The simulation of HPFC revealed in Fig. 47.5 is also carried out with $C_0 = 0.1$nF and $R_0 = 1.5$ kΩ. The SPICE generated frequency response has been depicted in Fig. 47.8.

Conclusion

A novel grounded impedance multiplier employing only two VDBAs has been described. This configuration can multiply any arbitrary impedance value Z with an electronically controllable multiplication factor. The proposed configuration is fit for monolithic integration due to the use of active elements only. The behavior of

47 New Active-Only Impedance Multiplier Using VDBAs 533

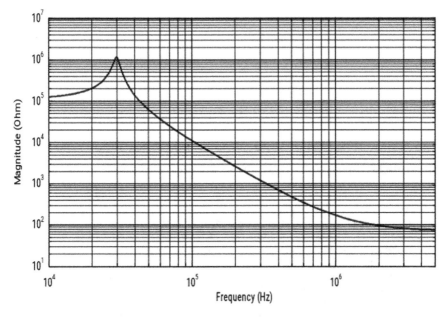

Fig. 47.7 Demonstration of proposed circuit as a grounded capacitance multiplier

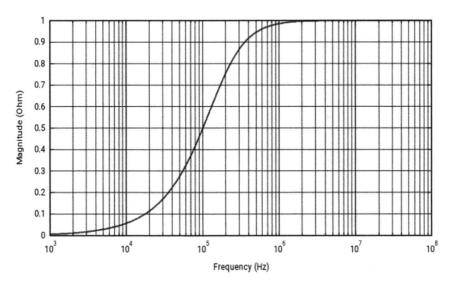

Fig. 47.8 Simulated frequency response of filter illustrated in Fig. 47.5

realized circuit configuration is verified by constructing an HPFC design example. The 0.18 μm TSMC technology has been used for executing simulations in PSPICE environment.

References

1. Khan IA, Ahmed MT (1986) OTA-based integrable voltage/current–controlled ideal C-multiplier. Electron Lett 22:365–366
2. Ahmed MT, Khan IA, Parveen T (1988) Wide range electronically tunable component multiplier. Int J Electron 65:1007–1011
3. Ahmed MT, Khan IA, Minhaj N (1995) Novel electronically tunable C-multipliers. Electron Lett 31:9–11
4. Surakampontorn W, Kumwachara K, Riewruja V, Surawatpunya C (1997) CMOS-based integrable electronically tunable floating general impedance inverter. Int J Electron 82:33–44
5. Myderrizi I, Zeki A (2014) Electronically tunable DXCCII-based grounded capacitancemultiplier. Int J Electron Commun (AEU) 68:899–906
6. Jantakun A, Pisutthipong N, Siripruchyanun M (2010) Single element based-novel temperature insensitive/electronically controllable floating capacitance multiplier and its application. In: Proceedings international conference on electrical engineering electronics computer, telecommunications and information technology (ECTI-CON 2010), pp 37–41
7. Yuce E (2007) On the implementation of the floating simulator employing a single active device. Int J Electron Commun (AEU) 61:453–458
8. Brinzoi P, Cracan A, Cojan N (2011) A new approach in designing electrically controlled capacitance multipliers. In: Proceedings 10th international symposium on signals circuits and systems(ISSCS 2011), pp 1–4
9. Ahmed MT, Khan IA, Minhaj N (1995) Novel electronically tunable C-multipliers. Electron Lett 31(1):9–11
10. Prommee P, Somdunyakanok M (2011) CMOS-based current-controlled DDCC andits applications to capacitance multiplier and universal filter. Int J Electron Commun (AEU) 65:1–8
11. Abuelmaatti MT, Tasadduq NA (1999) Electronically tunable capacitance multiplier and frequency-dependent negative- resistance simulator using the current controlled current conveyor. Microelectron J 30:869–873
12. Khan IA, Ahmed MT (1986) OTA-based integrable voltage/current-controlled ideal Cmultiplier. Electron Lett 22(7):365–366
13. Silapan P, Tanaphatsiri C, Siripruchyanan M (2008) Current controlled CCTA basednovel grounded capacitance multiplier with temperature compensation. In: Proceedings of Asia pacific conference on circuits and systems (APCCAS 2008), pp 1490–1493
14. Manhas PS, Pal K (2011) A low voltage active circuit for realizing floating inductance capacitance frequency dependent negative resistances and admittanceconverter. Arab J Sci Eng 36:1313–1319
15. Cataldo GD, Ferri G, Pennisi S (1998) Active capacitance multipliers using currentconveyors. In: Proceedings of IEEE ISCAS, vol 98. II pp 343–346
16. Pal K (1981) New inductance and capacitor floatation schemes using current conveyors. Electron Lett 17:807–808
17. Srivastava M, Prasad D, Roy A (2018) A purely active circuit simulator for realizing electronically tunable floating resistance. In: Springer international conference ICICCD-2017, AISC book series, vol 624, pp 429–441
18. Srivastava M, Prasad D, Laxya, Shukla AK (2016) Novel active circuit for realizing variable grounded passive elements. In: IEEE International conference on microelectronics and telecommunication engineering, 22–23 Sept 2016, Ghaziabad, India, pp 559–563

19. Srivastava M, Bhanja P, Mir SF (2016) A new configuration for simulating passive elements in floating state employing VDCCs and grounded passive elements. In: IEEE-international conference on power electronics, intelligent control and energy systems (ICPEICES- 2016), Delhi, India, pp 13–18
20. Srivastava Mayank (2017) New synthetic grounded FDNR with electronic controllability employing cascaded VDCCs and grounded passive elements. J Telecommun Electron Comput Eng 9(4):97–102
21. Gupta P, Srivastava M, Verma A, Singh A, Devyansi A, Ali A (2019) A VDCC based grounded passive element simulator/scaling configuration with electronic control. In: Advances in signal processing and communication, Lecture notes in electrical engineering, vol 526. Springer, pp 429–441
22. Biolek D, Senani R, Biolkova V, Kolka Z (2008) Active elements for analog signal processing; classification review and new proposals. Radioengg J 17(4):15–32